Sustainable Intensification of Crop Production

P. Parvatha Reddy

Sustainable Intensification of Crop Production

Springer

P. Parvatha Reddy
Entomology & Nematology
Indian Inst. of Horticultural Research
Bangalore, Karnataka, India

ISBN 978-981-10-2701-7 ISBN 978-981-10-2702-4 (eBook)
DOI 10.1007/978-981-10-2702-4

Library of Congress Control Number: 2016959392

Printed on acid-free paper

This Springer imprint is published by Springer Nature
The registered company is Springer Nature Singapore Pte Ltd.
The registered company address is: 152 Beach Road, #22-06/08 Gateway East, Singapore 189721, Singapore

Foreword

डॉ एन. के. कृष्ण कुमार
उप महानिदेशक (बाग. वि.)
Dr. N. K. Krishna Kumar
DEPUTY DIRECTOR GENERAL
(HORTICULTURE SCIENCE)

भारतीय कृषि अनुसंधान परिषद
कृषि अनुसंधान भवन-II
पूसा, नई दिल्ली -110 012

INDIAN COUNCIL OF AGRICULTURAL RESEARCH
KRISHI ANUSANDHAN BHAWAN-II
PUSA, NEW DELLHI-110 012

July 5, 2016

The green revolution (using high-yielding crop varieties, fertilization, irrigation, and pesticides) led to enormous gains in food production and contributed to world food security. In many countries, however, intensive crop production has had negative impacts on production, ecosystems, and the larger environment putting future productivity at risk. "Negative externalities" of intensification include land degradation, salinization of irrigated areas, over extraction of groundwater, the buildup of pest resistance, and the erosion of biodiversity. Agriculture intensification has also damaged the wider environment through, for example, deforestation, the emission of greenhouse gases, and nitrate pollution of water bodies. In order to meet the projected demands of a growing population expected to exceed nine billion by 2050, farmers in the developing world must double food production, a challenge complicated by the effects of climate change and growing competition for land, water, and energy. The total number of undernourished people who suffer from hunger around the world is estimated at 800 million.

The challenge, therefore, is to realize increased production while avoiding the more extreme of the effects detailed above. The development of more ecologically designed agricultural systems that reintegrate features of traditional agricultural knowledge and add new ecological knowledge into the sustainable intensification process can contribute to meeting this challenge. Sustainable intensification has been defined as producing more from the same area of land while reducing negative environmental impacts and increasing contributions to natural capital and the flow of environmental services. This alternative paradigm has been shown to work in many parts of the world and is biologically and ecologically as well as economically more efficient in producing the required outputs of goods such as edible and nonedible biological products and of water while at the same time taking care of other essential ecosystem services that regulate soil, crop, and ecosystem health; protect habitats and biodiversity; drive carbon, nutrient, and hydrological cycles as well as conserve stocks of carbon, nutrients, and water; and protect soils and landscapes from erosion and other forms of degradation.

During the past 50 years, ideas about how to achieve such "sustainable intensification" have matured from concepts to reality. Today we have at our disposal concepts and technologies of how intensive and highly productive and efficient agriculture can be combined in practice and on real farms with the enhancement of ecosystem services but without "trade-offs" or "unavoidable damages." There is a need for opening up the chances for farmers to receive payments not only for their produce but also for such services as, for example, the sequestration of carbon in the soils or the provision of clean water, using healthy and biologically active soils as medium for such ecosystem functions and services.

The information on sustainable intensification of crop production is very much scattered, and there is no book at present which comprehensively and exclusively deals with the above aspects in agriculture emphasizing on food security. The present book on *sustainable intensification of crop production (SICP)* outlines a new paradigm which aims to produce more from the same area of land, through increasing efficiency and reducing waste, while conserving resources, reducing negative impacts on the environment, and enhancing the provision of ecosystem services. The use of ecologically based management strategies can increase the sustainability of agricultural production while reducing off-site consequences. This book also highlights the underlying principles and outlines some of the key management practices and technologies required to implement SICP.

I compliment Dr. P. Parvatha Reddy for his meticulous contribution on a very potential topic of sustainable intensification of crop production. I am sure that this book will prove an invaluable source of reference for the policy makers, researchers, scientists, practicing farmers, and graduate students.

(N. K. Krishna Kumar)

Preface

The remarkable scientific and technological achievement by "the green revolution" during the 1960s is based largely on intensification of management on land already under agriculture, accomplished through the use of high-yielding crop varieties, chemical fertilizers and pesticides, irrigation, and mechanization. The green revolution led to enormous gains in food production and improved world food security. However, intensive crop production has had negative impacts on production, ecosystems, and the larger environment putting future productivity at risk. The world's population is expected to grow to over nine billion people by 2050; there will be a need to raise food production by 70 % globally and by almost 100 % in developing countries. An estimated 80 % of the required food production increases will thus need to come from land that is already under cultivation at a time when annual growth in crop productivity is decreasing from a rate of around 3–5 % a year in the 1960s to a projected 1 % in 2050. Increases in agricultural production will therefore have to come in the form of yield increases and higher cropping intensities. This increase must be achieved against a challenging backdrop including the decreasing availability of land and competition for water, resource degradation (e.g., poor soil fertility), energy scarcity (resulting in higher costs for input production and transport), as well as climate change where alterations in temperature, precipitation, and pest incidence will affect farmers' choice of crops to grow as well as their potential yields.

There is now widespread awareness of the importance of switching over to an ecosystem approach for intensifying crop production. The future of food and farming program has called for substantial changes throughout the world's food system including sustainable intensification to simultaneously raise yields, increase efficiency in the use of inputs, and reduce the negative impacts of food production. The International Assessment of Agricultural Knowledge, Science and Technology for Development (IAASTD) emphasized the need for a change in paradigm to encourage increased adoption of sustainable ecological agriculture and food systems and called for a shift from current farming practices to sustainable agricultural systems capable of providing significant productivity increases and enhanced ecosystem services.

Assessments in developing countries have shown how farm practices that conserve resources improve the supply of environmental services and increase productivity. A review of agricultural development projects in 57 low-income countries found that more efficient use of water, reduced use of pesticides, and improvements in soil health had led to average crop yield increases of 79 %. Another study concluded that agricultural systems that conserve ecosystem services by using practices such as conservation tillage, crop diversification, legume intensification, and biological pest control perform as well as intensive, high-input systems. Sustainable intensification of crop production (SICP), when effectively implemented and supported, will provide the "win-win" outcomes required to meet the dual challenges of feeding the world's population and conserving the resources of the planet. SICP will allow countries to plan, develop, and manage agricultural production in a manner that addresses society's needs and aspirations, without jeopardizing the rights of future generations to enjoy the full range of environmental goods and services.

The information on sustainable intensification of crop production is very much scattered, and there is no book at present which comprehensively and exclusively deals with the above aspects in agriculture emphasizing on ensuring food security. The present book on sustainable intensification of crop production comprehensively deals with important aspects of sustainable intensification. The book is divided into 11 parts comprising of 24 chapters which comprehensively deals with the following important aspects:

- Minimum soil disturbance (direct seeding, no-tillage, prevention of soil compaction)
- Permanent organic soil covers (crop residue management, cover crops, relay crops, intercrops)
- Species diversification (crop rotation, sequences, and associations; fodder crops/ pasture rotations and agro-forestry; integration of legumes in the rotation)
- Selection of suitable cultivar, planting time, age, and spacing (cultivars with good production potential, rooting abilities, good-quality seed material, planting at appropriate time, seed rates, and spacing)
- Balanced plant nutrition (increasing soil organic matter; balanced nutrient supply in adequate rates, including liming; placement of fertilizer where adequate)
- Agro-ecological pest management, including weed management
- Improved water management practices (micro-irrigation water harvesting, mulch, no-till, cover crops, regulation of irrigation water according to plant requirements, maintenance of aerobic soil compaction)
- Careful management of farm machinery
- Integrated crop–livestock management

The book is extensively illustrated with excellent-quality photographs enhancing the quality of publication. The book is written in lucid-style, easy-to-understand language along with adoptable recommendations involving sustainable eco-friendly practices.

This book will be of immense value to the scientific community involved in teaching, research, and extension activities. The material can be used for teaching postgraduate courses. The book can also serve as a very useful reference to policy makers. Suggestions to improve the contents of the book are most welcome (e-mail: reddy_parvatha@yahoo.com). The publisher, Springer India (Pvt.) Ltd., New Delhi, India, deserves commendation for their professional contribution.

Bangalore, India P. Parvatha Reddy
July 15, 2016

Contents

Part II Minimum Soil Disturbance

Part III Permanent Soil Organic Cover

About the Author

Dr. P. Parvatha Reddy obtained his Ph.D. degree jointly from the University of Florida, USA, and the University of Agricultural Sciences, Bangalore.

Dr. Reddy served as the director of the prestigious Indian Institute of Horticultural Research (IIHR) in Bangalore from 1999 to 2002 during which period the institute was honored with the "ICAR Best Institution Award." He also served as the head of the Division of Entomology and Nematology at the IIHR and gave tremendous impetus and direction to research, extension, and education in developing bio-intensive integrated pest management strategies in horticultural crops. These technologies are being practiced widely by the farmers across the country since they are effective, economical, eco-friendly, and residue-free. Dr. Reddy has about 34 years of experience working with horticultural crops and been involved in developing an F1 tomato hybrid "Arka Varadan" resistant to root-knot nematodes.

Dr. Reddy has over 250 scientific publications to his credit, which also include 30 books. He has guided two Ph.D. students at the University of Agricultural Sciences, Bangalore.

Dr. Reddy is serving as chairman of the Research Advisory Committee (RAC), Indian Institute of Vegetable Research, Varanasi, and senior scientific advisor of the Dr. Prem Nath Agricultural Science Foundation, Bangalore. He also served as a member of the RAC of the National Centre for Integrated Pest Management, New Delhi; member of the expert panel for monitoring the research program of the National Initiative on Climate Resilient Agriculture (NICRA) in the theme of horticulture including pest dynamics and pollinators; member of the RAC of the National Research Centre for Citrus, Nagpur; and member of the Project Directorate of Biological Control, Bangalore. He served as a member of the QRT to review the progress of the Central Tuber Crops Research Institute, Trivandrum, AICRP on tuber crops, AICRP on nematodes, and AICRP on betel vine. He is the honorary fellow of the Society for Plant Protection Sciences, New Delhi; fellow of the Indian Phytopathological Society, New Delhi; and founding president of the Association for Advancement of Pest Management in Horticultural Ecosystems (AAPMHE), Bangalore.

Dr. Reddy has been awarded with the prestigious "Association for Advancement Pest Management in Horticultural Ecosystems Award," "Dr. G.I. D'souza Memorial Lecture Award," "Prof. H.M. Shah Memorial Award," and "Hexamar Agricultural Research and Development Foundation Award" for his unstinted efforts in developing sustainable, bio-intensive, and eco-friendly integrated pest management strategies in horticultural crops.

Dr. Reddy has organized the "Fourth International Workshop on Biological Control and Management of *Chromolaena odorata*," "National Seminar on Hi-tech Horticulture," "First National Symposium on Pest Management in Horticultural Crops: Environmental Implications and Thrusts," and "Second National Symposium on Pest Management in Horticultural Crops: New Molecules and Biopesticides."

Part I
Introduction

Chapter 1
Sustainable Intensification—An Overview

Abstract Even though the intensive crop production practices adopted in achieving green revolution (using fertilizers, pesticides, and machines) led to enormous gains in food production and improved world food security, it had negative impacts on production, ecosystems, and the larger environment (causing environmental damage, pollution, reliance on fossil fuels) putting future productivity at risk. The food production in the developing world must be doubled, in order to feed a growing population that is expected to reach nine billion by 2050. This chapter outlines a new paradigm, Sustainable Crop Production Intensification (SCPI), which aims to produce more from the same area of land, through minimizing soil disturbance; enhancing and maintaining a protective organic cover; cultivating a wider range of plant species; using well-adapted, high-yielding varieties; enhancing crop nutrition; integrating pest (insect pests, diseases, nematodes, and weeds) management strategies; adopting efficient water management, using adequate and appropriate farm power and equipment; and integrating crop-livestock production. These ecologically based management strategies can increase the sustainability of agricultural production while reducing the off-site consequences.

Keywords Green revolution • Sustainable intensification • Crop production • Ecosystem services • Soil nutrition • Water management • Integrated pest management

1.1 Introduction

The world population is expected to reach over nine billion by 2050. Over 800 million people are suffering from hunger and malnutrition. In order to meet the food demand for the growing population, the food production has to be doubled by 2050 in the developing world. Hence, there is an urgent need and increased efforts to accelerate food production and productivity.

P.P. Reddy, *Sustainable Intensification of Crop Production*,
DOI 10.1007/978-981-10-2702-4_1

Table 1.1 Gaps between yields obtained by the farmer, on-farm and on-station

Crop	Area (M ha)	Total production (MMT)	Average yield (t/ha)	On-farm yield (t/ha)	On-station yield (t/ha)
Maize	2.01	6.16	3.01	6–8	9–12
Wheat	1.63	3.43	2.11	2.5–5	3.5–5.5
Sorghum	1.71	3.60	2.11	3.3	4.2

Source: CSA 07/08

1.1.1 Present Scenario

The shortage of food production in the developing countries has created a serious setback toward the Millennium Development Goals (higher productivity, less resource degradation) of achieving self sufficiency in food production. There is a wide gap between yields obtained by the farmer, on-farm and on-station (Table 1.1).

Some of the factors responsible for low yield in farmers' fields include:

- Agricultural inputs not affordable to all farmers.
- Climate change is an additional threat to agriculture (IPCC 2007).
- Conventional tillage is a major soil degrading factor.
- Cultivated soils have lost considerable carbon.
- Degradation of land resources.
  - In Asia 90 % of potential land is already used.
  - 1.3 Million ha of agricultural land are lost every year (urbanization).
  - Available land per person is declining.
- Delayed sowing.
- High soil erosion.
- Ineffective market, storage, and distribution chains.
- Inputs are not efficiently utilized.
- Low soil organic matter content (levels below 2 % are common).
- Most of the fodder resources are used by oxen.

1.2 Green Revolution

Green revolution by adopting improved seeds (high-yielding varieties/F1 hybrids), inputs (high doses of fertilizers, pesticides, and weedicides), heavy machinery, and copious water for irrigation was successful with good soils having abundant irrigation water supply but failed to address some social and environmental issues.

1.2.1 Positive Impacts

1.2.1.1 Increased Food Production

The green revolution, with the adoption of high-yielding varieties of seeds, chemical fertilizers, and irrigation, significantly increased wheat, rice, and maize production and productivity (2.5 times in wheat). From 1960 to 2000, the yields of wheat, rice, and maize in all developing countries increased by 208, 109, and 157 %, respectively.

1.2.1.2 Increased Nutrition

From 1960 to 1990, the number of undernourished people and malnourished children in the world fell significantly. Due to fall in food prices, the diet diversification even among poor people increased.

Other impacts of green revolution include:

- Adoption of improved package of practices for crop production
- Improvement in standard of living
- Increased employment opportunities (production of seeds, fertilizers, pesticides, and other inputs; transport and marketing; and food processing)
- Reduction in import of food grains
- Socioeconomic change

1.2.2 Negative Impacts

1.2.2.1 Loss of Biodiversity

The green revolution has significantly decreased the biodiversity of crops in the world today. The use of few selected genetically uniform crop varieties [particularly dwarf wheat varieties released from the International Maize and Wheat Improvement Center (CIMMYT), Mexico; and Taichung Native rice varieties released from the International Rice Research Institute (IRRI), Philippines], in very large areas, has led to mono-cropping. Genetic diversity was lost by replacing thousands of wild crop strains with a few monoculture strains. Ninety percent of all human food now comes from only 15 crop species and 8 livestock species. Before the green revolution, it is estimated that there were over 30,000 different variants of rice. Critics estimate that only ten modified rice variants are used today. Many valuable genetic traits bred traditional varieties over thousands of years were permanently lost.

1.2.2.2 Soil Quality Impact

Heavy doses of fertilizers were used for higher production. The soil quality is diminished due to increased reliance on synthetic fertilizers rather than natural fertilizers (FYM, compost, crop residues, green manures, and other organic amendments) which allowed replenishing of nutrients. Nitrate pollution of water bodies took place due to leaching of nitrogenous fertilizers.

1.2.2.3 Pesticide Hazard

Pesticides travel through food chain and accumulate in higher organisms (biomagnification). They persist in soil, air, and surface and groundwater and continue to poison them for a long time. In 1989, WHO and UNEP estimated that there were about one million human pesticide poisonings annually, and some 20,000 (mostly in developing nations) ended in death (Pimentel 1996). Long-term exposure to pesticides caused cancer in animals (e.g., chlorinated hydrocarbons).

1.2.2.4 Soil Salinization

Intensification of water use for irrigation has led to soil salinization. This has resulted in land toxicity due to accumulation of sodium ions in soil solution that adversely affected plant growth.

1.2.2.5 Increase in Pests and Disease Incidence

Due to mono-cropping, the pest and disease problems in crops increased. In order to manage pests, indiscriminate use of pesticides was adopted.

1.2.2.6 Soil Compaction

The use of heavy farm machinery for sowing; fertilizer, pesticide, and herbicide application; and harvesting resulted in soil compaction.

The environmental degradation makes the green revolution an overall inefficient, short-term solution to the problem of food insecurity. Hence, more sustainable and environment friendly agriculture system needs to be developed to both feed a growing world population without compromising the needs of future generations and to feed themselves.

Other negative impacts of green revolution include:

- Water depletion for irrigation
- Fossil fuels used for running heavy farm machinery

However, without the green revolution, much more land would have been converted for agriculture, destroying forests, wetlands, and other ecosystems.

1.3 Challenges and Solutions

1.3.1 Challenges

The challenge of meeting future demand for food in a sustainable manner is affected by the following challenges:

- Additional land for agriculture is limited (including irrigated land).
- Chemical pesticides pollute and kill pollinators, and pests evolve resistance.
- Climate change is an additional threat.
- Food production has to be doubled by 2050, in view of continuous population growth.
- GM crops show promise for social and environmental benefits, but questions linger about their impacts.
- Grazing and logging, as well as cropland agriculture, contribute to soil degradation.
- Higher production has to be achieved in eco-friendly manner.
- Increasing demand for bioenergy and renewable resources.
- Livestock pose benefits and harm for the environment and human health.
- Major changes needed to attain the MDGs (higher productivity, less resource degradation).
- More than 50 % of available water resources already used for irrigation.
- Much of the world's crop diversity has vanished.
- Salinization, waterlogging, and other soil degradation problems are leading to desertification.
- Soil erosion is a problem worldwide.

New technologies and innovation may help with these challenges for producing more food with less environmental impact.

1.3.2 Solutions

Developing new strategies that take advantage of ecological interactions within agricultural systems (Robertson 1997) along with policy support and adequate funding are crucial to the continuance of high-productivity sustainable agriculture in the future:

- More funding for seed banks can rebuild crop diversity.
- Green revolution advances have kept up with food demand so far. Improved distribution and slowed population growth would help further.

- Farming strategies, like no-till farming, contour farming, terracing, etc., help control erosion.
- Better grazing and logging practices exist that have far less impact on soils.
- Biocontrol and IPM offer alternatives to pesticides.
- Further research and experience with GM crops may eventually resolve questions about impacts and allow us to maximize benefits while minimizing harm.
- Ways are being developed to make feedlot agriculture and aquaculture safer and cleaner.
- Government laws, and government extension agents working with farmers, have helped improve farming practices and control soil degradation.

1.4 Sustainable Intensification—A New Paradigm for the Twenty-First Century

Sustainable intensification aims to produce more food from the same piece of land with less environmental impact. New technologies and innovation will help to overcome these challenges and help to achieve the aim. Sustainable agriculture is a profitable way of producing high-quality food and fiber that protects and renews the natural environment, builds local economies, and enhances the quality of life of farmers and farm workers.

Sustainable intensification maintains soil and water quality, reduces risk, uses integrated modern and traditional strategies, utilizes ecological processes (e.g., biological nitrogen fixation, natural predators), minimizes environmental hazards, and acknowledges local environmental and cultural conditions, resulting in increased productivity (Fig. 1.1) (Pretty et al. 2011).

The Sustainable Agriculture Research and Education Program (www.SARE.org) defines sustainable agriculture as an agricultural production and distribution system that:

- Achieves the integration of natural biological cycles and controls
- Minimizes adverse impacts on health, safety, wildlife, water quality, and the environment
- Optimizes the management and use of on-farm resources
- Promotes opportunity in family farming and farm communities
- Protects and renews soil fertility and the natural resource base
- Provides an adequate and dependable farm income
- Reduces the use of nonrenewable resources and purchased production inputs

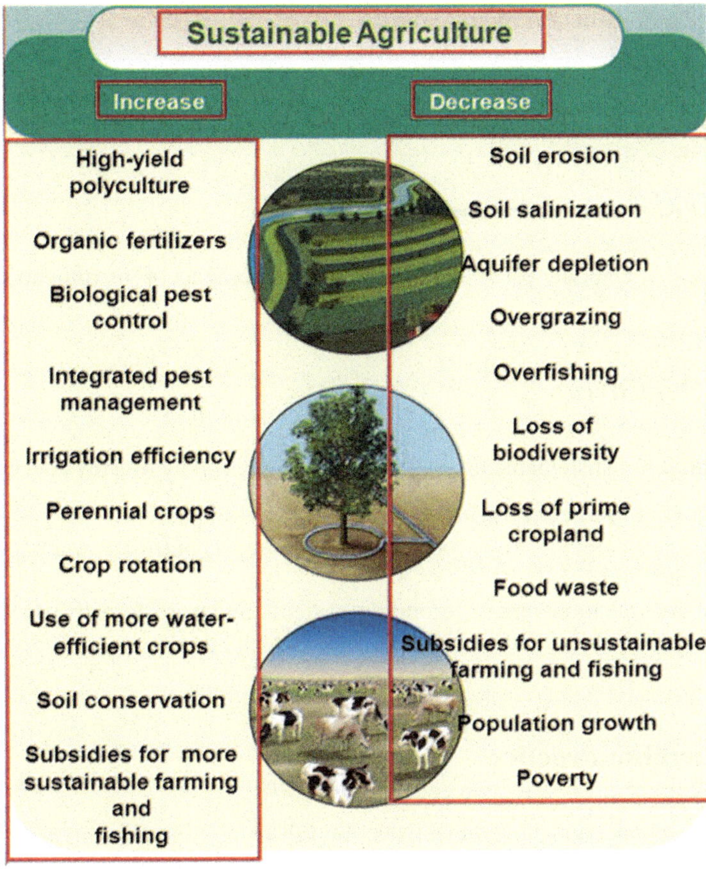

Fig. 1.1 Sustainable intensification—a new paradigm for the twenty-first century

1.4.1 Principles

The sustainable intensification is founded on a set of science-based economic, environmental, and fiscal and institutional principles.

1.4.1.1 Economic Sustainability

Sustainable livelihoods and improved well-being through growth and poverty reduction.

1.4.1.2 Environmental Sustainability

Target agricultural land, forests, water resources, protected areas, and biodiversity, so that opportunities and options of future generations are not degraded.

1.4.1.3 Fiscal and Institutional Sustainability

Must be realistic about cost and institutional requirements of instruments.

1.4.2 Objectives

The objectives of sustainable intensification as suggested by Murray (2012) include:

- Production of more food with less environmental impact
- Input use efficiency of key production inputs such as nutrients, water, pesticides, labor, land, and energy
- Use of natural biodiversity (cropping systems and agro-forestry) to overcome biotic, abiotic and economic stresses

1.5 Farming Practices

In order to achieve the above objectives, the following nine farming practices have to be adopted:

- Conservation tillage or direct sowing
- Retention of crop cover
- Agricultural biodiversity
- Selection of suitable cultivars
- Enhanced crop nutrition
- Efficient water management
- Ecological pest management
- Careful management of farm machinery
- Crop-livestock integration

1.5.1 Conservation Tillage

Conservation tillage is a generic term that includes many varied tillage systems that leave more than 30 % crop residue cover after planting. It can include no-till, minimum tillage systems such as strip tillage, ridge tillage, and mulch tillage that leaves

Fig. 1.2 Direct sowing

more than 30 % residue cover after planting (Fig. 1.2). Residue limit of 30 % was established as a result of the relationship between residue cover and inter-rill erosion. An increase from 0 to 30 % residue cover results in a 70 % reduction of inter-rill soil loss. Crop rotations and cover crops are central to this system approach.

On most soils, no-till yields should be similar to yields obtained with tillage. No-till should outyield tilled crops in areas where drought stress is a problem, due to the water conserved by the mulch cover.

1.5.1.1 No-Till/Strip-Till (>30 % Residue)

Soil is left undisturbed from harvest to planting except for strips up to one third of the row width. These strips may involve only residue disturbance or may include soil disturbance. Planting or drilling is accomplished using disk openers, coulters, row cleaners, in-row chisels, or rototillers. Weed control is accomplished primarily with herbicides. Cultivation may be used for emergency weed control. Other terms used to describe no-till include direct seeding, slot planting, zero-till, row-till, and slot-till.

1.5.1.2 Ridge-Till (>30 % Residue)

Soil is left undisturbed from harvest to planting except for strips up to one third of the row width. Planting is completed on the ridge with sweeps, disk openers, coulters, or row cleaners and usually involves the removal of the top of the ridge. Residue

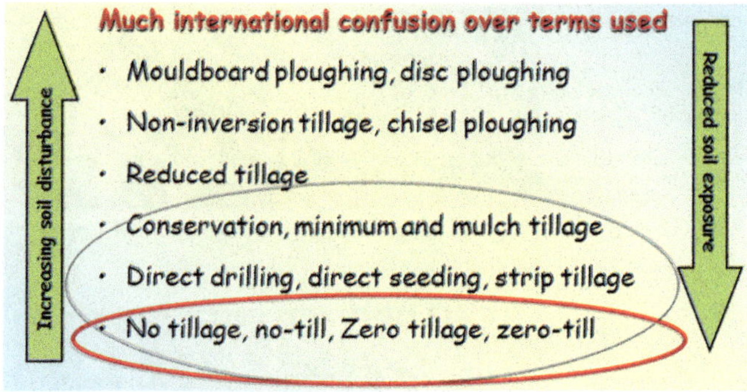

Fig. 1.3 Effect of different tillage systems on soil disturbance

is left on the surface between ridges. Weed control is accomplished with herbicides (frequently banded) or cultivation. Ridges are rebuilt during row cultivation.

1.5.1.3 Mulch-Till (>30 % Residue)

Full-width tillage involving one or more tillage trips which disturb the entire soil surface. Tillage tools such as chisels, field cultivators, disks, sweeps, or blades are used. Done prior to and/or during planting. Weed control is accomplished with herbicides or cultivation.

1.5.1.4 Reduced-Till (15–30 % Residue)

Full-width tillage involving one or more tillage trips that disturb the entire soil surface, done prior to planting. There is 15–30 % residue cover after planting or 500 to 1000 pounds per acre of small grain residue equivalent throughout the critical wind erosion period. Weed control is accomplished with herbicides or row cultivation.

The effect of different tillage systems on soil disturbance and soil exposure is presented in Fig. 1.3.

1.5.2 Retention of Crop Cover

The retention of crop residues (Fig. 1.4)/cover crops can be achieved by leaving the crop residues after harvest, planting cover crops after harvest, and adding organic amendments such as farm yard manure, compost, wood chips, straw, leaves, etc. as soil mulch. This helps in reducing the soil erosion (Table 1.2).

Fig. 1.4 Providing ground cover with crop residues

Table 1.2 Effect of soil cover on soil erosion

Cropping system	Annual loss of soil (t/ha)	Rainfall runoff (%)
No crop (bare soil)	41.0	30
Continuous maize	19.7	29
Continuous wheat	10.1	23
Maize, wheat, clover rotation	2.7	14
Continuous blue grass	0.3	12

Source: Missouri Experiment Station, Columbia, Missouri (based on 14 years data)

Growing cover crops, such as cereals (wheat, rye, and oats), legumes (hairy vetch, crimson clover, red clover, white clover, winter pea), and brassicas (mustard and radish) which provide quick soil cover, have an extensive root system and preferably survive winter. If the cover crops are killed without tillage and the main crop established with no-till methods, additional erosion protection will be provided by the resulting mulch.

The tine system results in short residue, well spread to maximize trash flow, while disk system produces long residue which is well spread.

Surface soil becomes more granular and friable with continuous residue management systems. Expected changes due to residue management systems include improved soil aggregate stability, increased water-holding and infiltration capacity, reduced soil erosion, increased organic matter and C sequestration, soil structure improvement, nitrate recapture, atmospheric nitrogen fixation, reduced water evaporation (increased water use efficiency), reduced soil temperature, weed smothering, nutrient release, and less surface ponding due to rainfall.

1.5.3 *Agricultural Biodiversity*

The components of agricultural biodiversity consist of herbivores, predators and parasites, pollinators, vegetation, soil mesofauna, soil microfauna, livestock, and earthworms. The functions of biodiversity include population regulation, biological control, predation, sources of natural enemies, disease suppression, pollination, genetic introgression, crop wild relatives, competition, allelopathy, biomass consumption, nutrient cycling, soil structure, and organic matter decomposition.

1.5.3.1 Ecosystem Services of Biodiversity

The following are the ecosystem services provided by biodiversity:

- Biological control: trophic (food web) dynamic regulations of populations
- Climate and gas regulation (global temperature, precipitation, other biologically mediated climatic processes at global/local levels; of atmospheric chemical composition)
- Erosion control and sediment retention (prevent loss of soil by wind, rain impact, runoff; storage of silt in ecosystem, in lakes and wetlands)
- Food production (the portion of gross primary production extractable as raw food or for processing of food)
- Genetic resources (sources of unique biological materials and products)
- Nutrient cycling (storage, cycling, processing, input of nutrients)
- Pollination (movement of floral gametes)
- Raw materials (the portion of gross primary production extractable as raw material)
- Refugee (habitat for local/transient populations)
- Resilience/disturbance regulation (ecosystem response to environmental fluctuation, mainly controlled by vegetation structure)
- Soil formation (processes of weathering of rock; soil buildup)
- Waste detoxification (recovery of mobile nutrients, removal/breakdown of excess or toxic nutrients/compounds, pollution control)
- Water regulation and supply (hydrological flow/regimes; water retention, storage, provisioning in the watershed: water supply in aquifers, surface water bodies; availability for consumption, irrigated agriculture, industry, transport)

1.5.3.2 Enhancement of Agricultural Biodiversity

Agricultural biodiversity can be enhanced by agro-forestry, composting, cover crops, crop rotations, green manures, intercropping, no-tillage, organic matter inputs, and windbreaks (Altieri 1994). The soil management strategies can also be used to enhance soil quality through biodiversity (Table 1.3).

Table 1.3 Soil management strategies to enhance soil quality through biodiversity

Management objective	Cropping system	Management strategies
Soil fertility improvement	Coffee/banana/maize/beans/in rotation with round potatoes	Application of FYM, incorporation of crop residues, house refuse, and weeds, *Sesbania*, *Grevillea*, and composting
	Maize/beans	Application of FYM, crop residues, and trash lines
Soil moisture conservation	Coffee/banana/maize/beans/in rotation with round potatoes	Crop residues, mulching, trash lines, *Setaria*, canopy cover
	Maize/beans	Self mulching, spreading/ incorporation of crop residues, trash lines
Soil erosion control	Coffee/banana/maize/beans/in rotation with round potatoes	Rainfall interception by trees canopy, mulching, trash lines, flower hedges, *Sesbania sesban*
	Maize/beans	Trash lines, crop canopy

1.5.4 Selection of Suitable Cultivars

There is a need to select suitable crop cultivars which can perform well under different agro-ecosystems, socioeconomic circumstances, farming practices, and resilient to climate change. The following aspects should be considered while selecting and developing crop varieties suitable for sustainable intensification:

- Adapted to various agro-ecosystems, changing production practices, farming systems, and less favored areas and resilient to climate change
- Elaborate root system to improve the root efficiency for mining the water and absorption of nutrients
- Enhanced carbon sequestration
- High input, water, and nutrient use efficiency
- High quality with good production potential
- Resistant/tolerant to biotic (pests, diseases, and nematodes) and abiotic (drought, heat, flood, salinity, and frost) stresses
- Varieties with higher nutritional value and desirable organoleptic properties

1.5.5 Balanced Crop Nutrition

1.5.5.1 Essential Nutrients of Plants

In today's world, because of concentration of people in cities, and waste filling landfills, nutrients are exported away from the soil. Loss of soil also occurs through erosion that comes as a result of extensive tillage. Nutrients should be managed in a way that meets crop needs without applying in excess which can lead to water quality problems.

Plants need macronutrients such as N, P, K, Ca, Mg, and S (>50 mg/kg in plant) and micronutrients such B, Cl, Cu, Fe, Mn, and Zn (<50 mg/kg in plant). The sources of nutrients include soil, organic manures, and inorganic fertilizers.

Nutrients can be applied to crop plants either in organic or inorganic form. Nutrients in inorganic fertilizers are dissolved into soil solution and are available immediately. Negative charges on soil surfaces hold cations (Ca, Mg, K, and Na) available for exchange with soil solution. Decomposition of organic matter releases nutrients, and hence they are slowly available. Nitrogen is easily lost via water through leaching or to air via denitrification or volatilization, and phosphorus is usually low because soil quickly fixes it. Hence, there is a need for supplemental application of N, P, and K, besides what is already present in the soil.

The main emphasis should be given for balanced use of mineral fertilizers, while the role of biofertilizers, organic manures, green manuring, and recycling of organic wastes should also be considered as supplementary. Biological nitrogen fixation through root nodule bacteria is an alternative method of replenishing soil nutrients. Manure utilization recycles nutrients back to the land.

Nutrient management means integration of on-farm nutrient sources (soil reserves and organic manures), with commercial inorganic fertilizers to meet the crop need, and minimizing nutrient loss. The development of nutrient management plan helps to decide on the suitable form, correct amount and proper timing of application of crop nutrients to obtain optimum yields, and less impact on quality of water. Best nutrient management strategies include soil testing, timing of applications, conservation tillage, buffer strips, cover crops, P-based nutrient management, and balance: inputs = exports. In future, there is an urgent need for judicious and integrated management of all sources of nutrients for promoting efficient and balanced use of plant nutrients.

1.5.6 Improved Water Management Practices

The fundamental question is how to increase water productivity in agriculture. For the region, it is important to shift from water-intensive crops (wheat, sugar beets, citrus) to less water-intensive crops (barley, fruit trees, etc.). This applies particularly to rain-fed crops, where there is untapped potential to be developed through research. Many technologies are available for improving the water use efficiency in agriculture, both rain-fed and irrigated:

- Adopt conservation tillage and planting on raised beds to use irrigation water more efficiently.
- Adopt water-saving technologies in rain-fed agriculture:
  - Improved small-scale and supplemental irrigation systems can increase productivity of rain-fed agriculture.

  - More efficient crop sequencing and timely planting can achieve significant savings in water use.
  - New crops, requiring less water, are now available.

- Installing drip irrigation systems and center pivot irrigation as an alternative to the traditional flood irrigation.
- Integrated water resources management: surface water, groundwater, catchment basin, and land-use planning.
- Use drip irrigation for irrigated crops, and use drought tolerant crops, for rain-fed agriculture.
- Various new precision irrigation systems are available for supplying water to plants only when needed.
- Wastewater can be treated and used for irrigation (mainly for peri-urban agriculture).

1.5.7 Agro-ecological Pest Management

The ecological goal of pest management strategy is to use naturally based controls such as farmscaping, beneficial overwintering sites, cover crops, and no-till production schemes, in order to increase number of natural enemies for long-term pest control. Leaving a residual pest population is important for natural enemies to have food in order to survive. Pest residues for natural enemies should be lower than the economic threshold. The pests should be suppressed but not to be eradicated. The pesticide treatments should be based on the need—"treat when necessary" driven by data.

Ecological pest management practices include (Fig. 1.5):

- Conservation tillage/residue management
- Crop nutrient management
- Conservation buffers
- Filter strips and crop rotation
- Other conservation practices

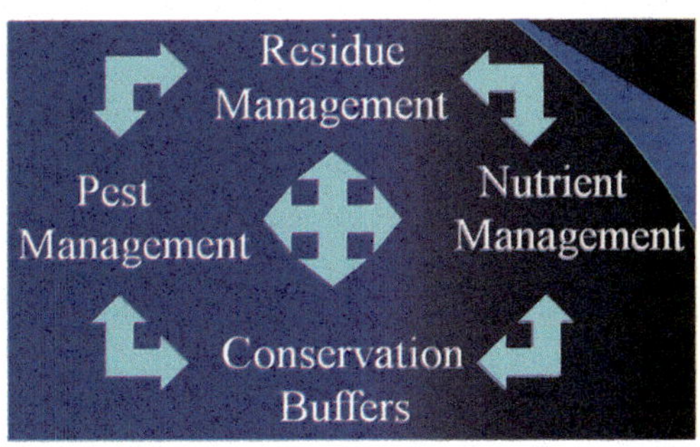

Fig. 1.5 Ecological pest management

1.5.7.1 Integrated Pest Management

Integrated pest management is an approach to pest control that combines biological, cultural, and other alternatives to chemical control with the judicious use of pesticides (Fig. 1.6). The objective of IPM is to maintain pest levels below economically damaging levels while minimizing harmful effects of pest control on human health and environmental resources.

1.5.7.2 Implementing IPM Theory

- Use cultural methods, biological controls, and other alternatives to pesticides to help delay the pesticide resistance.
- Use field scouting, pest forecasting, and economic thresholds to ensure that pesticides are only used for real pest problems.
- Match pesticides with site characteristics to minimize off-site environmental risks.

1.5.7.3 Cultural Controls

- Altering planting and harvest dates
- Altering seeding rates/crop spacing
- Certified seed that is free of pathogens and weed seed
- Cover crops
- Crop rotation
- Pest-resistant varieties
- Sanitation practices such as cleaning tillage and harvesting equipment

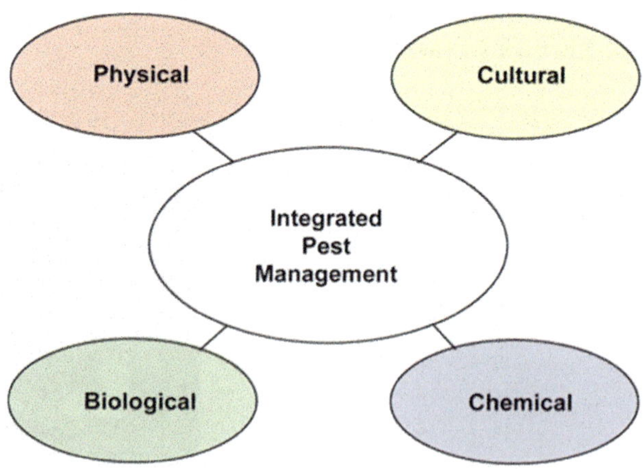

Fig. 1.6 Integrated pest management

- Seedbed preparation, fertilizer application, and irrigation schedules that help plants outgrow pests
- Trap crops

1.5.7.4 Biological Controls

- Predators—free-living animals that eat other animals (insects)
- Parasitoids—insect parasites of other insects
- Pathogens—disease-causing microorganisms

1.5.8 Careful Management of Farm Machinery

Agriculture mechanization refers to usage of any improved tool, implemented or machined by the farm workers to improve the efficiency of any farm operation to increase crop productivity. It also helps to reduce drudgery or stress which adversely affects human mental faculties leading to errors, imprecision, hazards, and eventually loss of efficiency if done manually. The timeliness of operations has assumed greater significance in obtaining optimal yields from different crops. For example, the sowing of wheat in Punjab is done up to the first fortnight of November. A delay beyond this period by every 1 week leads to reduction in yield of about 375 kg/ha. The timeliness is also important for farm operations like weeding, irrigation, and harvesting. To ensure timeliness of various operations, it is necessary to use farm machinery which has high output capacity. Farm mechanization has helped to increase area under cultivation and the intensity of cropping.

The quality and precision of operations are equally significant for realizing higher yields. The various farm operations need a high degree of precision to increase the efficiency of the inputs and reduce the losses.

Farm mechanization is gaining importance because the output of work done per hour is several times more than human labor. This also entails reduction in the total labor requirement. The ever-increasing wages for the human labor, their nonavailability at time of need, and the raising cost of maintaining draft animals also speak in favor of mechanization,

1.5.8.1 Mechanization of Farming

- Irrigation was the first operation that was mechanized.
- Tillage was the next operation that was mechanized.
- Power thresher has been developed to thresh bulk production of rice, and it is estimated that more than 60 thousand are in use in the country.
- Entire maize shelling is done by power corn sheller.

Fig. 1.7 CA planters for small-scale farmers

- Rice transplanter, reaper, and combine harvester became cost-effective for farmer using these during sowing/planting and harvesting time (Fig. 1.7).

1.5.9 Integrated Crop-Livestock Production

Integrated crop-livestock production is a promising approach for enhancing overall productivity and profitability through recycling of the farm by-products and efficient utilization of available resources (Fig. 1.8). This integrates natural resources into farming system to achieve maximum replacement of off-farm inputs. Crop residues are substantial component of livestock diets. It can generate rural employment opportunities to the farming sector throughout the year and provide better economic and nutritional security. This can go a long way to uplift rural life through increased income.

The advantages of integrated system include increased productivity, profitability, and soil fertility; provide balanced diet, recycling of nutrients, generation of employment, and money flow round the year; and solve energy crisis.

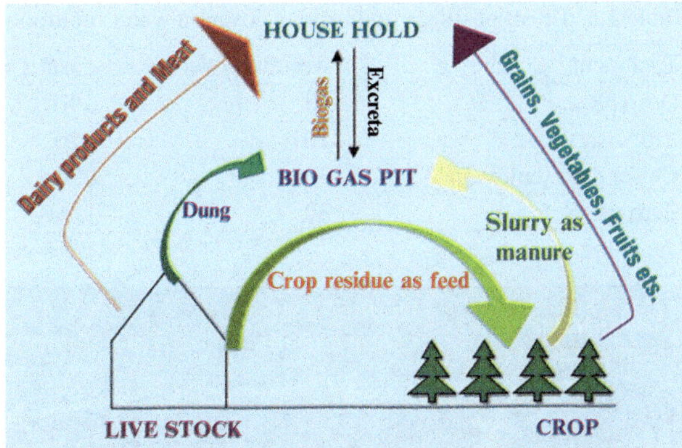

Fig. 1.8 Integrated crop-livestock production

Table 1.4 Effect of tillage on water erosion

Tillage type	Soil loss (t/ha/yr)	
	Brazil (Sidiras)	Paraguay (Venialgo)
Conventional tillage	68.2	22.9
Vertical tillage	55.0	–
Zero tillage	6.9	0.5

1.6 Benefits

The several benefits of sustainable intensification are as follows:

- Resembles a natural ecosystem.
- Soil is continuously covered.
- Builds soil organic carbon.
- Fertilizer requirement is reduced.
- No turning of the soil, less use of energy (30–40 % less use of energy).
- Helps in pest and weed control.
- Yields are comparable to conventional agriculture.
- Higher yields especially in drought years.
- CA adds up to 1 mm soil per year.
- Organic matter increase at about 0.1–0.2 % per year until reaching a saturation.
- Production costs can be reduced.
- Different rooting systems for more efficient use of soil nutrients.
- Soil structure more stable.
- Drudgery is reduced.
- Planting area can be expanded with existing resources.
- Soil erosion and degradation stopped/reversed (Table 1.4).
- Less labor requirement.
- Machinery use efficiency is increased.

Table 1.5 Effect of tillage and residue cover on water infiltration

Treatment	Residue cover (%)	Water runoff (%)	Soil loss (t/ha)
Deep plow, disk	0	90	28.5
Rip, disk	10	70	6.7
Chisel plow, cultivator	30	34	1.6
Zero tillage	80	6	1.0

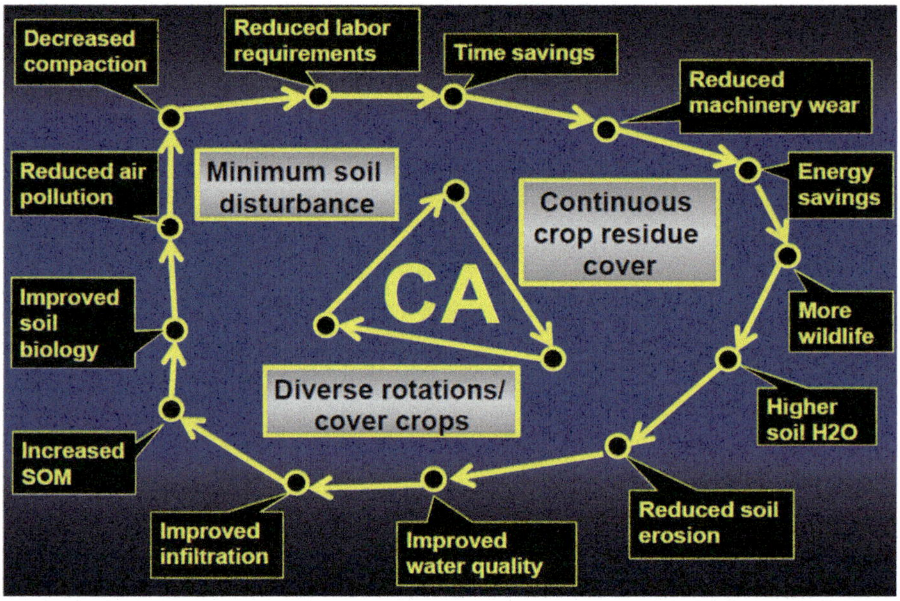

Fig. 1.9 Benefits of conservation agriculture

- Mitigation through emission reductions (fuel, NOx, CH_4).
- Mitigation through carbon sequestration up to 0.2 t of C/ha/yr.
- Adaptation through better drought tolerance.
- Adaptation through better water infiltration (less flooding) (Table 1.5).

The overall benefits of sustainable intensification are summarized in Fig. 1.9.

1.7 Conclusions

The biological and environmental consequences of green revolution are increasingly apparent. There is a need to integrate features of traditional agricultural knowledge with more ecologically designed agricultural knowledge into the intensification process. There is an indication of the renewed interest in ecologically designed systems such as agro-forestry, intercropping, and integrated crop-livestock systems. There is also an increasing interest in integrated nutrient—organic matter management and pest management approaches. The adoption of "precision agriculture," an emerging technology that facilitates application of inputs differentially across fields to match crop demands, allows a technological step toward increased efficiency.

Capacity building for farmers through technical support and providing on-hand learning opportunities through farmer field schools to improve the skills of all stakeholders are the need of the hour. This will require coordinated efforts of extension workers, researchers, and policy makers to enhance the adaptation of sustainable intensification. The SI agricultural system can only be able to feed more than nine billion people by 2050 in a sustainable way.

References

Altieri MA (1994) Biodiversity and pest management in agro-ecosystems. Haworth Press, New York

IPCC (2007) Climate change 2007: synthesis report. An assessment of the Intergovernmental Panel on Climate Change. Geneva, Switzerland

Murray WJ (2012) Sustainable crop production intensification. In: Barney ST (ed) Proceedings of the international scientific symposium on biodiversity and sustainable diets united against hunger, FAO Headquarters, Rome, Italy, pp, 66–74

Pimentel D (1996) Green revolution, agriculture and chemical hazards. Sci Total Environ 188(Suppl):S86–S98. doi:10.1016/0048-9697(96)05280-1

Pretty J, Toulmin C, Williams S (2011) Sustainable intensification in African agriculture. Int J Agric Sustain 9(1):5–24

Robertson GP (1997) Nitrogen use efficiency in row crop agriculture: crop nitrogen use and soil nitrogen loss. In: Jackson L (ed) Ecology in agriculture. Academic Press, New York, pp 347–365

Part II
Minimum Soil Disturbance

Chapter 2
Conservation Tillage

Touch the earth lightly, use the earth gently, nourish the life of all the world in our care.

Shirley Erina Murray (1992)

Abstract Conservation tillage is defined as a form of non-inversion tillage system that conserves natural or other resources, reduces loss of soil or moisture, and retains protective amounts of residue mulch (more than 30 %) on the soil surface. It holds great potential for the sustainability of agricultural productivity and the environment by preserving soil quality, reducing production costs, conserving soil and water, reducing herbicide, and weeding labor input costs and greenhouse gas emissions. Conservation tillage practices are technically appropriate, economically viable, socially acceptable, and environmentally non-degrading.

Keywords No-till • Strip-till • Ridge-till • Crop residue mulch • Reduced-till • Conventional till

2.1 Introduction

Intensive tillage is any form of tillage that degrades and fractures the natural soil structure. It destroys or disturbs the ecosystems of soil fauna so important for nutrient cycling. Tillage moves the soil downslope via tillage erosion. Intensive tillage loosens the soil and buries the crop residue, allowing the soil to dry, setting up the system for severe erosion with the next high-intensity rainfall event. The purposes of intensive tillage include:

- Aeration is improved due to loosening of the soil.
- Crop residues are incorporated into soil.
- Enhanced root growth and crop productivity.
- Favorable environment is provided for seed placement, germination, and crop growth.
- Fertilizer and lime can be incorporated into the soil.

P.P. Reddy, *Sustainable Intensification of Crop Production*,
DOI 10.1007/978-981-10-2702-4_2

- Pests and disease can be controlled. Burying infested stalks controls maize borer, while plowing infested stubble of wheat controls Hessian fly.
- Preparation of seedbed, which is required especially when planting costly, small-seeded crops such as clover and lettuce.
- Seedbed should be free of clods that prevent proper seedling emergence.
- Soil is loosened; soil crusts are broken, thereby increasing water infiltration and storage.
- Soil should compact enough around the seed for good soil–seed contact.
- Soil surface is smoothened.
- Weed control.

2.1.1 Negative Aspects of Soil Tillage

- Brings rocks to surface
- Costs money (fuel, equipment, maintenance)
- Destroys soil tilth
- Increases erosion
- Increases evaporation losses
- Increases runoff
- Promotes soil crusting
- Reduces biological activity (e.g., earthworms)
- Reduces organic matter content
- Takes time

2.2 Conservation Tillage

The portion of previous crop residues that are left on soil surface is called conservation tillage. The intensity of tillage and surface residues is indirectly related, i.e., the more the intensity, the less the residues, while the less the intensity, the more the residues (Fig. 2.1). Conservation tillage is any tillage system with primary objective of reducing soil and water loss. It is also called no-tillage, minimum tillage, reduced-tillage, or crop residue management, depending on the relative quantity of crop residues and different levels of incorporation (top removed, root stubbles left; top and root stubbles left; and nothing removed). It is a generic term which includes many varied tillage systems that leave more than 30 % crop residue cover after planting. It can include no-till, minimum tillage systems, mulch tillage, strip tillage, and ridge tillage, as long as these systems leave more than 30 % residue cover after planting.

Residue limit of 30 % was established as a result of the relationship between residue cover and inter-rill erosion. An increase from 0 to 30 % residue cover results in a 70 % reduction of inter-rill soil loss (Figs. 2.2 and 2.3).

Fig. 2.1 Conservation tillage with more than 30 % residue cover

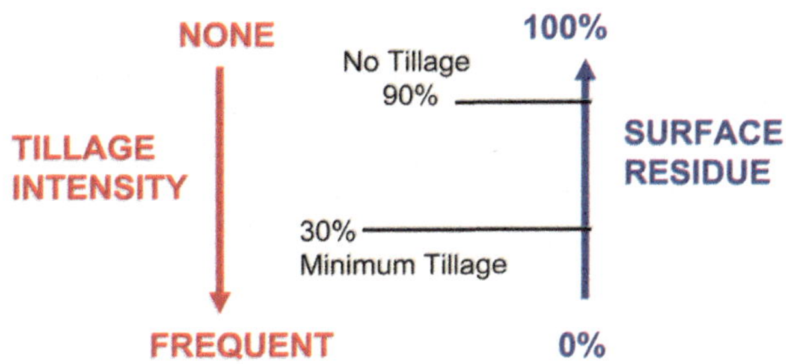

Fig. 2.2 Effect of tillage intensity on surface residue

2.3 Area

Worldwide, the conservation tillage is being currently practiced in 125 million hectares. The major countries adopting conservation tillage include the USA, Argentina, Brazil, Australia, Canada, Russia, China, Paraguay, and Kazakhstan (Table 2.1).

The major continents that are following conservation agriculture include South America, North America, and Australia and New Zealand. The continent-wise area under conservation agriculture is presented in Table 2.2.

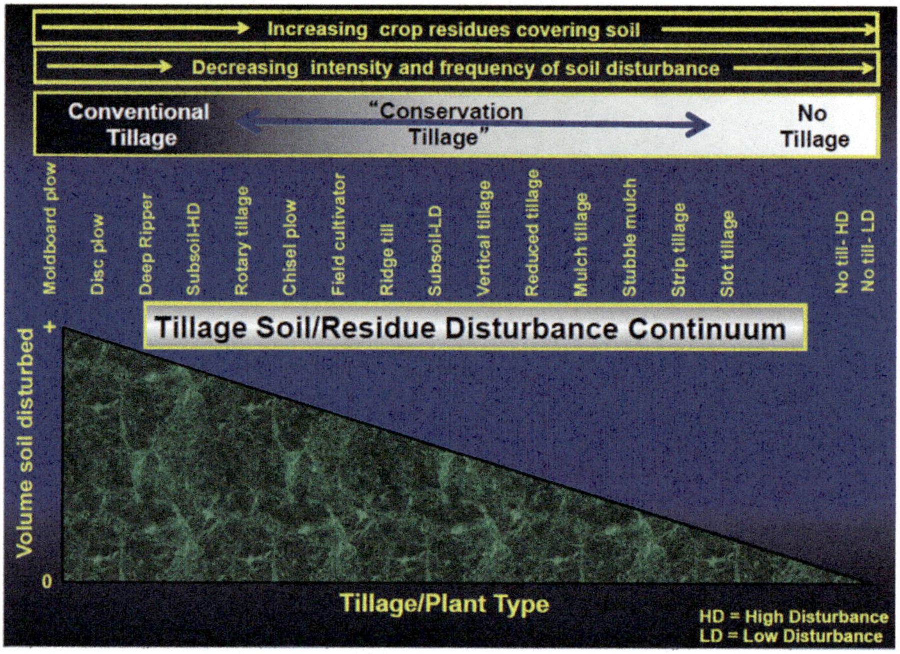

Fig. 2.3 Effect of tillage on soil disturbance

2.4 Tillage Systems

2.4.1 Goals

The three goals of a tillage system are improving soil physical properties, crop residue management, and weed control.

2.4.1.1 Soil Physical Properties

The soil physical properties such as structure, temperature, and moisture are modified with tillage. The preplant tillage of seedbed loosens soil, increases aeration, and provides a suitable medium for plant growth.

2.4.1.2 Crop Residue Management

After harvesting, the crop residues such as leaves, stalks, and roots should not be incorporated into the soil, but allowed to cover the soil surface. The quantity of crop residues produced depends on the crop, its growth, and harvesting type. The amount of crop residues left on the surface soil depends on different tillage systems.

Table 2.1 Extent of adoption of conservation agriculture worldwide (countries with >100,000 ha)

Country	CA area (ha)	Country	CA area (ha)
USA	26,500,000	South Africa	368,000
Argentina	25,553,000	Venezuela	300,000
Brazil	25,502,000	France	200,000
Australia	17,000,000	Zambia	200,000
Canada	13,481,000	Chile	180,000
Russia	4,500,000	New Zealand	162,000
China	3,100,000	Finland	160,000
Paraguay	2,400,000	Mozambique	152,000
Kazakhstan	1,600,000	United Kingdom	150,000
Bolivia	706,000	Zimbabwe	139,300
Uruguay	655,100	Colombia	127,000
Spain	650,000	Others	409,440
Ukraine	600,000	**Total**	**124,794,840**

FAO (2011)

Table 2.2 Area under conservation agriculture by continent

Continent	Area (ha)	% of total
South America	55,464,100	45
North America	39,981,000	32
Australia and New Zealand	17,162,000	14
Russia and Ukraine	5,100,000	4
Asia	4,723,000	3
Europe	1,351,900	1
Africa	1,012,840	1
World total	**124,794,840**	100

2.4.1.3 Weed Control

The preplant tillage destroys young weed seedlings and reduces weed pressure during growing period. The post-plant tillage destroys and buries emerging weed seedlings.

2.4.2 Different Types

The five types of tillage systems include:

- No-till/strip-till (>30 % residue)
- Ridge-till (>30 % residue)

Fig. 2.4 Strip tillage farming system

- Mulch-till (>30 % residue)
- Reduced-till (15–30 % residue)
- Conventional till/intensive-till (<15 % residue)

2.4.2.1 No-Till (>30 % Residue)

- Other terms used to describe no-till include direct seeding, slot planting, zero-till, row-till, and slot-till.
- Soil is left undisturbed from harvest to planting except for strips up to one third of the row width (Fig. 2.4) (MDA 2011). These strips may involve only residue disturbance or may include soil disturbance.
- Crops are planted and drilled in the narrow tilled strip (10–30 cm) using disk openers, coulters, row cleaners, in-row chisels, or rototillers.
- Weed control is accomplished primarily with herbicides. Cultivation may be used for emergency weed control.
- It leaves about 50 % crop residue on the soil surface.

2.4.2.2 Ridge-Till (>30 % Residue)

- Soil is left undisturbed from harvest to planting except for strips up to one third of the row width.
- Planting is completed on the ridge with sweeps, disk openers, coulters, or row cleaners and usually involves the removal of the top of the ridge (Fig. 2.5) (MDA 2011).

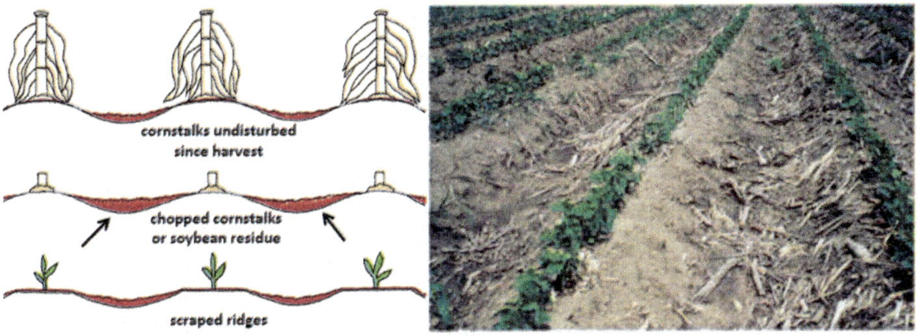

Fig. 2.5 Ridge-tillage farming system

Fig. 2.6 Mulch-tillage farming system

- Crop residues (65 %) are left on the surface between ridges.
- Weed control is accomplished with herbicides (frequently banded) or cultivation.
- Ridges are rebuilt during row cultivation.

2.4.2.3 Strip-/Mulch-Till (>30 % Residue)

- Full-width tillage involving one or more tillage trips which disturb the entire soil surface. Sub-tillage equipments such as chisels, field cultivators, disks, sweeps, or blades are used.
- The primary mulch tillage with chisel plow loosens the soil, but do not invert it. It leaves 50–80 % residue cover over the soil (Fig. 2.6) (MDA 2011).
- Light disking later reduces crop residue cover to 30–50 %
- Done prior to and/or during planting.
- Weed control is accomplished with mulch-till.

Fig. 2.7 Conventional tillage (primary) farming system

2.4.2.4 Reduced-Till (15–30 % Residue)

- Full-width tillage involving one or more tillage trips which disturb the entire soil surface.
- Done prior to planting. There is 15–30 % residue cover after planting or 500–1000 pounds per acre of small grain residue equivalent throughout the critical wind erosion period.
- Weed control is accomplished with herbicides or row cultivation.

2.4.2.5 Conventional Till/Intensive-Till (<15 % Residue)

- It consists of primary and secondary tillage. The soil is broken and crop residues are buried in primary tillage which is carried out by plow that inverts the soil (Fig. 2.7). The operation that breaks up the soil into finer particles to prepare a fine seedbed is accomplished in secondary tillage carried out by disking, and harrowing.
- Full-width tillage which disturbs the entire soil surface and is performed prior to and/or during planting. There is less than 15 % residue cover after planting.
- Generally involves plowing or intensive (numerous) tillage trips.
- Weed control is accomplished with herbicides or row cultivation.

In summary, the comparison of different types of tillage systems is presented in Table 2.3.

Table 2.3 Comparison of different types of tillage systems

System	Typical field operations	Major advantages	Major disadvantages
No-till	Plant into undisturbed surface	Maximum erosion control	No incorporation. Increased dependence on herbicides. Some limitations with poorly drained soils, especially with heavy residue. Slow soil warming
	Post-emergent weedicide spray as needed	Soil moisture conservation. Minimum fuel and labor costs	
Ridge-till	Chop stalks (on furrow irrigation)	Excellent erosion control if on contour	No incorporation
	Plant on ridges	Well adapted to wide range of soils	Narrow row soybeans and small grains not well suited. No forage crops
	Cultivate for weed control and to rebuild ridges	Excellent for furrow irrigation. Ridges warm up and dry out quickly	Machinery modifications required
		Low fuel and labor costs	
Strip-/mulch-till	Fall strip-till	Clears residue from row area to allow preplant soil warming and drying	Cost of preplant operation. Strips may dry too much, crust, or erode without residue. Not suited for drilled crops
	Plant on cleared strips. Post-emergent weedicide spray as needed	Injection of nutrients directly into row area	
		Well suited for poorly drained soils	Potential for nitrogen fertilizer losses
Reduced-till/chisel plow	Fall chisel	Less erosion than from cleanly tilled systems and less wind erosion than fall plow or fall disk because of rough surface. Well adapted to poorly drained soils	Little erosion control
	One or two spring diskings or field cultivations	Good to excellent incorporation	High soil moisture loss. Medium to high labor and fuel requirements
	Plant		
	Cultivate		
Moldboard plow	Fall or spring plow; one or two spring diskings or field cultivations; plant; cultivate	Suited for poorly drained soils. Excellent incorporation. Well-tilled seedbed	Major soil erosion. High soil moisture loss. Timeliness considerations. Highest fuel and labor costs

2.4.3 Effect on Crop Yield

On most soils, no-till yields should be similar to yields obtained with tillage. No-till should outyield tilled crops in areas where drought stress is a problem (5–15 % increase in yield), due to the water conserved by the mulch cover (Table 2.4).

Table 2.4 Effect of tillage on grain yield and net returns in different crops

	Grain yield (t/ha)		Net returns (Rs/ha)	
Crop	No-tillage	Conventional tillage	No-tillage	Conventional tillage
Pigeon pea (early)	1.15	0.90	13,900	8181
Maize (*kharif*)	2.60	2.35	2350	358
Maize (*Rabi*)	4.25	3.66	8213	4215
Pea	2.28	1.95	21,840	11,850
Lentil	1.70	1.50	21,800	1750
Chickpea	2.00	1.86	23,200	–
Lathyrus	1.45	1.31	12,050	20,830
Mung bean (summer)	0.45	0.39	3738	10,030

2.5 Benefits and Limitations

2.5.1 Benefits

- Most of the environmental benefits of no-tillage are due to the mulch cover at the soil surface.
- Soil improvement with no-till takes years. Continuous no-till is recommended.
- No-till affects many other aspects of crop production (nutrient, weed and pest dynamics, residue distribution) that needs to be integrated into a system approach. Crop rotations and cover crops are central to this system approach.
- Helps in early planting.
- Reduces soil erosion by water and wind to the extent of 40–50 % and thereby reduces soil degradation (Fig. 2.8) (Table 2.5).
- Organic matter is doubled in topsoil (soil fertility increased and soil erosion reduced).
- Soil structure improvement.
- Reduces soil compaction.
- Increases C sequestration.
- Increases fertilizer use efficiency.
- Reduces trips over field, fuel saving, and less compaction and saves time.
- Reduces cost of cultivation.
- In conservation tillage, lateral root spread and total root mass are higher as compared to conventional plowing (Fig. 2.9).
- Reduces fuel consumption and thereby CO_2 emission is also reduced.
- Reduces labor costs.
- Healthy soil.
- Improves air/water quality.
- Atmospheric nitrogen fixation.
- Reduces CO_2 release from soil.
- Nitrate recapture.

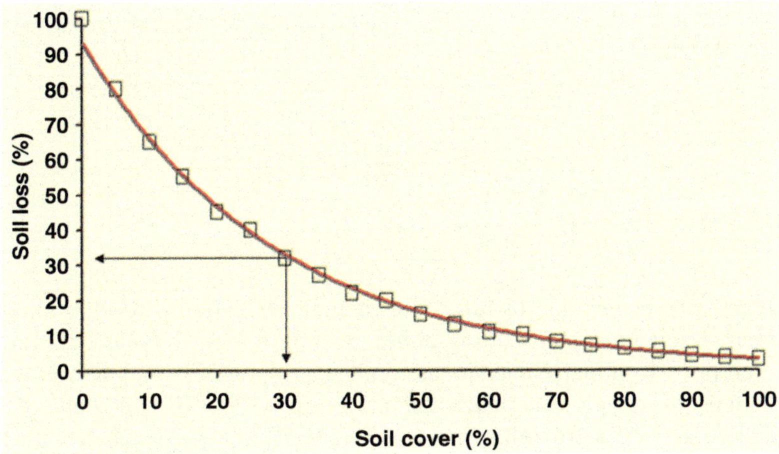

Fig. 2.8 Effect of soil cover on soil loss

Table 2.5 Effect of soil surface cover on soil erosion

Soil surface cover (%)	Reduction in soil erosion (%)
10	30
30	65
50	83
70	91

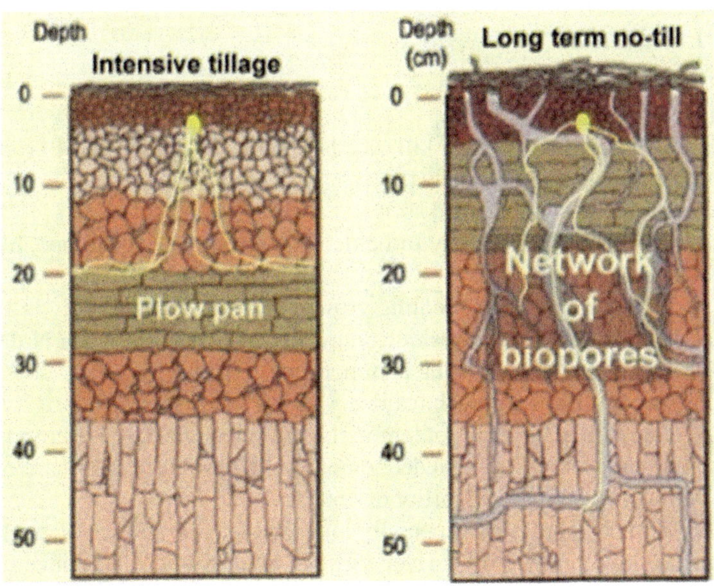

Fig. 2.9 Total root mass and lateral root spread are greater in conservation tillage compared to conventional plowing

Table 2.6 Effect of soil surface cover on water evaporation

Soil surface cover (%)	Water evaporation (%)
0	100
20	78
40	67
60	61
80	58

- Soil remains cooler due to reflecting of light by the crop residue mulch and increased soil moisture. Cutting and spreading the cut biomass of sun hemp intercropped with maize and spreading the cut biomass as mulch improved soil moisture conservation at maize harvest (+1.63–1.94 %) and yield of maize (12.0 %) as well as of following wheat (13.8 %) crop compared with the no mulching (Singh et al. 2011). Cooler soil benefits production in warm climates and conserves soil moisture especially in topsoil (less evaporation and improved water-holding capacity) (Table 2.6) (Schertz 1994).
- Surface water quality improvement (erosion and runoff reduced).
- Pest management (favors beneficials by providing habitat for predators).
- Allows several crops per season.
- Benefits earthworms.
- Increases wildlife habitat.
- Yield enhancement.

2.5.2 Limitations

- Often weed pressure is severe.
- Herbicides have to be used to kill weeds. Herbicide-tolerant GM crops can be used to overcome weed problems.
- Delay in drying of waterlogged soils.
- In north, soil warming is likely to be delayed during spring. Hence, planting is delayed.
- Seed germination and early seedling growth are reduced especially in spring.
- Availability of N from crop residues may be affected resulting in N deficiency. Hence, C/N ratio of crop residue is important.
- Difficult to plant through much residue.
- Surface irrigation may be a problem in the presence of excessive crop residues.
- Soil acidity may increase due to decomposition of residues and release of organic acids, which can affect availability of nutrients.
- Risk of toxins is increased (especially injurious to early growth stages) by leaving crop residue on the soil surface through allelopathy (e.g., maize on maize, wheat on wheat).

- Pest and disease problems may increase. Crop residues may provide habitat or hiding places for crop pests (cutworms). Cool temperature is likely to aggravate disease problems. Dependence on pesticides for pest and disease management is increased (increases pesticide use by 15–40 %).

2.6 Integration with Other Practices

There is a need for integration of conservation tillage with other farming practices in order to increase crop and livestock production and productivity in sustainable manner. The conservation tillage can be integrated with eco-friendly pest and weed management strategies, including biological, physical, and chemical methods to decrease the threat of pesticide resistance. The conservation tillage farming can be made more appealing to the farmers by integrating with growing a diverse selection of economically viable crops. Likewise, there is a need to combine conservation tillage with crop rotation to produce them sustainably.

2.7 Conclusions

In the future, the significant transformation of intensive agriculture into a sustainable and productive endeavor can take place through the conservation tillage production system. The efforts of agricultural scientists (soil scientists, agricultural engineers), farmers, government personnel, machinery manufacturers, and members of industry associations can successfully diffuse this technology. The appropriate approach can include the following aspects that need to be emphasized:

- Based on cropping systems, adapted crops, and climate, tillage management regions for the country should be established.
- By using a wide array of crop production systems and crop diversity across tillage management regions that are compatible with smallholder farm operations, an adaptable model of conservation tillage-planting systems should be developed.
- Demonstration of aggressive, farmer-oriented, on-farm, participatory methodologies and practices should be taken up to sensitize all parties (extension personnel, farmers, government and nongovernment operators, researchers, and support service providers) to play their roles.
- Government policies should encourage training for farmers, financial subsidy for purchase of equipment, and support for local manufacturers.
- Procurement of suitable conservation tillage equipment regionally or nationally by tapping the financial resources.

- To capture agro-forestry practices such as environmental protection, soil management techniques, and socioeconomic and other concerns of smallholder farmers, a system approach to multidisciplinary and multi-sector tillage research network should be organized.

References

FAO (2011) CA Adoption Worldwide. FAO-CA website available online at: http://www.ao.org/ag/ca/6c.html

MDA (2011) Conservation Practices, Minnesota Conservation Funding Guide, Minnesota Department of Agriculture. Available at: http://www.mda.state.mn.us/protecting/conservation/practices/constillage

Murray SE (1992) "Touch the earth lightly". Worship & Song, No. 3129, Genesis 1:26–30; 28:10–17

Schertz DL (1994) Conservation tillage – A national perspective. In: Bauer PJ, Busscher WJ (eds) Proceedings of the southern conservation tillage conference for sustainable agriculture, pp 1.5. Columbia, South Carolina

Singh R, Sharma AR, Dhyani SK, Dube RK (2011) Tillage and mulching effects on performance of maize (*Zea mays*) – wheat (*Triticum aestivum*) cropping system under varying land slopes. Indian J Agric Sci 81(4):330–335

Chapter 3
Agricultural Soil Compaction

Abstract The soil type, crops grown, and weather conditions determine the effects of soil compaction on plant growth and yield. Soil compaction can significantly reduce yields due to N losses, decreased K availability, and lowered root respiration because of reduced soil aeration, under wet conditions. Water infiltration and storage, root growth, and soil volume explored by roots and crop yields are reduced due to excessive soil compaction. Axle load and soil moisture content are directly correlated with soil compaction. Soil compaction can be reduced through crop rotations and management, direct seeding, drainage improvement, and introduction of deep rooting plants, using in-row or bent-leg subsoilers that minimally disturb soil surface and using smallest vehicle possible for farm operations.

Keywords Soil compaction causes • Symptoms • Measurement • Effects • Management

3.1 Introduction

The reduction of soil volume (a simple reduction in pore space) due to external factors is called soil compaction. Soil compaction is defined as increase in soil bulk density or decrease in soil volume and porosity (Fig. 3.1) due to mechanical stress on soil (e.g., from traffic of farm machinery). It can also occur due to natural reconsolidation of soil. There are two types of compaction, namely, surface compaction and subsoil compaction. The compaction that occurs in the surface "plow layer" is called surface compaction, while the compaction that occurs as a result of a surface load below the plow layer is called subsoil compaction.

3.2 Causes

At present, increase in the size of farm equipment used to carry out various farm operations increases the risk of soil compaction. The agricultural soil compaction can take place due to frequent movement of farm machinery. Factors responsible for

P.P. Reddy, *Sustainable Intensification of Crop Production*,
DOI 10.1007/978-981-10-2702-4_3

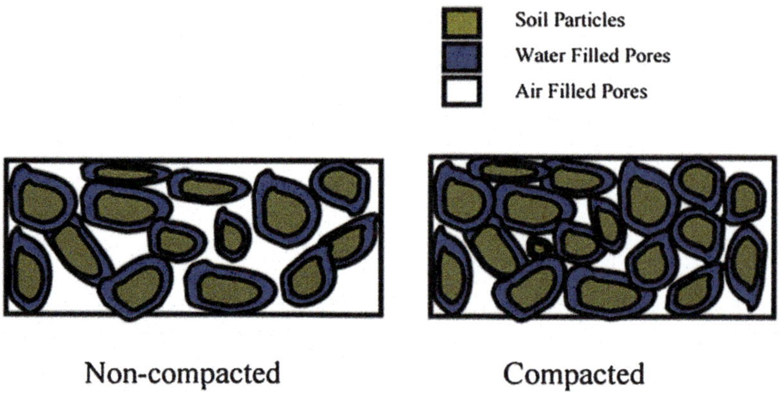

Fig. 3.1 Effect of soil compaction on pore space

compaction due to vehicular traffic include weak soil (soil density and moisture content effect) and excessive loads (size of vehicles, tire size, and number of passes are directly related to compaction). Soil tillage operations are also responsible for soil compaction.

Contact pressure of tires (expressed in psi) causes surface soil compaction. This can be reduced by increasing the tire width and height. The narrower the tires, the more the surface soil compaction, while the wider the floatation tires, the lesser the surface soil compaction (Fig. 3.2).

Axle load (expressed in tons) is responsible for subsoil compaction; the higher the axle (or wheel) load, the deeper the stress transmitted into the soil (Fig. 3.3). The subsoil compaction is reduced when axle load is reduced.

The soil compaction due to natural soil reconsolidation is more in well-graded soils compared to poorly graded soils. The raindrop impact is certainly a natural cause of soil compaction which causes soil crust (usually less than 1.25 cm thick at the soil surface) that may prevent seedling emergence.

The plow pans (hard pans caused by severe soil compaction) are caused by moldboard plow and disk plow and harrow just below the tillage level.

Soil moisture content also affects soil compaction (Fig. 3.4). Maximum soil compaction occurs at a moisture content approximating field capacity.

Significant livestock trampling resulting from livestock grazing in agricultural land is also a major cause of soil compaction. This is not affected whether the grazing is continuous or short term; however it is affected by the intensity of grazing.

3.3 Symptoms

The symptoms of soil compaction include ponding of water, degradation of soil structure (clod formation) (Fig. 3.5), more runoff, and resistance to tillage implements. The symptoms on plants include deficiency of nutrients (nitrogen and

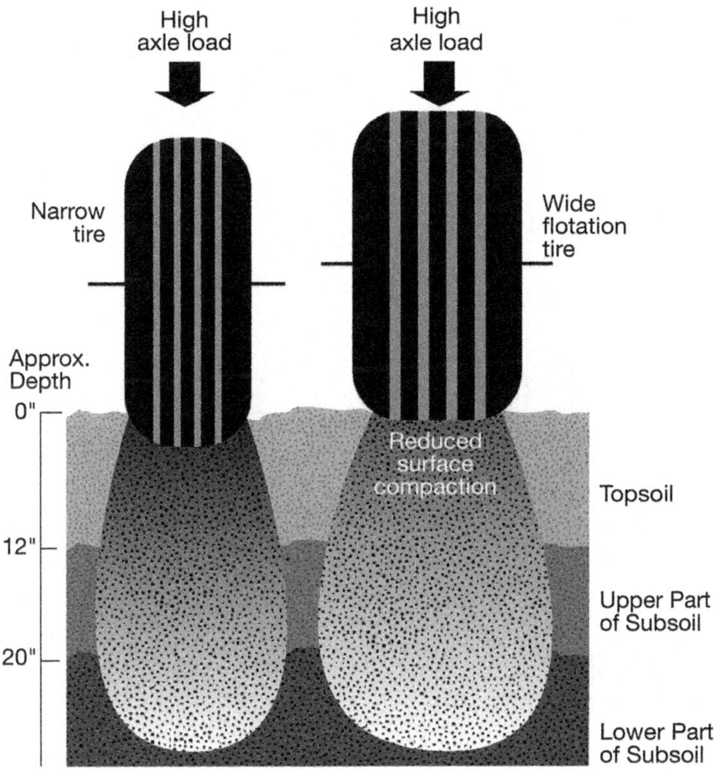

Fig. 3.2 Effect of contact pressure of tires on surface soil compaction

phosphorus) (Fig. 3.6), uneven and stunted growth (Fig. 3.7), reduced root growth (Fig. 3.8), and decrease in yield.

3.4 Effects

The following effects are observed due to soil compaction:

- Affects water-holding capacity of soil, water infiltration, water redistribution over landscapes, and roots ability for extraction of water (slow root growth).
- Enhanced erosion of soil.
- Poor soil drainage.
- Reduction in uptake of nitrogen and other nutrients. Results in increased denitrification rates because of limited aeration, increase in ammonia volatilization losses in surface-applied liquid manure (due to reduced infiltration), and reduced uptake of phosphorus and potassium (due to inhibition of root growth).

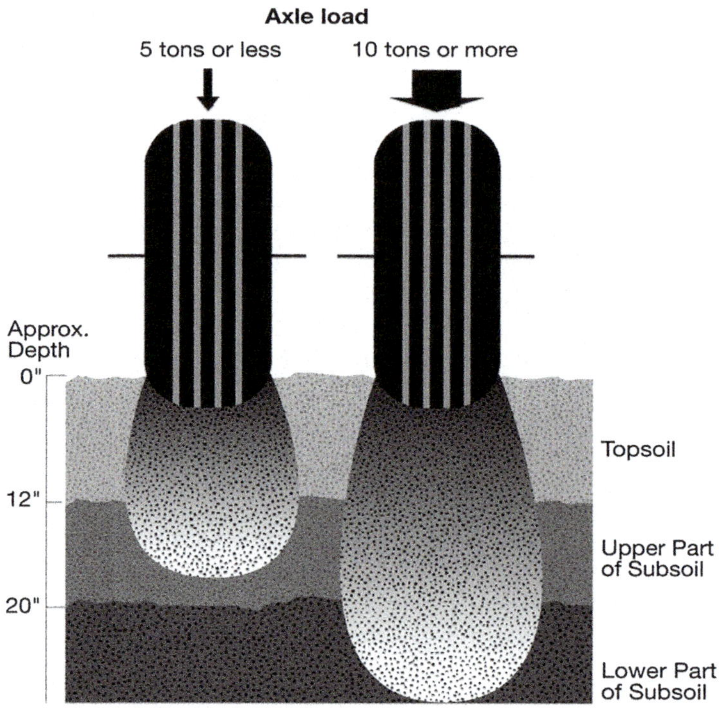

Fig. 3.3 Effect of axle load on subsoil compaction

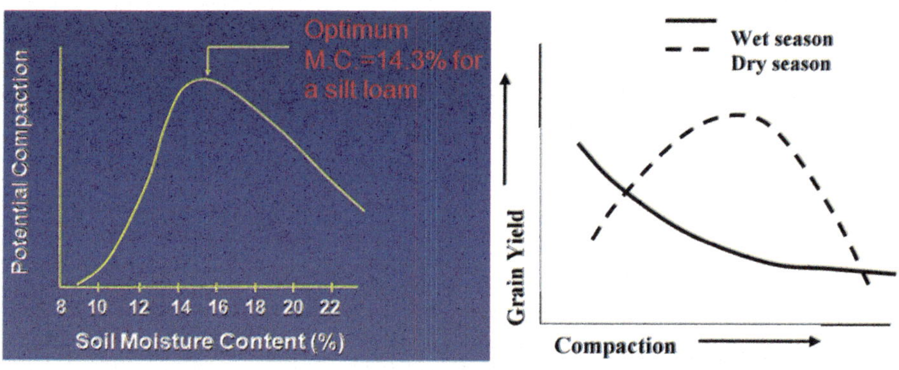

Fig. 3.4 Effects of soil moisture on soil compaction (*left*) and crop yield (*right*) (Soane and van Ouwerkerk 1994)

- Resistance to root penetration. Reduced root and shoot growth (Fig. 3.9).
- Soil is dense with low porosity.
- More energy is required for tilling soil.
- Decreased soil oxygen (oxygen is required for root respiration and active uptake).
- Unfavorable environment for soil organisms (especially earthworms).

Fig. 3.5 Clod formation due to soil compaction

Fig. 3.6 Nitrogen (*left*) and phosphorus (*right*) deficiency symptoms on maize due to soil compaction

- The raindrop impact is certainly a natural cause of soil compaction which causes soil crusting that may delay seedling emergence.
- Yield reduction (Table 3.1) (Fig. 3.10). The average yield loss of 15 % was noticed in tilled soils due to compaction during the first year.

3.5 Measurement

The agricultural soil compaction is directly affected by change in bulk density, porosity, and soil volume. Soil cone penetrometer can be used as a diagnostic tool to measure soil compaction (Fig. 3.11). The compaction categories based on penetrometer readings in field surveys are presented in Table 3.2.

Fig. 3.7 Uneven crop stands and stunting of growth of maize plants due to soil compaction

Fig. 3.8 Pancake root mass due to soil compaction

3.6 Management Strategies

The management strategies (avoiding or remediation) can be developed by understanding the causes of soil compaction. Avoiding soil compaction is easier, while remediation strategies are costly and may not correct the problem entirely.

3.6.1 Avoiding Compaction

Limiting the surface soil compaction as far as possible and avoiding subsoil compaction altogether should be the aim of compaction management.

Fig. 3.9 Effect of soil compaction on root and shoot development

Table 3.1 Effect of soil compaction on crop yield

Crop	Yield reduction in different categories of soil compaction		
	Moderate	Severe	Extreme
Wheat and soybean	<5 %	5–10 %	15–25 %
Tobacco and maize	5–10 %	10–20 %	30–50 %

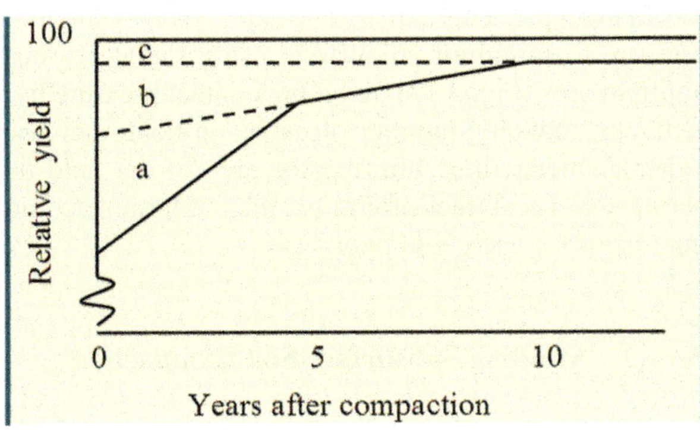

Fig. 3.10 Effect of soil compaction on yield (**a**) yield losses due to compaction of the topsoil, (**b**) yield losses due to compaction of the upper subsoil, (**c**) yield losses due to compaction of the lower subsoil) (Hakansson and Reeder 1994)

Fig. 3.11 Soil cone penetrometer used for measuring soil compaction

Table 3.2 Compaction categories based on field survey

Categories	% of penetrometer readings
Little or none	<30 % of penetrometer readings in field were 300 psi or greater
Slight	30–50 % of the penetrometer readings were 300 psi or greater
Moderate	50–75 % of the penetrometer readings were 300 psi or greater
Severe	75–100 % of the penetrometer readings were 300 or higher

3.6.1.1 Avoiding Surface Soil Compaction

The surface compaction can be avoided by reducing the contact pressure of tires to less than 35 psi. This can also be achieved by using floatation tires, reducing tire air pressure to minimum allowable, using radial-ply instead of bias-ply tires, using multiple tires (Fig. 3.12), reducing contact pressure by replacing singles by doubles or tracks, reducing number of passes in field, increasing footprint length by using larger diameter tires, limiting the area of the field that is impacted by providing permanent wheel tracks for movement of farm machinery traffic, and proper ballasting tractor for each field operation.

3.6.1.2 Avoiding Subsurface Soil Compaction

The subsurface compaction can be avoided by increasing the number of axles and reducing the axle load below 10 tons (Fig. 3.13).

Fig. 3.12 Multiple tires reducing contact pressure

Fig. 3.13 Effect of axle load on soil compaction

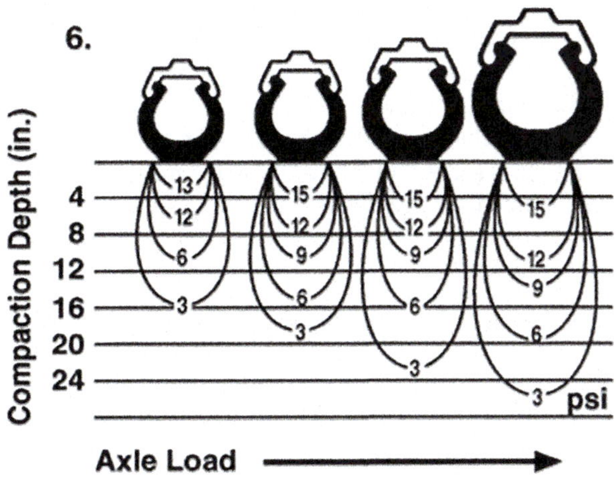

3.6.2 Avoiding Plow Pans

The plow pans can be avoided by using no-tillage, not driving a tractor wheel in the furrow, using a field cultivator instead of disk harrow, and using a chisel instead of moldboard plow.

3.6.3 Enhancing Soil Resistance

The soil resistance to compaction can be increased by building a soil ecosystem that has a permanent macropore system, increasing its organic matter content, and using cover crops with deep root systems that serve to reduce or remediate the effects of soil compaction (Fig. 3.14).

Fig. 3.14 Effect of taproot of canola (*left*) versus fibrous root of pea (*right*) on soil compaction (Photo: R Dunn)

3.6.4 Remediation Through Farm Machinery

The soil compaction can be remediated through tillage by subsoilers and chisel plow and recommending more than 30 % residue cover after planting which reduces erosion of soil and increases the soil quality. Tillage implements used should not reduce the soil residue covers below 30 %. Hence, subsoilers which have narrow shanks, coulters cut through residue in front of the shank, leave the soil in a condition that is ready to be planted, leave most residues at the soil surface, and do not create much surface disturbance (Fig. 3.15).

In conclusion, the soil compaction can be managed by using the following recommendations:

- Increase soil organic matter content (Fig. 3.16) and soil biota.
- Allow traffic only when soil moisture is low.
- Adopt conservation tillage system including cover crops.
- Use controlled traffic systems (research shows 10–15 % yield increase from controlled traffic).
- Use smallest vehicle possible for the job.
- Direct seeding.
- Crop rotations and management (alternating maize and soybean with fibrous and tap root systems, respectively).
- Introduction of deep rooting plants.
- Use in-row or bent-leg subsoilers that minimally disturb soil surface.
- Drainage improvement.

3.7 Conclusions

The soil structure, soil quality, and productivity everywhere in the world are threatened by soil compaction. The strength of the soil and soil compaction are directly correlated. The soil strength is reduced by higher soil moisture content. The

Fig. 3.15 The modern subsoiler for remediating subsoil compaction and retaining surface residue cover

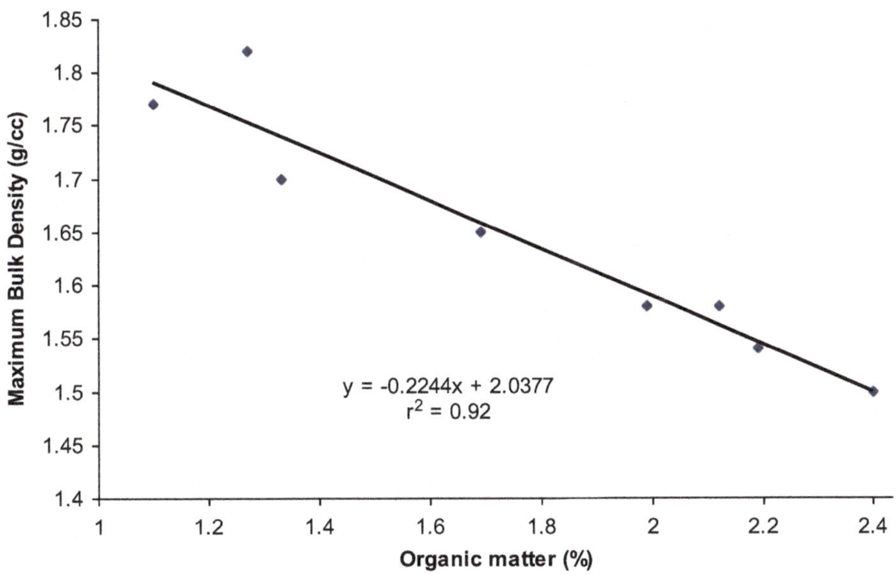

Fig. 3.16 Effect of soil organic matter on soil compaction

conservation agricultural practices include controlling wheel/track loads, using low tire inflation pressure, avoiding unnecessary field traffic, and subsoiling/chiseling to remove hardpan developed due to traffic and puddling. The interaction effects of landscape position and land management practices on soil compaction are needed to avoid site-specific compaction and to optimize soil management practices using precision agriculture principles and technologies.

References

Hakansson I, Reeder RC (1994) Subsoil compaction by vehicles with high axle load – extent, persistence and crop response. Soil Tillage Res 29:277–304

Soane BD, van Ouwerkerk C (eds) (1994) Soil compaction in crop production. Elsevier, ISBN 0-444-88286-3

Part III
Permanent Soil Organic Cover

Chapter 4
Cover/Green Manure Crops

Abstract For any annual cropping system to be sustainable, the cover/green manure crops can be considered as the backbone. They are important in sustainable farming because of nitrogen fixation, nutrient enhancement, enhancement of organic matter, improvement of soil structure, rooting action, soil and water conservation, soil microbial activity, and weed suppression. This chapter summarizes different aspects of cover/green manure intercrops such as types of cover/green manure crops, nutrient potential, benefits, and limitations. The role of cover/green manure crops in crop rotation systems in enhancing the production is also discussed.

Keywords Soil erosion • Fertility management • Soil quality • Microbial activity • Weed suppression • Pest management

4.1 Introduction

Cover/green manure crops are planted between main crops which are valuable tools for agricultural production and productivity (Fig. 4.1). The major fertility management tools for organic farmers include cover/green manure crops. They are cultivated in order to improve the physical, chemical, and biological properties of the soil and to cover the soil surface. Historically cover crops revolve around legumes.

The cover crops are multifunctional and mainly grown for nutrient, soil, water, and pest management (Fig. 4.2) (Magdoff and Weil 2004). The ideal cover/green manure crop should fix N from air, produce huge quantities of biomass within a short period, develop deep roots, absorb minerals in soil in large quantities, be easy to cultivate, not compete with main crop if grown in association, be easy to work in, provide palatable animal fodder, be tolerant to pests and diseases, suppress weeds, germinate and emerge quickly, tolerate adverse soil and climatic conditions, be competitive, and be with acceptable cost of cultivation. The suitability of various cover crops for different purposes is presented in Table 4.1.

In Iowa, winter cover crops grow between harvest and planting of annual cash crops such as corn and soybean. Rye cover crop is grown in spring. Some of the cover/green manure crops that can be grown in different seasons are presented in Fig. 4.3 and Table 4.2.

P.P. Reddy, *Sustainable Intensification of Crop Production*,
DOI 10.1007/978-981-10-2702-4_4

Fig. 4.1 Bean cover crop in maize

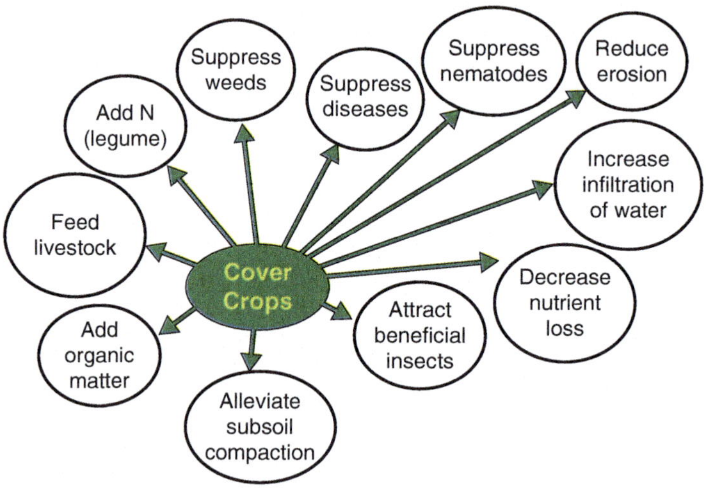

Fig. 4.2 Different functions of cover crops (Magdoff and Weil 2004)

4.2 Cover/Green Manure Crops

Both legumes and cereals can be used as cover crops. The legume cover crops include alfalfa, sweet clover (annual), winter pea, buckwheat, cowpea, crimson clover, flat pea, hairy vetch, red clover, white clover, etc. The cereal (grass) cover crops include ryegrass (annual), barley, rye, orchard grass, sorghum, Sudan grass, oats (spring), timothy, winter wheat, etc. The mixed legume and cereal cover crops can also be grown (perennial seeding—medium red clover, or white clover + annual or perennial ryegrass, or orchard grass, or timothy; annual seeding—spring oats + mammoth red clover, small grain + crimson clover, small grain + hairy vetch, barley + crimson clover).

Table 4.1 Suitability of various cover crops for different purposes

Purpose/objective	Suitable cover crops
N fixation	Winter pea, soybean, sun hemp, cowpea, lentil, vetches, clovers, chickling vetch
Nutrient cycling	Ryegrass (annual), small grains, brassicas, sunflower, sugar beet
Weed suppression	Sorghum–Sudan grass, brassicas, buckwheat, cereal rye
Biofumigation	Brassicas, sorghum–Sudan grass, sesame, sun hemp
Bio-plow (compaction reduction)	Alfalfa, sorghum–Sudan grass, brassicas, sugar beet, sweet clover, sunflower, radish, canola, turnip (and hybrids), soybean
Grazing	Ryegrass (annual), small grains, brassicas, sorghum–Sudan grass, clovers, turnips, rape, radish, lentils, oat, triticale
Residue cycling	Canola, rape, radishes, turnips, mustards

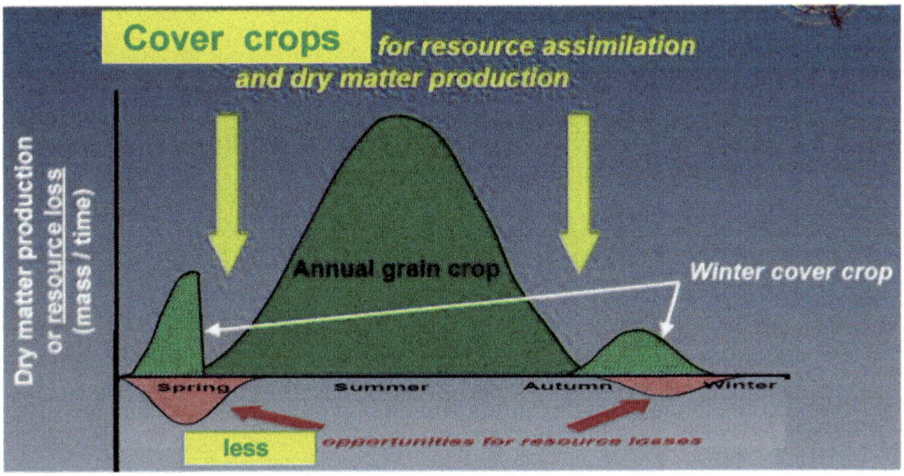

Fig. 4.3 Cover crops that can be grown in different seasons

Table 4.2 Cover/green manure crops for different seasons

Spring–summer species		Fall–winter species		Summer species	
Legumes	Nonlegumes	Legumes	Nonlegumes	Legumes	Nonlegumes
Mucuna	Pearl millet	Hairy vetch	Cereal rye	Cowpea	Sorghum
Pigeon pea	Rye	Crimson clover	Annual ryegrass	Soybean	Sorghum–Sudan grass
Jack bean				Velvet bean	
Sun hemp		Subterranean clover	Wheat	Sun hemp	Foxtail millet
Lablab			Barley	Buckwheat	Pearl millet
Forage peanut		Austrian winter pea	Oats		Japanese millet
Tephrosia		Soybean			Browntop millet
Leucaena					
Creeping indigo					

Table 4.3 Different types of cover/green manure crops

Green leaf manuring crops	In situ green manure crops
Gliricidia maculata	*Crotalaria juncea, C. mucronata, C. anagyroides*
Pongamia glabra	*Sesbania aculeata, S. rostrata*
Calotropis gigantea	*Phaseolus mungo, P. trilobus*
Tephrosia purpurea	*Melilotus indica, M. alba*
Indigofera teysmannii	*Vigna radiata, V. unguiculata*
Sesbania speciosa	*Cyamopsis tetragonoloba*
Ipomoea carnea	*Trifolium alexandrinum*
Cassia tora	*Lathyrus sativus*
Leucaena leucocephala	

4.2.1 Different Types

The cover/green manure crops are of two types, namely, green leaf manuring crops and in situ green manure crops.

4.2.1.1 In Situ Green Manure Crops

Green manure legume cover crops are grown for a short duration of about 60 to 80 days and incorporated into soil at the same site. This system is practiced in rice-growing regions of southern and northern parts of India in order to generate nutrient resources on site.

4.2.1.2 Green Leaf Manuring Crops

The foliage and tender parts of green leaf manuring crops are collected from nearby forests, shrubs, trees, and degraded lands and incorporated into the soil 15 to 30 days before planting of the main crops.

The different types of cover/green manure crops are presented in Table 4.3.

4.3 Nutrient Potential

Green manure cover crops such as *Sesbania aculeata* and *Crotalaria juncea* have larger accumulation of macro- and micronutrients in view of higher production of biomass and good nutrient content compared to other crops. Water hyacinth is another potential crop with higher nutrient content which can contribute 7 tons of dry matter to the soil. The availability of different nutrients in various commonly grown cover/green manure crops is presented in Table 4.4.

Table 4.4 Nutrient content in cover/green manure crops

Crop	Dry matter (t/ha)	Nutrient content						
		Major nutrients (kg/ha)			Micronutrients (g/ha)			
		N	P_2O_5	K_2O	Zn	Fe	Cu	Mn
Sesbania rostrata	**5.00**	**131.0**	**18.5**	**62.5**	**200**	**9840**	**180**	**1050**
S. speciosa	3.00	119.4	7.2	39.0	150	1440	132	330
Crotalaria juncea	**5.25**	**150.2**	**47.3**	**93.9**	**262**	**2467**	**100**	**2205**
Eichhornia crassipes	**7.00**	**198.1**	**63.0**	**125.3**	**350**	**3290**	**133**	**2940**
Trianthema spp.	2.50	58.5	7.5	28.7	75	4980	47	500
Parthenium hysterophorus	4.00	106.4	35.2	51.6	280	1880	76	640
Gliricidia maculata	3.60	125.6	7.9	46.8	108	1980	68	540

Singh et al. (1992) and Palaniappan (1992)

The effect of green manure cover crops on weed control, nitrogen fixing rate, biomass production, and consumption is presented in Table 4.5.

4.4 Benefits and Limitations

4.4.1 Benefits

The benefits of cover/green manure crops include:

- Prevention of soil erosion since soil particles are bound together by vigorous root system of cover/green manure crops (Fig. 4.4).
- Enhancement of soil tilth and structure.
- Cover soil surface which prevents water evaporation from soil resulting in conservation of soil moisture.
- Control insect pests resulting in savings on costs of insecticides. Provide habitat for beneficials by providing food sources (pollen and nectar) (Table 4.6).
- Management of crop diseases resulting in savings on costs of fungicides. Cover crops such as sorghum and ryegrass roots secrete toxins that control crop diseases due to allelopathic effects. Green manure cover crops can significantly reduce *Verticillium* wilt and scab of potato and *Phytophthora* in alfalfa due to the presence of indigenous strains of antibiotic producing *Streptomyces* species.
- Protect from nematode damage resulting in savings on costs of nematicides. Some cover crops such as triticale, cereal rye, forage rapeseeds, mustards, and oil seed radish have nematicidal effects (Table 4.7). Cover crops such as sorghum and ryegrass roots secrete toxins that control nematodes (Table) due to allelopathic effects.
- Suppression of weeds resulting in savings on costs of herbicides. Cover crops such as sorghum and ryegrass roots secrete toxins that control weeds (Table 4.8) due to allelopathic effects. Sweet clover (yellow blossom), sunflowers, black

Table 4.5 Effects of green manure cover crops on biomass production, nitrogen fixing rate, weed control, and consumption

Green manure cover crop	Fresh biomass produced (t/ha)	N fixation rate (kg/ha)	Weed control	Consumption
Cowpea/black bean (*Vigna unguiculata*)	≤35 (FAO 2011)	73–354 (Silva and Uchida 2000); 80 (Bunch 2003)	Good	Edible tender pods and dried beans
Rice bean (*Vigna umbellata*)	≤33 (Ecocrop 2007c)	80 (Bunch 2003)	Good	Edible tender pods and dried beans
Jack bean (*Canavalia ensiformis*)	40–50 (Ecocrop 2007a)	240 (Bunch 2003)	Good	Edible tender pods (dried beans are toxic)
Lablab bean (*Lablab purpureus*)	25–40 (Ecocrop 2007b)	220 (FAO 2011); 130 (Bunch 2003)	Good	Edible tender pods for garden varieties; edible dried beans for field varieties
Peanut/groundnut (*Arachis hypogaea*)	13.17 (Weerasinghe and Lathiff 2011)	72–124 (Silva and Uchida 2000)	Fair	Seeds readily consumed

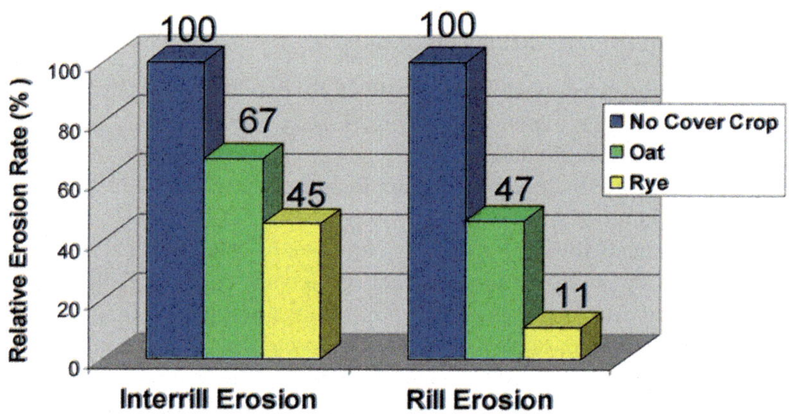

Fig. 4.4 Effect of cover crop on erosion rate in no-till soybean (Kaspar et al. 2001)

Table 4.6 Beneficial insects attracted to common cover crop species

Cover crop	Beneficial insects
Buckwheat	Parasitic wasps, ladybugs, tachinid and hover flies, lacewings
Clovers	Parasitic wasps, big eyed bugs, minute pirate bugs, ladybugs, tachinid flies, and aphid midges
Hairy vetch	Minute pirate bugs, ladybugs, predatory and parasitic wasps
Cereals	Ladybugs

Table 4.7 Root-knot nematode (*Meloidogyne chitwoodi*) management in potato using cover crops

Cover crop	Root-knot larval population/250 g soil	Culled tubers (%)
Wheat stubble (control)	3246	82
Sudan grass (cv. Trudan 8)	246	76
Rape seed (cv. Humus)	6	13

Ingham (1993)

Table 4.8 The allelopathic effects of cover crops on weed suppression

Cover crop	Weeds suppressed	Reference
Hairy vetch	Lambs quarters, yellow foxtail, yellow nut sedge, pitted morning glory	Teasdale and Daughtry (1993) and White et al. (1989)
Crimson clover	Pitted morning glory, wild mustard, Italian ryegrass	White et al. (1989)
Cereal rye	Lambs quarters, redroot pigweed, common ragweed	Barnes and Putnam (1986) and Masiunas et al. (1995)
Wheat	Morning glory, prickly sida	Liebl and Worsham (1983)
Velvet bean	Yellow nutsedge, chickweed	Hepperly et al. (1992) and Fujii et al. (1992)
Sorghum–Sudan grass	Annual ryegrass	Forney and Foy (1985)

mustard, white mustard, and rapeseed have also allelopathic effect. The weed suppression is also due to crop residue cover which prevents sunlight required for germination of small-seeded weed species.

- Stabilization of soil temperature (Fig. 4.5).
- Addition of organic matter (Table 4.9) improves soil fertility.
- Legume cover crops add of N (40–80 kg/ha) (Tables 4.10 and 4.11) and micronutrients to the soil, resulting in savings on fertilizers. Legume cover crops fix atmospheric N through *Rhizobium* (Fig. 4.6). There is no need to apply off-field organic matter.
- Deep-rooted cover crops mine leached nutrients from deeper soil layers (e.g., legumes). They break through compacted subsoils, decrease soil compaction, and act as biological plow.
- Improves soil drainage.
- Recycle nutrients.
- Increase in microbial activity. Organic matter serves as a substrate for the growth of microorganisms. Microbial breakdown of organic matter releases nutrients (Table 4.12) for plant growth.
- Useful in reclamation of saline/alkaline soils.
- Great affinity with mycorrhizal fungi which mobilize P and make it easily available to plants.
- Increase water infiltration in soil.

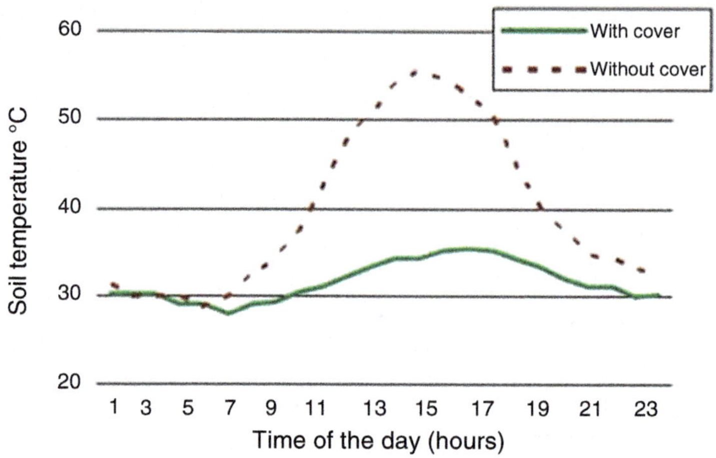

Fig. 4.5 Effect of cover crop *Mucuna* on temperature fluctuations at a soil depth of 3 cm in a cotton crop

Table 4.9 Effect of summer and winter cover/green manure crops on biomass production and accumulation of organic carbon (average of three seasons)

Cover/green manure crops	Dry matter[a] (kg/ha)	Organic carbon kg/ha	Organic carbon % of D.M.[a]
Summer species[b]			
Pigeon pea	9153	5153	56.30
Black-seeded Mucuna	7500	3911	52.15
Jack bean	7703	3863	50.15
Winter species[c]			
Oilseed radish	4771	2219	46.52
White lupine	4012	1925	47.97
Black oats	3680	1818	49.40
Hairy vetch	2942	1395	47.43
Black oats + hairy vetch	5440	2641	48.55
Black oats + white lupine	4259	2069	48.57

Derpsch and Florentín (1992)
[a]Dry matter—for summer species, data was taken from tissue analysis from Brazil
[b]Sown in association with corn and dry matter determined at approximately 7 months
[c]Sown after cotton and dry matter determined at approximately 4 months

Table 4.10 Nitrogen content in different legume cover crops

Legume cover crop	Tops (%N)	Roots (%N)
Beans	93	7
Peas	89	11
Clovers	68	32
Alfalfa	58	42

Table 4.11 Biomass and nitrogen production by legume cover crops

Legume cover crop	Biomass (tons/acre)	Nitrogen (lbs/acre)
Sweet clovers	1.8	120
White clover	1.0–3.0	80–200
Red clover	1.0–2.5	70–150
Winter peas	2.0–3.0	90–150

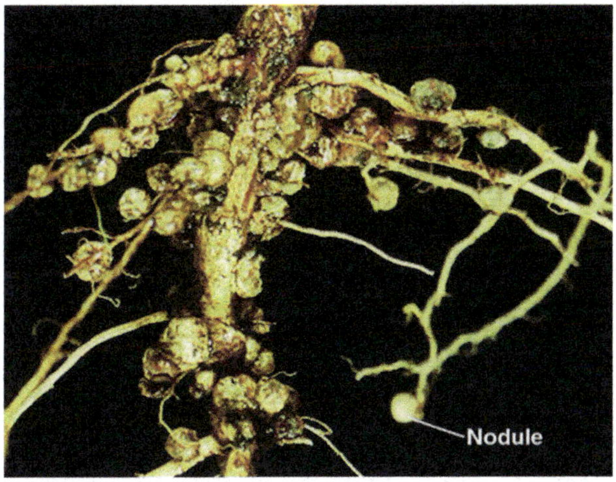

Fig. 4.6 Legume roots fix atmospheric nitrogen through *Rhizobium* nodules

Table 4.12 Effect of cover crops on biomass production and release of nutrients

Cover crop	Biomass (tons/acre)	Nutrients released (lbs/acre)		
		N	P	K
Clovers	4.2	110	20	130
Winter peas	4.2	150	20	160
Rye	5.6	90	20	110

- Can be easily grown on residual soil moisture or with less rainfall because of increased water use efficiency.
- Humus (decomposed organic matter) binds soil particles in stable aggregates.
- Produce large biomass due to fast and vigorous growth which results in nutrient mobilization from soil to vegetation and weed suppression. Initial fast growth provides quick soil cover for effective soil protection.
- Erosion and leaching losses of N are avoided (Kaspar et al. 2007) (Fig. 4.7) and thereby reduce pollution of water.
- Produce more foliage than wood resulting in easy decomposition of soil organic matter.
- Plant material has low C/N ratio leading to increased nutrient availability to the next main crop.

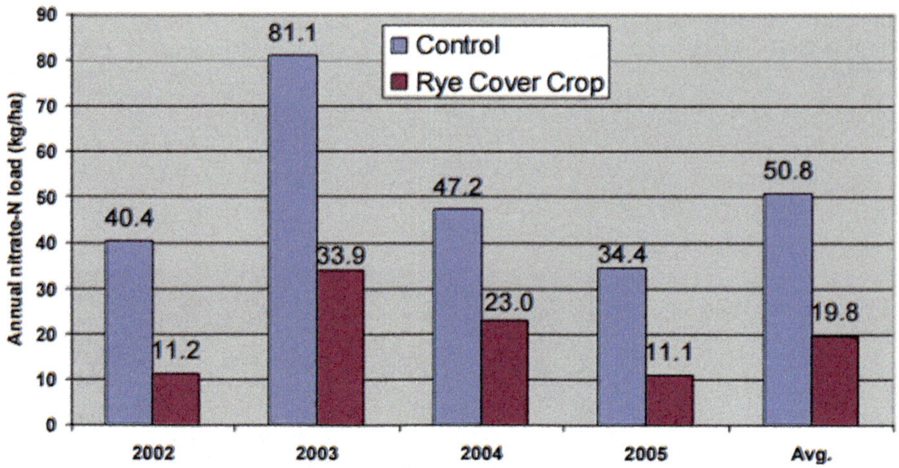

Fig. 4.7 Effect of rye cover crop on reduction of nitrate leaching in drainage water (Kaspar et al. 2007)

Table 4.13 Influence of green manure cover crops on yield and quality of rice

Treatment	Grain yield (t/ha)	Straw yield (t/ha)	Amylase content (%)	Crude protein content (%)
Control	5.373	7.248	25.2	7.8
FYM	5.934	7.717	26.4	8.3
Sesbania rostrata	6.374	7.717	26.4	8.3
Crotalaria juncea	6.142	8.098	27.2	8.3
CD (P = 0.05)	0.156	0.221	0.8	0.31

Hemalatha et al. (2000)

- Easy to terminate and cover the soil surface.
- Serves as fodder for livestock.
- Increase in yields, quality of the produce (Table 4.13) and economic returns.

4.4.2 Limitations

- Competition with main crops. Main crop yield may be affected (corn grain yields 4 to 8 bu/ac lower following rye cover crop).
- Costly to establish.
- Deficit of nitrogen may occur at the beginning of the growing period when decomposition of cover crops is taking place.
- Difficulty of planting into the crop residues.
- Elimination may be difficult.

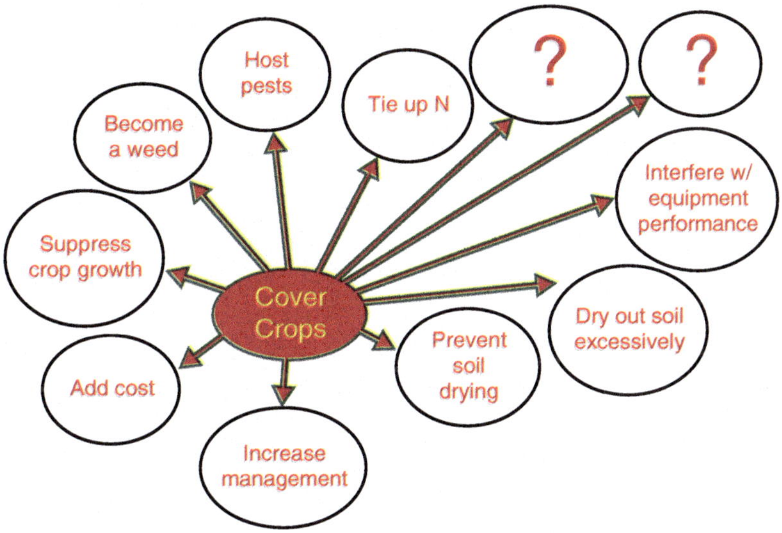

Fig. 4.8 Limitations of cover crops (Magdoff and Weil 2004)

- Not cash crops, income is lost.
- Possibility of becoming weeds (e.g., vetch).
- Require management skills.

In conclusion, the limitations of cover/green manure crops are depicted in Fig. 4.8. (Magdoff and Weil 2004).

4.5 Integration with Crop Rotation

The cover/green manure crops can be included into crop rotation systems during periods of the year when no cash crop is being grown. A rye cover crop can fit into corn/soybean rotation that provide ground cover in the fall and winter and scavenge residual nitrogen. Rye can also be used as mulch during spring to conserve moisture and to suppress early-season weeds for the following soybean crop. In order to provide nitrogen for corn the following spring, hairy vetch can be planted behind soybeans (Sullivan 2003).

4.6 Conclusions

A significant contribution has been made to sustainable agriculture by cover/green manure crops. The favorable outcomes due to cover/green manure crops include enhancing agro-biodiversity, high flexibility, and low input costs, increasing soil organic carbon, managing soil erosion, managing pests, and reducing risk. These benefits from cover/green manure crops make agriculture eco-friendly and economically viable system for many end users. By using cover/green manure crop residue as animal feed can fetch additional income to the farmer. The scope of cover/green manure crops can be expanded by integrating them with cropping system research.

References

Barnes JP, Putnam AR (1986) Evidence for allelopathy by residues and aqueous extracts of rye (*Secale cereale*). Weed Sci 86:384–390

Bunch R (2003) Adoption of green manure and cover crops: reversing degradation. LEISA Mag 19(4):16–18

Derpsch R, Florentín M (1992) La mucuna y otras plantas de abono verde para pequeñas propiedades. Asunción, Paraguay, MAG, Miscelánea No. 22, 44 pp

Ecocrop (2007a) *Canavalia ensiformis*. Food and Agriculture Organization of the United Nations. http://ecocrop.fao.org/ecocrop/srv/en/cropView?id = 609

Ecocrop (2007b) *Lablab purpureus*. Food and Agriculture Organization of the United Nations. http://ecocrop.fao.org/ecocrop/srv/en/cropView?id = 1311

Ecocrop (2007c) *Vigna umbellata*. Food and Agriculture Organization of the United Nations. http://ecocrop.fao.org/ecocrop/srv/en/cropView?id = 2152

FAO (2011) Grassland species profiles. *Vigna unguiculata* (L.) Walp. http://www.fao.org/ag/AGP/AGPC/doc/Gbase/data/pf000090.htm

Forney DR, Foy CL (1985) Phytotoxicity of products from rhizospheres of a sorghum-Sudan grass hybrid (*Sorghum bicolor* x *Sorghum sudanese*). Weed Sci 33:597–604

Fujii Y, Shibuya T, Yasuda T (1992) Allelopathy of velvet bean: its organic production—cover crops for organic farms 18 discrimination and identification of LDOPA as a candidate of allelopathic substances. Japan Agric Res Q 25:238–247

Hemalatha M, Thirumugan V, Balasubramanian R (2000) Effect of organic sources of nitrogen on productivity and quality of rice (*Oryza sativa*) and soil fertility. Indian J Agron 45(3):564–567

Hepperly P, Aguilar EH, Perez R, Diaz M, Reyes C (1992) Pigeon pea and velvet bean allelopathy. In: Rizvi SJH, Rizvi V (eds) Allelopathy: basic and applied aspects. Chapman and Hall, London, pp 357–370

Ingham RE (1993) Biology and control of nematodes in potato. Report to the Agriculture Research Foundation, Oregon Potato Commission, pp 18–36

Kaspar TC, Radke JK, Laflen JM (2001) Small grain cover crops and wheel traffic effects on infiltration, runoff, and erosion. J Soil Water Conserv 56:160–164

Kaspar TC, Jaynes DB, Parkin TB, Moorman TB (2007) Rye cover crop and gama grass strip effects on NO_3 concentration and load in tile drainage. J Environ Qual 36:1503–1511

Liebl RA, Worsham AD (1983) Inhibition of pitted morning glory (*Ipomoea lacunose* L.) and certain other weed species by phytotoxic components of wheat (*Triticum aestivum* L.) straw. J Chem Ecol 9:1027–1043

Magdoff F, Weil RR (2004) Soil organic matter in sustainable agro-ecosystems. CRC Press, Boca Raton

Masiunas JB, Weston LA, Weller SC (1995) The impact of rye cover crops on weed populations in a tomato cropping system. Weed Sci 43:318–323

Palaniappan SP (1992) Green manuring: nutrient potentials and management. In: Tendon HLS (ed) Fertilizers, organic manures, recyclable wastes and biofertilizers – components of integrated plant nutrition. Fertilizer Development and Consultant Organization, New Delhi, pp 52–71

Silva JA, Uchida R (eds) (2000) Plant nutrition management in Hawaii's soils, approaches for tropical and subtropical agriculture. College of Tropical Agriculture and Human Resources, University of Hawaii, Manoa

Singh B, Singh Y, Sadana US, Meelu OP (1992) Effect of green manure, wheat straw and organic manure on DTPA on extractable Fe, Mn, Zn and Cu in a calcareous sandy loam soil at field capacity and under water-logged conditions. J Indian Soc Soil Sci 40:114–118

Sullivan P (2003) Overview of cover crops and green manures. In: Williams P (ed) Fundamentals of sustainable agriculture, pp 1–16. ATTRA, National Center for Appropriate Technology, Fayetteville. http://www.attra.ncat.org/attra-pub/PDF/covercrop.pdf

Teasdale JR, Daughtry CST (1993) Weed suppression by live and desiccated hairy vetch. Weed Sci 41:207–212

Weerasinghe P, Lathiff MA (2011) Influence of groundnut residue on the growth and yield of subsequent rice crop in non-calcic brown soils. Regional Agricultural and Research and Development Centre, Aralaganwila. http://www.goviya.lk/agri_learning/Paddy/Paddy_Research/Paddypdf/A18.pdf

White RH, Worsham AD, Blum U (1989) Allelopathic potential of legume debris. Weed Sci 37:674–679

Chapter 5
Intercropping

Abstract The increased diversity in an agro-ecosystem can be brought about by intercropping (cultivating two or more crops at the same time in the same field). The main benefits of intercropping which increase crop yields include higher growth rate; better utilization of resources; ecological balance; increase soil conservation through greater ground cover; better lodging resistance for crops susceptible to lodging; insurance against crop failure; reduced damage by pests, diseases, and weeds; improvement in the quality and quantity of products; and greater financial stability. The main types of intercropping are mixed intercropping, relay intercropping, row intercropping, and strip intercropping. By using the plants from Leguminosae family in intercropping, the soil fertility can be increased due to the increasing amount of biological nitrogen fixation. Intercropping minimizes environmental impacts of agriculture through reduced fertilizer and pesticide requirements.

Keywords Agro-biodiversity • Intercropping • Mono-cropping • Sustainable agriculture • Crop mixtures

5.1 Introduction

A large portion of food production mainly comes from mixed cultures (Govinden et al. 1984) through several farming systems developed in tropical countries (Plucknett and Smith 1986). The key strategy for sustainable agriculture is restoring on-farm biodiversity through diversified farming systems that mimic nature (Scherr and McNeely 2008). The correct assembly of on-farm biodiversity can provide agro-ecosystems capable of maintaining soil fertility, regulating natural protection against pests, and sustaining productivity (Thrupp 2002). The on-farm biodiversity can be increased by intercropping (Malézieux et al. 2009).

Conventional mono-cropping system is responsible depletion of the natural resources and causes environmental pollution. In view of these and other problems with monoculture farming becoming more apparent, "sustainability" is the need of the hour, and interest in intercropping is gaining importance as part of the solution. Intercropping can be defined as simultaneous growing of two or more different

P.P. Reddy, *Sustainable Intensification of Crop Production*,
DOI 10.1007/978-981-10-2702-4_5

Fig. 5.1 Intercropping of The Three Sisters (maize, beans, and squash)

crops (e.g., grain + legume—nitrogen-fixing plant) at the same time in the same field or in alternate rows. Increasing the productivity per unit area is the objective of intercropping. Intercropping systems increases diversity in an agricultural ecosystem, ecological balance, and more utilization of resources, improves the quantity and quality of products, and reduces damage by pests, diseases, and weeds. In developing countries, intercropping is a common practice in agriculture system. In intercropping, the subsidiary crops are grown in between two widely spaced rows of main crops. The most common intercropping includes cereals and legumes, both for forage and for grain. The cereal–legume intercropping provides the nitrogen to the system by the fixation of atmospheric nitrogen by the legume. Examples of intercropping include green gram or red gram–cotton, sugarcane–soybean, and green gram or black gram–maize.

The history of intercropping, i.e., growing of "The Three Sisters," namely, maize, beans, and squash dates back to 5000 years. The spreading squash with big leaves covers the soil surface conserving moisture, suppresses weeds, and deters pests. The maize acts as a support to beans which contribute nitrogen to the soil for the growth of other crops. Together, "The Three Sisters" provide a balanced food (Fig. 5.1). At optimum plant densities, the maize yields were considerably higher in the mixture than in a pure stand. Considerable yield reductions in bean and squash are noticed when grown in mixture. The Land Equivalent Ratio (LER) was considerable higher (1.6) for the whole mixture than any of the pure stands (Table 5.1) (Amador 1980).

Table 5.1 Effect of intercropping on yield

Crop	Yield (pounds/acre)	
	Pure stand	Intercrop
Maize	1096	1533
Beans	544	98
Squash	383	71

5.2 Different Types

There are four types of intercropping systems, namely, mixed intercropping, relay intercropping, row intercropping, and strip intercropping (Vandermeer 1992; Sullivan 2003).

5.2.1 *Mixed Intercropping*

Simultaneous growing of two or more crops mixed together and then sown either in lines or they are broadcasted is called mixed intercropping. Since this system is not scientific based, it causes problem in performing all agricultural operations including harvesting of crops. In areas where climatic hazards (flood, drought, frost, etc.) are common and frequent, the mixed-intercropping system is followed. There is always the threat of crop failure. The crops may mature either together or at different times, even though they are sown at the same time.

This type of intercropping system is used for grass–legume in pastures in India (Fig. 5.2), forage sorghum with silage corn in Oregon, and corn, bean, and squash in Central America. The simultaneous growing of corn (16,000 seeding rate per acre equivalent to 67 % of the sole corn rate) and soybean (seeding rate of 135,000 per acre equivalent to 67 % of the sole bean rate) for silage in the same rows showed increased production and the highest profit per unit area in Canada.

5.2.2 *Relay Intercropping*

Simultaneous growing of two or more crops during part of the life cycle of each is called relay intercropping. When the first crop has reached its reproductive stage but before it is ready for harvest, a second crop is planted so that the first crop is harvested to make room for the full development of the second. In relay cropping, a legume crop, or relay crop, is seeded directly into the established first crop (e.g., alfalfa and red clover can be sown as relay crops into winter wheat and fall rye in the spring) (Fig. 5.3).

Some of the examples of relay intercropping include cassava, cotton, sweet potato, and sesban with corn, chickpea, lentil, and wheat with upland rice.

Fig. 5.2 Mixed intercropping of corn with climbing bean

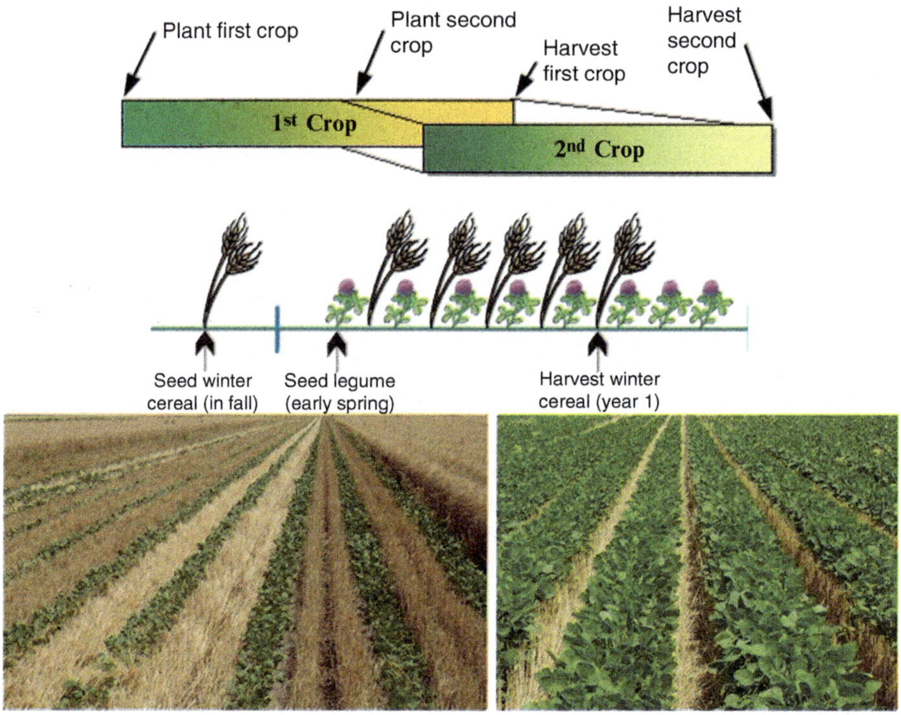

Fig. 5.3 Relay intercropping of winter cereal crop (wheat) with early-spring seed legume (soybean)

Relay intercropping of black medic (a self-seeding legume) can be taken up under a flax crop. The black medic forms low-growing living mulch under the flax crop canopy (Fig. 5.4). The black medic continues to grow even after the flax is harvested and set seed until the first killing frost.

The shallow (wheat)- and deep (cotton)-rooted crops can be relay intercropped in order to mine the nutrients both in shallow and deep soil layers (Fig. 5.5).

Fig. 5.4 Relay intercropping of black medic (a self-seeding legume) under a flax crop

Fig. 5.5 Relay intercropping of wheat with cotton

Table 5.2 Some crop combinations under relay intercropping in the USA

First crop	Second crop	Planting comments
Rice	Soybean	Seeding may be done by aerial broadcast
Small grain	Soybean or other legumes	The legume may be over seeded into the small grain. In seeding the first crop, skipping every third or fourth row facilitates planting of the second crop. Weeds, however, may be a problem in the skipped row
Maize, sorghum or millet	Soybean or other legume	The second crop may be banded between rows if seeded at the last cultivation of the first crop

Fig. 5.6 Row intercropping of corn with climbing bean

Some crop combinations under relay intercropping in the USA are presented in Table 5.2.

5.2.3 *Row Intercropping*

Simultaneous growing of two or more crops (with at least one crop planted in rows) either in alternate rows or one crop alternated with multiple rows of another crop is called row intercropping (Fig. 5.6).

The annual crops like corn, rice, and pineapple are commonly grown as intercrops between the rows of the perennial crops (coconut, mango, citrus, sapota, etc.) especially during the initial years after planting. This cropping system effectively utilizes the large vacant spaces in between rows and suppresses the weed growth during the early stage of the perennial crops.

In case of annual crops, the component crops are grown simultaneously. The pigeon pea and sorghum are normally planted together in India. After the harvest of sorghum plants, the pigeon pea plants start flowering. Similarly, velvet bean or cowpea is planted in standing maize crop at the last cultivation in the USA. The beans or peas use the corn stalks as natural trellis when the corn plants mature. Later, both the crops are harvested together.

Fig. 5.7 Strip intercropping of one broomcorn row with two bush bean rows

5.2.4 Strip Intercropping

Simultaneous growing of two or more crops in different strips which are wide enough to permit independent cultivation but narrow enough for the crops to interact agronomically is called strip intercropping (Fig. 5.7). There are two types of strip-intercropping systems: contour strip cropping and field strip cropping. The strips with uniform width across the exact contour of the field are called contour strip cropping, while the strips with uniform width across the general slope of the land are called field strip cropping.

In an intercropping experiment with hot pepper and snap bean, snap bean strip intercropped with pepper gave higher biomass and produced higher yields (more available light favored snap bean) than snap bean row intercropped with pepper. In view of improved microclimate, the growth of strip-intercropped snap bean was even better than in the mono-crop (Kleinhenz 1992).

The alternating strips of wheat, corn, and soybean 6 rows wide each; oat, corn, and soybean; and 6 rows of corn with 12 rows of soybean are some of the examples of strip intercropping in the USA.

5.3 Experiences

The intercropping of spring wheat, canola, and field peas grown in all possible combinations (wheat–canola, wheat–pea, canola–pea, and wheat–canola–pea) gave higher yields, suppressed weeds, and improved grain quality as compared to monocropping (Anon 2004) (Table 5.3).

The yield increase took place (LERs greater than 1) in about two-thirds of the intercropping treatments. In pesticide-free plots (PFPs), only about half of the intercrops had higher yield, while in conventional plots, 75 % had higher yield. The monocultures used about 10 % more land than the intercrops as indicated by the overall average of LERs (1.11). The canola–pea (CP) intercrop had the highest LERs and most consistent higher yields (100 % of the time under conventional management) (Fig. 5.8).

Table 5.3 Effect of intercropping on land equivalent ratio and yield

Intercrops	Land equivalent ratio (LER) Pesticide free	Conventional	Average	Frequency of occurrence of high yield Pesticide free (%)	Conventional (%)	Average (%)
Wheat–canola	1.13	1.04	1.09	40	80	60
Wheat–pea	0.87	1.10	0.99	20	40	30
Canola–pea	1.19	1.22	1.21	67	100	84
Wheat–canola–pea	1.16	1.12	1.14	60	80	70
Average	1.09	1.12	1.11	47	75	61

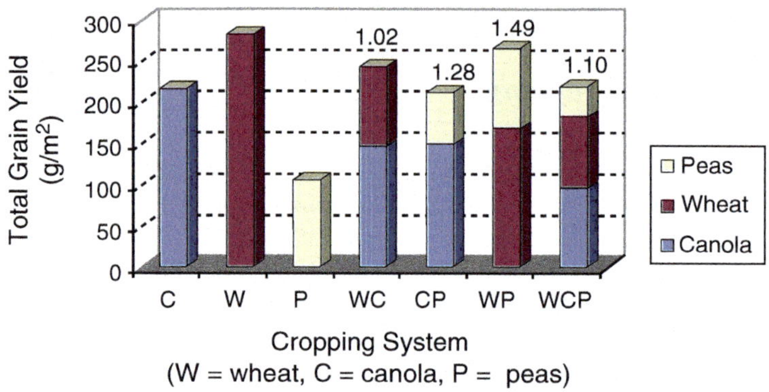

Fig. 5.8 Effect of intercropping on grain yield and LERs. All intercrops had LERs greater than 1

The intercrops suppressed weeds better than mono-crops. The number of crops in the intercropping and the weed biomass are directly correlated. As the number of crops increased in the mix, the weed biomass decreased. Weed suppression was maximum in wheat-canola (WC) and minimum in wheat-pea (WP) which had problems with weeds and lodging (Fig. 5.9).

The nitrogen and protein content increased in grain when peas were included in intercropping. This gave an added advantage, as the wheat with high protein fetched the premium price (6 to 20 cents per bushel higher) (Fig. 5.10).

The intercropping of sweet corn and peas gave higher yield over mono-cropping of two crops. At higher corn densities, the pea yields were reduced (Table 5.4) (Francis and Decoteau 1993).

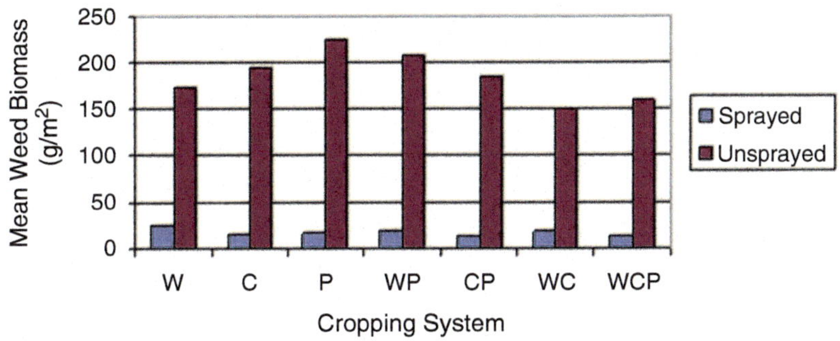

Fig. 5.9 Effect of intercropping on weed biomass

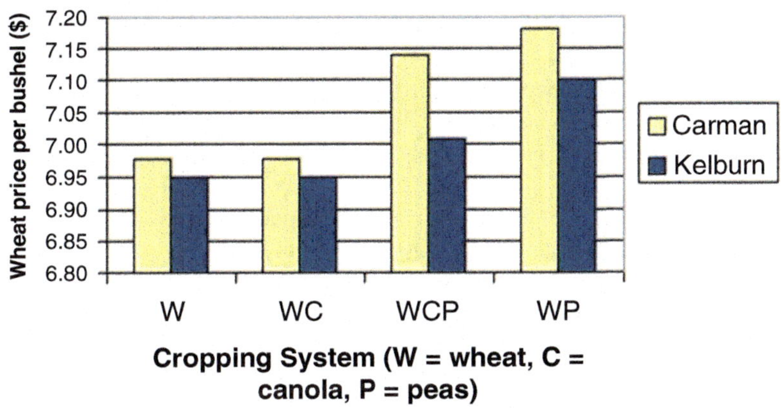

Fig. 5.10 Effect of intercropping on wheat yield and grain price at two locations

Table 5.4 Effect of intercropping seed rates on land equivalent ratio and yield

Seed rates	Yield (pounds/acre)		LER
	Corn	Peas	
Corn	5600	***	***
Peas	***	1200	***
Corn + peas (low corn)	4200	800	1.24
Corn + peas (medium corn)	4600	800	1.32
Corn + peas (high corn)	5000	500	1.26

5.4 Benefits and Limitations

5.4.1 Benefits

- Growth resources are efficiently utilized (land, sunlight, and vertical space). Even though plants grow in the same land, but resource partitioning and niches for crops are different.

Table 5.5 Effect of intercropping on yield in different regions

Region	Yield (kg/ha)		
	Solo maize	Intercrop (maize + cowpea)	% Increase
Humid (Ibadan)	1850	1980	+7
Subhumid (Ile-Ife)	2100	2450	+ 17
Semiarid (Zaria)	1865	2050	+ 10
Arid (ICRISAT)	1570	1754 (Millet + groundnut)	+ 12

Aina and Adejuwon (2004)

- Enhanced soil fertility and fixation of significant amount of N in soil. The intercropping of plants from Leguminosae family and nonlegume plants increase the nitrogen content in nonlegume plants (Eskandari et al. 2009). The Rhizobium bacteria with legume roots fix atmospheric nitrogen which is added to the soil to increase the soil fertility.
- Greater yield stability over different seasons and economic stability to farmers (Table 5.5).
- Increase of biodiversity in agro-ecosystems. Two or more crops can be taken up at a time.
- Reduce nitrate leaching (wheat acts as a scavenger crop).
- Time, labor, and equipment can be used efficiently.
- Enhanced carbon sequestration.
- Provides continuous ground surface cover and protecting the soil from raindrop impact, wind, and water erosion.
- Provision of shelter and physical support (one crop to the other crop).
- Beneficial to the small farmers with limited means.
- Addition of huge biomass (thereby organic matter) to the soil.
- Suppression of pests, diseases, and weeds (4–8 months in a year) which may be due to three reasons.

  - According to the disruptive crop hypothesis, the subsidiary crop species disrupt the pest's ability to attack the host crop species effectively by interfering with host finding ability of insect pests through volatiles, visual, etc.
  - As per the natural enemies hypothesis, more number of natural enemies (predators and parasitoids) in intercrop, e.g., the weedy intercrop, White Dill (Queen Anne's lace), attracts several natural enemies; sunflower intercropped with bell pepper encourages biological control by attracting minute pirate bug (*Orius* spp.) and other beneficial, provides refuge and food resources for natural enemies, and provides perches for birds which prey on insect pests.
  - According to the trap crop hypothesis, insect pests are diverted away from the main crop and are attracted by the trap intercrop which is more attractive [e.g., the strip intercrop alfalfa attracts Lygus bugs away from cotton in California (Altieri 1994), flea beetles from collard to cruciferous intercrops with strong chemical attractants, and stem borers from maize and sorghum to molasses grass which produces volatiles that attract parasitic wasps in Kenya (Khan et al. 1997)]. Other examples of trap intercrop systems used successfully for pest management in agriculture are presented in Fig. 5.11 and Table 5.6.

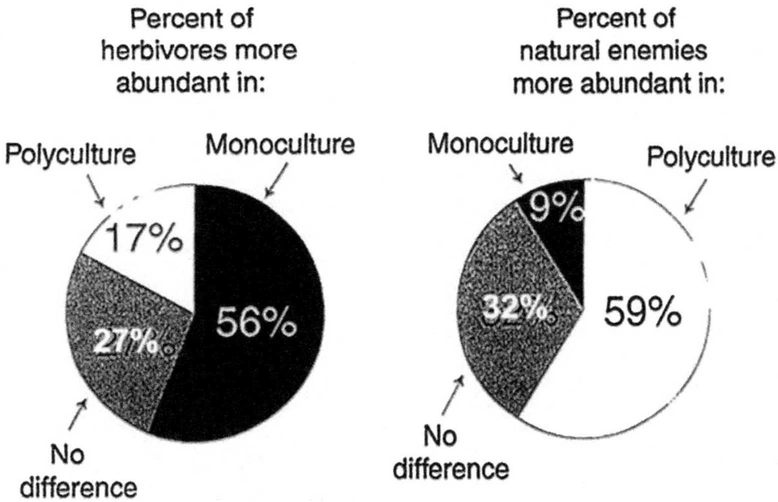

Fig. 5.11 Effect of intercrops on insect pests and their natural enemies (Andow 1991)

Table 5.6 Effect of trap intercropping systems on pest management

Pests managed	Trap intercropping systems	Country
Lygus bugs, *Lygus hesperus, L. elisus*	Cotton–alfalfa	California, USA
Cotton boll weevil, *Anthonomus grandis*	Cotton–cotton	USA
		Nicaragua
Stink bugs, *Nezara viridula, Euschistus* spp., *Acrosternum hilare, Piezodorus guildinii*	Soybean–snap bean	USA
		Brazil
		Nigeria
Mexican bean beetle, *Epilachna varivestis*	Soybean–snap bean	USA
Bean leaf beetle, *Cerotoma trifurcata*	Soybean–snap bean	USA
Colorado potato beetle, *Leptinotarsa decemlineata*	Potato–potato	USSR
		Bulgaria
Blossom beetle, *Meligethes aeneus*	Rape–rape	Finland
	Cauliflower–marigold	
Pine shoot beetle, *Tomicus piniperda*	Pine trees–pine logs	Great Britain
Spruce bark beetle, *Ips typographus*	Spruce–spruce tree logs	Europe

Hokkanen (1991)

5.4.2 Limitations

- Decrease in yield because of adverse competition effect for water, light, and soil nutrients.
- Main crop growth may be affected due to allelopathic effect.
- Overextraction of nutrients is possible.
- Free use of machines for intercultural operations may be obstructed.

- The systems are little understood and more complex.
- Likely to get less benefits out of intercropping for small farmers with inadequate resources.
- Increase in evapotranspiration loss of water.
- Intercrop competition is significant since both the crop species may aggressively pursue resources from the same niche.
- Harvesting of one crop component may cause damage to the other crops.
- Some pests may cause more damage in certain cases.
- Higher relative humidity in the air may favor diseases.
- Shortage of trained personnel.
- Requires a greater level of management.

5.5 Conclusions

The promotion of biodiversity, reduced use of agrochemicals (inorganic fertilizers and pesticides), and sustainable yields year-round is provided by multiple cropping systems followed by smallholders in traditional agriculture. The reduction of pest incidence and more efficient use of nutrients, water, and solar radiation in intercropping had a potential to increase yields in the range of 20–60 % (Altieri 1999). Intercropping is a potential solution where the capital is in short supply, production must be sustainable without expensive fossil fuels, pollution control, and rural unemployment exists.

The economic and ecological efficiency of the agricultural practices in a diversity of different countries and biomes can be increased by widespread implementation of intercropping systems. It appears reasonable to continue research on intercropping in view of its multiple advantages that it can offer, particularly in the search for sustainable agricultural systems and environmental problems with the present farming systems. Hence, more research is required to better understand how intercrops function and to develop intercropping systems that are compatible with current farming systems.

References

Aina PO, Adejuwon JO (2004) Agronomic adaptations to climate extremes (AF23). Regional Workshop on Assessments of Impacts and Adaptations to Climate Change (AIACC), Dakar, Senegal

Altieri MA (1994) Biodiversity and pest management in agroecosystems. Food Products Press, New York

Altieri MA (1999) The ecological role of biodiversity in agroecosystems. Agr Ecosyst Environ 74:19–31

Amador MF (1980) Behavior of Three Species (Corn, Beans, Squash) in Polyculture in Chontalpa, Tabasco, Mexico. CSAT, Cardenas, Tabasco, Mexico

Andow DA (1991) Control of arthropods using crop diversity. In: Pimentel D (ed) CRC handbook of pest management in agriculture, vol I, 2nd edn. CRC Press, Boca Raton, pp 257–284
Anon (2004) Agronomic benefits of intercropping annual crops in Manitoba. University of Manitoba
Eskandari H, Ghanbari A, Javanmard A (2009) Intercropping of cereals and legumes for forage production. Notulae Scientia Biologicae 1:7–13
Francis R, Decoteau DR (1993) Developing an effective southern pea and sweet corn intercrop system. Hort Technol 3(2):178–184
Govinden N, Arnason JT, Philogene BJR, Lambert JDH (1984) Intercropping in the tropics: advantages and relevance to the small farmer. Can J Dev Stud 5:213–232
Hokkanen HMT (1991) Trap cropping in pest management. Annu Rev Entomol 36:119–138
Khan ZR, Ampong-Nyarko K, Chilishwa P, Hassanali A, Kimani S, Lwande W, Overholt WA, Pickett JA, Smart LE, Wadhams LJ, Woodcock CM (1997) Intercropping increases parasitism of pests. Nature 388:631–632
Kleinhenz V (1992) Intercropping with legumes in the tropics. Asian Vegetable Research and Development Center, Tainan
Malézieux E, Crozat Y, Dupraz C, Laurans M, Makowski D, Ozier-Lafontaine H, Rapidel B, de Tourdonnet S, Valantin-Morison M (2009) Mixing plant species in cropping systems: concepts, tools and models. A review. Agron Sustain Dev 29:43–62
Plucknett DL, Smith NJH (1986) Historical perspectives on multiple cropping. In: Francis CA (ed) Multiple cropping systems. Macmillan, New York, pp 20–39
Scherr SJ, McNeely JA (2008) Biodiversity conservation and agricultural sustainability: towards a new paradigm of 'ecoagriculture' landscapes. Philos Trans Royal Soc B 363:477–494
Sullivan P (2003) Intercropping principles and production practices. Retrieved October 5, 2010 from http://attra.ncat.org/attra-pub/intercrop.html
Thrupp LA (2002) Linking agricultural biodiversity and food security: the valuable role of agrobiodiversity for sustainable agriculture. Int Aff 76:283–297
Vandermeer JH (1992) The ecology of intercropping. Cambridge University Press, Cambridge

Chapter 6
Crop Residue Management

Abstract In developing countries, the key in promoting physical, chemical, and biological attributes of soil health in agricultural systems is mainly through the retention of crop residue in fields in view of scarcity of alternative source of organic amendments. The environmental and soil health benefits of crop residue management include reduced soil erosion, enhanced nutrient cycling, higher soil moisture retention, improved soil quality, and increased soil organic matter and carbon storage. The decreased soil temperature, nitrogen immobilization, and waterlogging are some of the limitations of crop residue retention on soil surface. The use of crop residues as feed for livestock can limit the amount of residue retained on the soil surface, but partial retention (more than 30 % soil cover) is recommended.

Keywords Animal fodder • Conservation agriculture • Crop residue • Soil health • Soil properties • Soil quality

6.1 Introduction

The material that is left after the harvesting of crop is called crop residue. It is a byproduct of agriculture-based enterprise. Crop residue consists of, e.g., leaves, stems, stalks, cobs, husk, etc. (Fig. 6.1). The permanent soil cover is one of the main pillars of conservation agriculture, and crop residues form one of the integral parts of the soil surface cover. In order to facilitate direct planting in the crop residue-covered soil surface (up to 10 tons ha^{-1}), the zero-till seed-cum-fertilizer drill/planters like happy seeder, turbo seeder, rotary-disc drill, and easy seeder have been developed.

The crop residues on soil surface helps in conserving soil moisture (Jat et al. 2009) and nutrients, input use-efficiency, suppressing weeds, conserving resources, improving soil health, enhancing yield (Mandal et al. 2004), moderating soil temperature, and adaptation to climate change (Gupta and Seth 2007). It is possible to grow a second crop in dryland ecosystems (where only a single crop is grown in a year) with residual soil moisture under crop residue-covered soil, as evidenced by the research conducted at Central Research Institute for Dryland Agriculture (CRIDA), Hyderabad (Fig. 6.2).

P.P. Reddy, *Sustainable Intensification of Crop Production*,
DOI 10.1007/978-981-10-2702-4_6

Fig. 6.1 Retention of crop residues on soil surface

Fig. 6.2 Growing of horse gram during post-rainy season with maize residues under conservation agriculture in rain-fed condition (Courtesy: Srinivas Rao, CRIDA, Hyderabad)

Retention of crop residues differentiates conservation agriculture from conventional agriculture, and all conservation systems include at least a certain level of surface residue cover. Retaining of at least 30 % crop residues on soil surface is an essential component of conservation agriculture. Proper management of crop residue can control soil erosion and enhance the efficiency of irrigation. Field residue can also be used as animal fodder, for preparation of compost and farmyard manure, as bedding material for livestock, for generation of biogas and bioenergy, as substrate for mushroom production, for mulching the soil, and as industry raw material. The crop residues form more than one half of all dry matter in the global harvest. The amounts of nutrients present in crop residues are often higher than the quantities applied as fertilizers in most of the developing countries. Soil fertility and crop production can be sustained by proper management of crop residues.

The organic matter is rapidly lost through accelerated oxidation due to hot and moist conditions in tropical countries. The retention of crop residues prevents the rapid losses of organic matter. The crop residue material decomposes slowly (low nitrogen content, high C/N ratio) and will provide a long-term protection compared to quickly decomposing material (INFONET BIOVISION 2010). Crop residue management helps in attaining sustainable agricultural production and environmental preservation in a wide range of cropping systems and soils.

6.2 Generation

The various crop residues are produced in a huge quantity in several countries. In the USA alone, crop residues to the extent of 488 (10^6 Mg) are produced during 2001 (Table 6.1).

India produces 500 million tons of crop residues annually (cereal crops, 352 Mt; fiber crops, 66 Mt; oilseeds, 29 Mt; pulses, 13 Mt; and sugarcane, 12 Mt) (MNRE 2009). They are rich in plant nutrients such as N, P, K, S, and carbon. The crop residues can be used for improving soil health, increasing productivity, reducing pollution, enhancing sustainability, and increasing resilience in conservation agriculture.

6.3 Benefits and Limitations

6.3.1 Benefits

The ecosystem services provided by crop residue retention as surface mulch on cropland are presented in Fig. 6.3. Retention of on-site crop residues improves soil physical (structure, infiltration rate, plant available water capacity), chemical (nutrient cycling, cation exchange capacity, soil reaction), and biological (soil organic carbon sequestration, microbial biomass C, activity and species diversity of soil biota) properties. Off-site crop residues in the form of mulches improve quality of water and air through reduction in water and wind erosion, nonpoint sedimentation, source pollution, and transport of pollutants into the water bodies and aquatic ecosystems. The use of crop residues in combination with no-tillage in conservation agriculture enhances agronomic productivity and profitability.

The benefits of crop residue management are as follows:

- Improvement in soil quality.
- Improves soil tilth.
- Reduced water and wind soil erosion.

Table 6.1 Crop residue production in the USA (10^6 Mg)

	Crop residue production (10^6 Mg)	
Species	USA (2001)	World (2001)
Cereals	367	2802
Legumes	82	305
Oil crops	20	108
Sugar crops	14	373
Tuber crops	5	170
Total	**488**	**3758**

Lal (2005)

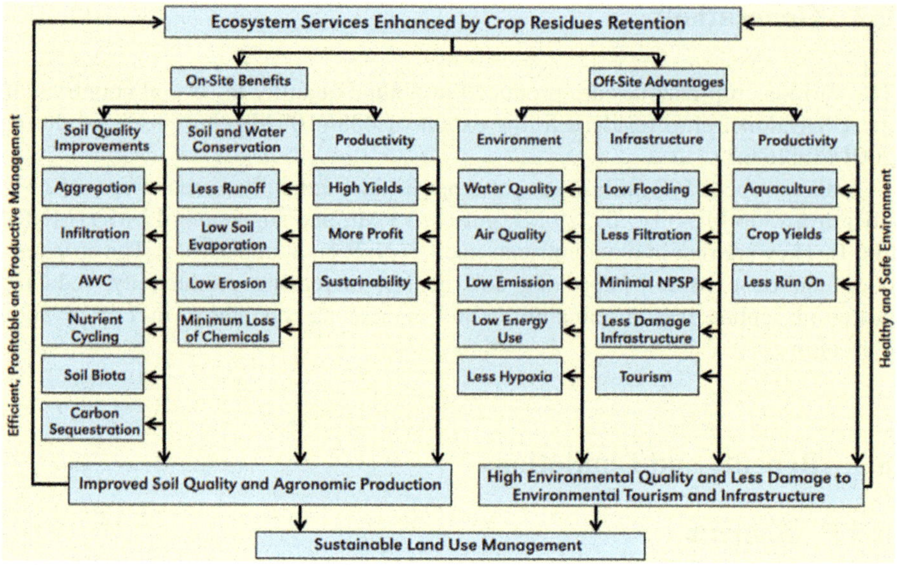

Fig. 6.3 Ecosystem services provided by crop residues

- Moderation of soil temperatures.
- Nutrient cycling and enhanced mineralization of N. Nutrients accumulate at the soil surface.
- The fertility of soil is increased because of addition of organic matter to the soil as it decomposes.
- Weed growth is reduced due to shading.
- Microorganism activity is increased.
- Enrichment of soil organic carbon pool.
- Reduced root and foliage diseases (anthracnose, *Colletotrichum lindemuthianum*; *Fusarium*-type root rot; and root-knot nematode, *Meloidogyne* spp. on beans).
- Increased arthropod diversity.
- Reduced evaporation and increased water infiltration. Moisture retention in soil longer and water storage is increased (Fig. 6.4).
- Increased yields (Fig. 6.5).

6.3.2 Limitations

- Decreased soil temperature
- Host some pathogens in wet climates if not managed properly (*Rhizoctonia*-type bean root rot—significantly more).
- Nitrogen immobilization.
- Waterlogging.

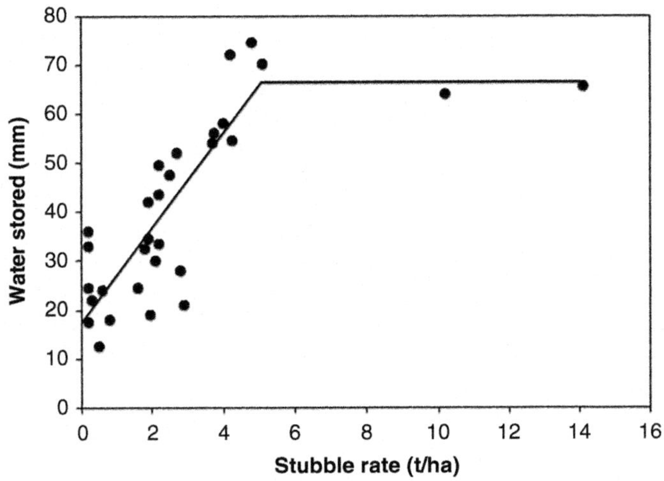

Fig. 6.4 Effect of crop residues on water storage (Cornish 1987)

6.4 Management Systems

Crop residue management includes several tillage systems including no-till, ridge-till, mulch-till, and reduced-till.

6.4.1 No-Till/Strip-Till

Other terms used to describe no-till include direct seeding, slot planting, zero-till, row-till, and slot-till. Soil is left undisturbed from harvest to planting except for strips up to 1/3 of the row width. These strips may involve only residue disturbance or may include soil disturbance. Crops are planted or drilled in the narrow-tilled strip (10–30 cm) using disk openers, coulters, row cleaners, in-row chisels, or roto-tillers. Weed control is accomplished primarily with herbicides. Cultivation may be used for emergency weed control. It leaves about 50 % crop residue on the soil surface.

Fig. 6.5 Effect of crop residue mulch on wheat yield (*left*) and biomass production (*right*)

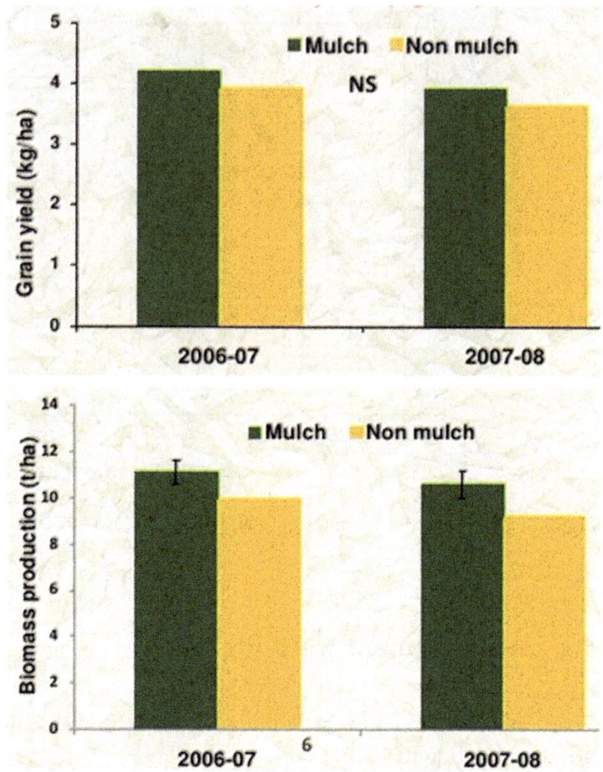

6.4.2 *Ridge-Till*

Soil is left undisturbed from harvest to planting except for strips up to 1/3 of the row width. Planting is completed on the ridge with sweeps, disk openers, coulters, or row cleaners and usually involves the removal of the top of the ridge. Crop residues (65 %) are left on the surface between ridges. Weed control is accomplished with herbicides (frequently banded) or cultivation. Ridges are rebuilt during row cultivation.

6.4.3 *Mulch-Till*

Full-width tillage involving one or more tillage trips which disturb the entire soil surface. Subtillage equipments such as chisels, field cultivators, disks, sweeps, or blades are used. The primary mulch tillage with chisel plow loosens the soil, but do not invert it. It leaves 50–80 % residue cover over the soil. Light disking later reduces crop residue cover to 30–50 %. It is done prior to and/or during planting. Weed control is accomplished with mulch-till.

6.4.4 *Reduced-Till*

Full-width tillage involving one or more tillage trips which disturb the entire soil surface. It is done prior to planting. There is 15–30 % residue cover after planting or 500–1000 pounds per acre of small grain residue equivalent throughout the critical wind erosion period. Weed control is accomplished with herbicides or row cultivation.

6.5 Impact on Soil Quality

Retention of crop residues on surface has a profound effect on physical (soil structure, runoff, erosion, soil compaction, soil temperature, and moisture content), chemical (soil organic carbon, soil pH, and cation exchange capacity), and biological (soil biodiversity and soil microbial biomass) properties of soil (Fig. 6.6). It also affects soil quality and nutrient cycling (Fig. 6.7). The plant biomass is produced

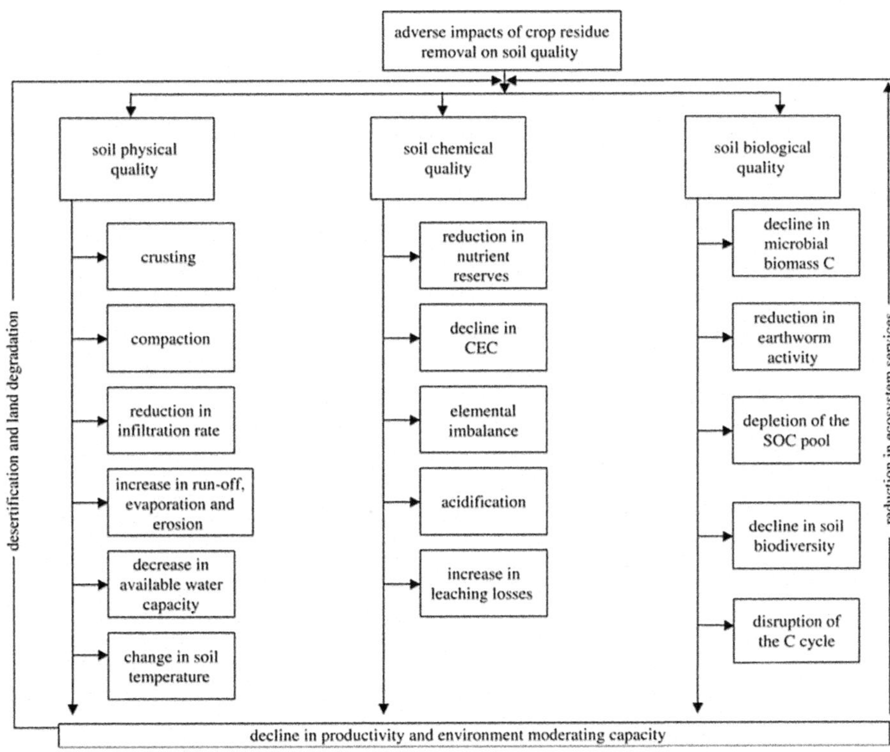

Fig. 6.6 Effect of crop residues on physical, chemical, and biological properties of soil (McGill and Cole 1981)

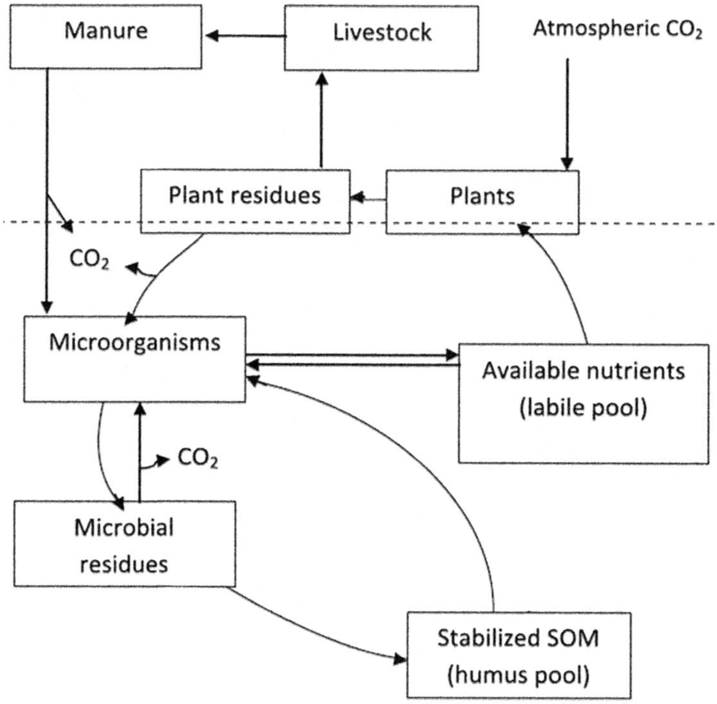

Fig. 6.7 Decomposition of crop residues and manure by soil microorganisms

when the crop plants capture carbon from the atmosphere through photosynthesis. The livestock consume part of the crop residues as feed in integrated crop-livestock farming systems and return to the soil in the form of farmyard manure that contain some available nutrients. The microorganisms that are present in the soil decompose crop residues and manure and release nutrients such as C, N, S, and P to the soil. These nutrients are taken up by crop plants and the nutrient cycling continues (McGill and Cole 1981).

6.6 Model Plan for Management

A model plan given in Table 6.2 may be used as a guideline for managing crop residues at local and regional scales.

6.7 Conclusions

The effective solution to challenges of food, water, and energy insecurity, issues related to climate change, and degradation of natural resources is through sustainable intensification, with crop residues as an integral component that ensures a

Table 6.2 Model plan for managing crop residues at local and regional scales

Query	Response	Crop residues management options
1. Can crop residues be used for conservation agriculture? If the answer is no, move to query 2	Yes	Retain it on soil surface
		Use drill for sowing with residues (e.g., happy seeder)
		Follow conservation agriculture for all crops in rotation
2. Can it be used as fodder? If the answer is no, move to query 3	Yes	Leave stubbles in field
		Enrich fodder with supplements (e.g., urea and molasses)
		Use manure in conservation agriculture
3. Can it be used for biogas generation? If the answer is no, move to query 4	Yes	Leave stubbles in field
		Adopt community biogas plant (e.g., KVIC design modified by IARI)
		Use slurry in conservation agriculture
4. Can it be used for composting? If the answer is no, move to query 5	Yes	Leave stubbles in field
		Adopt modern composting technique (e.g., IARI model)
		Use compost in conservation agriculture
5. Can it be used for biofuel generation? If the answer is no, move to query 6	Yes	Leave stubbles in field
		Install biofuel plant
		Use liquid slurry in conservation agriculture
6. Can it be used for electricity generation? If the answer is no, move to query 7	Yes	Leave stubbles in field
		Install biomass energy plant (e.g., KPTL model)
		Use ash in conservation agriculture
7. Can it be used for gasification? If the answer is no, move to query 8	Yes	Leave stubbles in field
		Install biomass gasifier (e.g., CIAE model)
		Use ash in conservation agriculture
8. Can it be used for biochar making?	Yes	Leave stubbles in field
		Install biochar kiln (e.g., IARI model)
		Use biochar in conservation agriculture

strong natural resource base. Crop residues are a prerequisite for conservation agriculture besides their use as livestock feed, fuel, and industrial raw material. Retaining crop residues on soil surface improves water infiltration and storage that is valuable for crop growth. For harnessing the full potential of the valuable crop residue resources for sustainability and resilience of agriculture, the cooperation of all the stakeholders, viz., farmers, supply and value chain service providers, researchers, extension agents, policy makers, civil servants, and consumers, is required to understand and spread the technology.

References

Cornish P (1987) Effects of wheat residues and tillage on the water balance of a red earth soil. IVth Australian Agronomy Conference, La Trobe. www.agronomy.org.au

Gupta RK, Seth A (2007) A review of resource conserving technologies for sustainable management of the rice–wheat cropping systems of the Indo-Gangetic plains. Crop Prot 26:436–447

INFONET-BIOVISION (ed) (2010) Mulching. Biovision, Zuerich. URL Accessed 19 June 2012

Jat ML, Gathala MK, Ladha JK, Saharawat YS, Jat AS, Kumar V, Sharma SK, Gupta RK (2009) Evaluation of precision land leveling and double zero-till systems in the rice–wheat rotation: water use, productivity, profitability and soil physical properties. Soil Tillage Res 105:112–121

Lal R (2005) World crop residues production and implication of its use as a biofuel. Env Int 31:575–586

Mandal KG, Misra AK, Hati KM, Bandyopadhyay KK, Ghosh PK, Mohanty M (2004) Rice residue management options and effects on soil properties and crop productivity. Food Agric Environ 2:224–231

McGill WB, Cole CV (1981) Comparative aspects of cycling of organic C, N, S, and P through soil organic matter. Geoderma 26:267–286

MNRE (2009) Ministry of New and Renewable Energy Resources, Govt. of India, New Delhi. www.mnre.gov.in/biomassrsources. Accessed on Dec 2012

Part IV
Species Diversification

Chapter 7
Crop Rotation

Abstract Growing a series of different types of crops in sequence in the same field is called crop rotation. The cereal–legume (including green manure crops) crop rotation helps to replenish nitrogen through biological N fixation. The benefits of crop rotation include enhancement of biodiversity; reduction of soil erosion; increasing soil fertility and crop yield; management of insect pests, diseases, nematodes, and weeds; and improvement of soil structure. The primary drawbacks and challenges include higher management skills are required, limited markets for alternative crops, need for extra equipment and storage, soils and climate suitability for production of alternative crops, and some rotation crops may have allelopathic effect.

Keywords Soil erosion • Soil fertility • Crop yield • Soil structure • Pests • Diseases

7.1 Introduction

The soil and crop management practices increase or sustain soil productivity. Sustaining or increasing soil productivity depends in part on soil and crop management practices that maintain or increase soil organic matter and productivity. Crop rotation is one of the components of crop management practices (along with cover crops and green manures) that produce high residues and maintain surface residue cover resulting in greater soil organic C and N, which may improve soil productivity.

Since ancient times, farmers learned from experience that cultivating the same crop, season after season, on the same land resulted in lower yields, and by growing a sequence of crops over several years, the productivity of the land can be increased significantly. The crop rotations enhanced soil organic matter, soil fertility, and soil tilth.

Irish potato famine is linked to one crucial farming mistake, i.e., growing same crop in the same piece of land year after year. The land is depleted of nutrients, and the soil is eroded. The pathogen responsible for late blight of potato remained in the soil and multiplied. The famine could have been prevented by following the simple cultural practice of crop rotation to divert the spread of late blight disease.

P.P. Reddy, *Sustainable Intensification of Crop Production*,
DOI 10.1007/978-981-10-2702-4_7

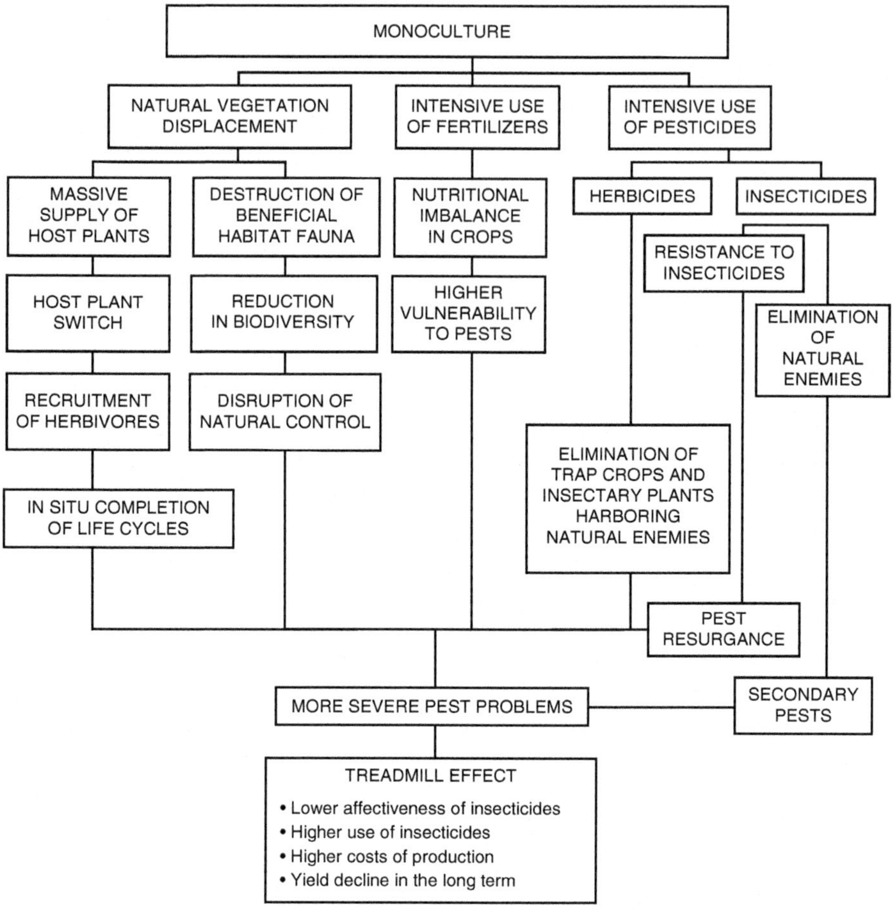

Fig. 7.1 The problems associated with monoculture

7.2 Impact of Monoculture

The problems associated with monoculture (growing of the same crop year after year in same land) are as follows (Fig. 7.1):

- Leads to soil degradation.
- Root development is decreased.
- Reduction in biological activity of the soil (earthworm population is decreased).
- Soilborne diseases and specific crop pests are increased.
- Proliferation of specific weeds (generally of the same class).
- Allelopathic toxins are accumulated.
- Labor distribution throughout the year is inadequate.
- Due to continuous exploration of the same soil zone, plant nutrients are decreased.

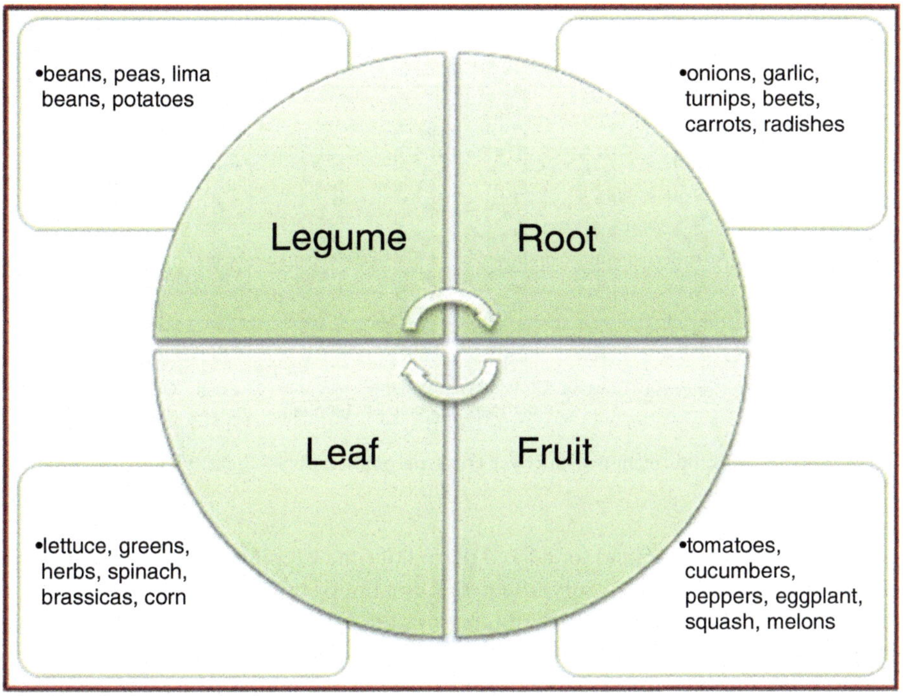

Fig. 7.2 Different types of crops grown under crop rotation [leafy vegetables (need nitrogen), fruiting crops (need phosphorus), rooting crops (need potassium), and legumes (soil builders)]

- Crop diversification is reduced.
- Reduction in yields.
- Economic risk is increased.

7.3 Crop Rotation

The crop rotation is an ancient practice which was followed in Rome, Africa, and Egypt. It prevents depletion of specific nutrients and buildup of soilborne pathogens and pests specific to one group of crops. Crop rotation builds up soil organic matter and nutrients which are used for plant growth. The crops belonging to the same family should not be grown continuously in the same field. The crops belonging to different families should be grown in succession. The replenishment of N through the use of green manure crops is the traditional component of crop rotation. Crop rotation improves soil fertility and structure by growing shallow and deep-rooted crops in succession.

Growing of different types of crops (root, leaves, fruit, flowers, legumes) in succession is called crop rotation (Fig. 7.2). Crop rotation means sowing of crop in a

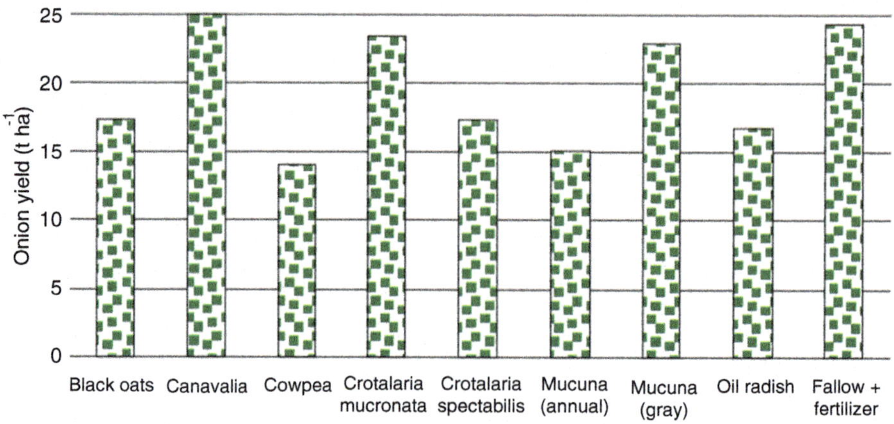

Fig. 7.3 Effect of crop rotation with cover crops on onion bulb (4–8 cm) yield (Calegari et al. 1993)

regular order on a same land for a fixed period of time keeping in view that the fertility of soil may not be adversely affected. Growing of crops in a planned sequence on the same field without repetition (crops in the same species or family) is called crop rotation. In recent times, the interest in crop rotations has increased due to their role in improving soil health and yields, reducing insect and disease risks, and diversifying farm income. For maximum profitability and production, a diverse range of crops should be utilized in crop rotation.

7.3.1 Principles

- Crops with tap roots should be followed by crops with a fibrous root system as this helps in the proper and uniform use of nutrients from the soil. In addition, roots do not compete with each other for the uptake of nutrients.
- Legumes should be grown after nonlegumes as they fix atmospheric nitrogen into the soil and add more organic matter to the soil.
- Exhaustive crops (crops which need more inputs like more fertilizer, irrigation, insecticide, etc.) should be followed by less exhaustive crops, which require less care (i.e., potato or maize should be followed by leguminous crops).
- Selection of crop should be demand-based (i.e., crops needed by the market should be chosen as it can be sold at a higher price).
- Crop selection also depends on land type, irrigation facilities, soil, and climatic considerations. Financial constraints of the farmer should also be kept in mind.

The crop rotation of onion with different cover crops gave higher onion bulb yields (Fig. 7.3). The highest onion yield was obtained when rotated with Canavalia, followed by *Crotalaria mucronata*, and gray *Mucuna.*

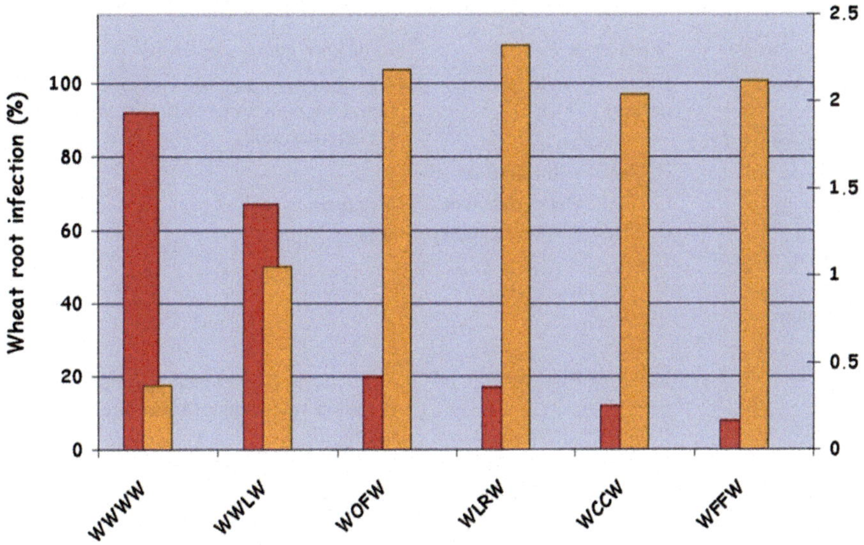

Fig. 7.4 Effect of crop rotation on fungal root diseases in wheat (W, wheat; L, lupines; O, oats; F, fallow) (Reis 1983)

7.4 Benefits

Crop rotation has several agronomic, socioeconomic, and environmental benefits compared to monoculture cropping (Cook and Ellis 1987) that include the following.

7.4.1 Agronomic Benefits

7.4.1.1 Breaking Life Cycle of Pests

The crop rotation plays a crucial role in reducing the pest (insect pests, diseases, nematodes, and weeds) damage by including nonhost plants in rotation that interrupts their life cycle. The crops from the same family should not be planted in sequence because they share the same pests (e.g., wheat and barley, tomato and eggplant). Hence, there is a need to rotate crops belonging to different families in sequence. The rotation system of families may include Brassicaceae (cole crops), followed by Asteraceae (lettuce, cut flowers), followed by Solanaceae (tomatoes, potatoes, peppers, eggplants), and followed by Cucurbitaceae (squashes, cucumbers, melons). At least 2-year rotation should be followed to manage foliar diseases, while 4-year rotation for the management of soilborne pathogens (Fig. 7.4). *Phytophthora capsici* infection in Solanaceous and Cucurbit crops can be controlled with 3–5-year crop rotation with Fabaceae (Leguminosae), Brassicaceae

Table 7.1 Insect pests, diseases, and nematodes controlled (partially or entirely) by crop rotation

Insect pests/crops	Diseases/crops	Root-knot nematodes/crops
Alfalfa weevil (alfalfa)	Bacterial blight (wheat, barley)	*Meloidogyne incognita* (velvet bean, rapeseed, common vetch, castor bean, French marigold)
Corn rootworm (corn)		
Colorado potato beetle (potato, tomato)	Bacterial wilt (alfalfa)	
	Granville (bacterial) wilt (tobacco, potato, eggplant)	*M. javanica* (velvet bean, rapeseed, sesame, common vetch, French marigold)
Wheat stem maggot (wheat)		
Pepper maggot (pepper)		
Squash bug (cucurbits)	Black shank (tobacco)	*M. arenaria* (velvet bean, sesame, common vetch, castor bean, French marigold)
Hessian fly (wheat)	Black dot (potato)	
	Northern corn leaf blight (corn)	*M. hapla* (French marigold)
	Smut (corn)	
	Stem rot (peanuts)	
	Verticillium wilt (potato, sunflower)	
	White mold (peanuts, potato, soybean)	

(Cruciferae), Apiaceae (Umbelliferae), Alliaceae (Amaryllidaceae), Asteraceae (Compositae), Chenopodiaceae, and Poaceae (Gramineae).

Crop rotations can also be used to manage insect pests. Some of the common features of crop pests that can be controlled by rotations include less mobility, limited host range, source in the field itself, and not capable of surviving long periods without a living host (Flint and Roberts 1988).

The plant parasitic nematodes can be controlled by using crop rotation. Inclusion of crops such as rapeseed, mustard, velvet bean, sorghum-Sudan grass, and sun hemp (antagonistic crops) in rotation gives effective control of nematodes.

Crop rotations can also be used to suppress weeds in crop plants. Rotation of spring lettuce with summer cover crops, such as soybean or cowpea and Japanese millet (vigorous and fast growing), and then fall broccoli will have less weed competition to subsequent lettuce crops due to little or no weed seed spill.

The cover crops such as Sudan grass, sorghum, rye, barley, wheat, and oats can be used in rotation with crop plants to suppress weeds through allelopathy, shading, and competition. The leachates and residues of hairy vetch, crimson clover, and other legumes can also suppress weeds when used in rotation with main crops.

Some of the insect pests, diseases, and nematodes controlled (partially or entirely) by crop rotation are presented in Table 7.1.

Table 7.2 Amount of nitrogen fixed by different legume crops

Legume	Rate of N fixation (kg/ha/year)
Centrosema pubescens	84–395
Desmodium uncinatum	178
Leucaena leucocephala	110–600
Stylosanthes guianensis	94–290
Trifolium semipilosum	80

Haque and Jutzi (1984)

7.4.1.2 Improving Structure of Soil

Crop rotation with shallow and deep-rooted crops helps to explore different soil profiles for water and nutrients (which contributes to enhancement of yield) and thereby improve physical properties of the soil such as tilth and bulk density.

7.4.1.3 Supply of Nutrients

Crop rotation helps to build up organic matter in soil. Each crop feeds on certain nutrients and then returns plant food elements for subsequent crops by leaving organic matter in the soil. Inclusion of legume crops in rotation enhances nitrogen levels in the soil for the benefit of subsequent crops.

The crops that are included in crop rotation should improve soil quality and prevent degradation of the environment. The use of rotational organic farming practices over a long period (8 years) was responsible for increasing soil organic carbon (2 %), total soil nitrogen (22 %), soluble phosphorus, exchangeable potassium, and soil pH (Clark et al. 1998).

(i) Nitrogen

The use of legumes in crop rotation fixes atmospheric nitrogen through their symbiotic relationship with *Rhizobium* spp. associated with their root system. This biological process adds nitrogen to soil by nodule excretion (direct) or by decomposition of root nodules and tissues (indirect). The inclusion of legumes in crop rotation normally contribute 84 to 395 kg N/ha (Table 7.2) (Haque and Jutzi 1984).

The nitrogen is mineralized during an extended period of time, and the surpluses do not run off into groundwater unlike N in conventional fertilizers. Depending on the environmental conditions, about 40–75 % of the total N contained in a legume cover crop is available for subsequent crops.

Crop rotation enhances the availability of soil N, reduces the amount of N fertilizer applied, and minimizes the potential risk of N leaching to groundwater. During conditions of wet weather, rotations with alfalfa, corn, oat, and soybean increase the soil nutrient pool thereby reduce the fertilizer N input and minimize the leaching of excess N.

(ii) Phosphorus and Potassium

The inclusion of deep-rooted cover crops in rotations helps for better absorption and transport of phosphorus and potassium from the deep soil profile to the soil surface, where plant roots have better access to them.

The combination of crop rotation and nutrient management can reduce or eliminate the need for purchased fertilizer. For example, in corn/alfalfa rotation, the manure is applied during the corn rotation, resulting in efficient use of the N and a buildup of P and K levels, while during the alfalfa phase of the rotation (manure is not applied), the forage crop will utilize the soil P and K levels that were built up during the corn phase of the rotation.

7.4.1.4 Reduction of Soil Erosion

The use of cover crops in crop rotation can reduce soil erosion by water and runoff. The integration of crop rotation along with no-till and reduced-till reduces impact of raindrop, sediment detachment, sediment transport, surface soil runoff, and soil loss (Unger and McCalla 1980).

Crop rotations accompanied with strip cropping system (perennial grasses and legumes or row crops of small grain and annual legume crops) can be more effective in controlling soil erosion. The two cultivated strips should have different planting and harvesting dates, and the sequence of crops should form a dense fibrous root system to hold the soil and retard the erosion.

The soil erosion can be reduced by farmers following the long-term crop rotation practice. In the Palouse region of Washington State, the organic farm contained 15 cm more topsoil as compared to the conventional farm.

The most suitable rotational crops for reducing soil erosion include legumes and grasses credited with the following main benefits:

- High level of soil surface cover reduces soil erosion.
- Enhancement of soil organic matter to form more stable soil aggregates to prevent soil erosion.
- Legumes fix atmospheric nitrogen.

7.4.1.5 Timeliness of Planting Operations

The workload during planting season of a good crop mix is spread over for several weeks that help in timely planting to obtain higher crop yields. In the corn and soybean rotation system, the two-third area of corn should be planted by the end of May, and soybeans can be planted in one-third of the acreage during the first 2 weeks of June, that should result in 90 % or more of their yield potential.

7.4.2 *Environmental Benefits*

7.4.2.1 Biodiversity Enhancement

Crop rotations have positive effects on above- and below-ground microbial communities that have favorable effects on soil organic matter and soil fertility. The increased crop rotational diversity basically alters the microbial community structure and activity, with positive interactions on aggregate formation and soil organic matter accrual. The sustainability of agricultural ecosystems can be enhanced by following legume-based crop rotations that support diversity of soil microbial communities.

7.4.2.2 Reduced Greenhouse Gas Emissions

The legume-based crop rotations can decrease nitrogen fertilizer use by up to 100 kg N per ha per year, thereby significantly reducing related greenhouse gas (GHG) emissions of nitrous oxide which has a global warming potential of 310 times higher than CO_2. The GHG emissions from the manufacturing process and transportation can also be decreased by reduced use of synthetic fertilizers.

7.4.2.3 Reduced Water Pollution

Limiting the use of synthetic fertilizers in crop rotations reduces groundwater pollution by leaching of nitrates. The diversified crop rotations used for the management of crop pests reduce the dependence on pesticides; thereby groundwater pollution from pesticides is reduced.

7.4.3 *Socioeconomic Benefits*

7.4.3.1 Distribution of Work Load

The workload is evenly distributed throughout the year for planting and harvesting operations under crop rotation systems.

7.4.3.2 Increased Margins

Crops when grown in rotation need lesser inputs, which represent a significant expenditure item. For example, the corn–soybean rotation needs less nitrogen fertilizers (since soybean fixes N) and no soil insecticide (needed for control corn rootworm larvae in continuous corn), thereby saving US$ 63 per hectare.

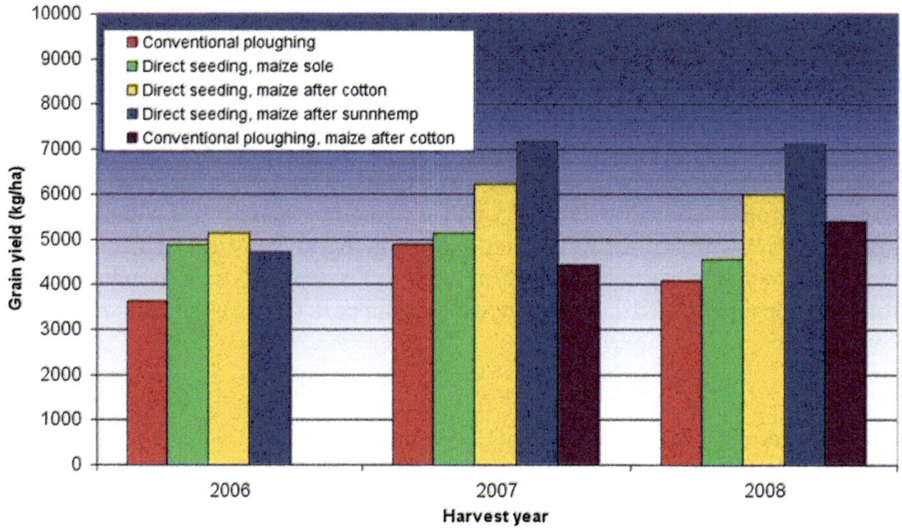

Fig. 7.5 Effect of crop rotation, direct seeding, and conventional plowing on maize grain yield

Table 7.3 Effect of crop rotation on maize corn yield (Morrow plots, University of Illinois)

Crop rotation	Maize yield (bu/A)
Continuous corn	139
Corn/soybeans	145
Corn/2-year alfalfa	154
Corn/corn/3-year alfalfa	153[a]
Corn/corn/3-year alfalfa	148[b]

[a]First-year corn yield
[b]Second-year corn yield

Crop rotations also increase crop yields (10–25 %) through improvement in soil properties and reducing weed and insect population (Fig. 7.5). For example, in corn–soybean rotation, 5–20 % increases in corn yields were observed as compared to the continuous corn (Table 7.3).

7.4.3.3 Improving the Economic Security of Farms

The farmers can overcome the drop in the price of some crops, since several crops are grown under crop rotation systems. By foreseeing the behavior of markets and using crops with increasing prices in crop rotations, the farmers can increase their farm income.

7.4.3.4 Stopping Tillage

As per the principle of conservation agriculture, the tillage operations can be curtailed to save on fuel and labor costs.

7.4.3.5 Organizing Open Field Days

By organizing open field days to demonstrate the advantages of crop rotation systems for leading farmers' group visits to farms, the relationship between consumers and farmers can be strengthened.

7.4.4 Other Benefits

The other benefits from rotation are as follows:

- More than one successful crop per year.
- Efficient nutrient cycling and utilization.
- Soil compaction is reduced due to deep and extensive root systems in some annuals and most perennial crops in rotation resulting in improved soil condition.
- Water use efficiency is increased.
- Biofumigation effect from canola or mustard for disease control.

7.5 Limitations

- Increased management skills are required.
- Uncertainty about markets and income potential. Limited markets for alternative crops.
- Soils and climate suitability for production of alternative crops.
- Additional crops produced need extra equipment and storage.
- Raising of livestock for utilization of forages produced.
- Some rotation crops may have allelopathic effect (Fig. 7.6).

7.6 Integration with Tillage Systems

The integration of maize-cover crops rotation with conservation tillage system (direct seeding) gave higher maize yields as compared to fallow and conventional tillage systems (Fig. 7.7). The combination of rotation with common vetch + direct seeding gave maximum maize yield, followed by hairy vetch, blue lupine + direct seeding, serradella + direct seeding, and *Lathyrus sativus* + direct seeding.

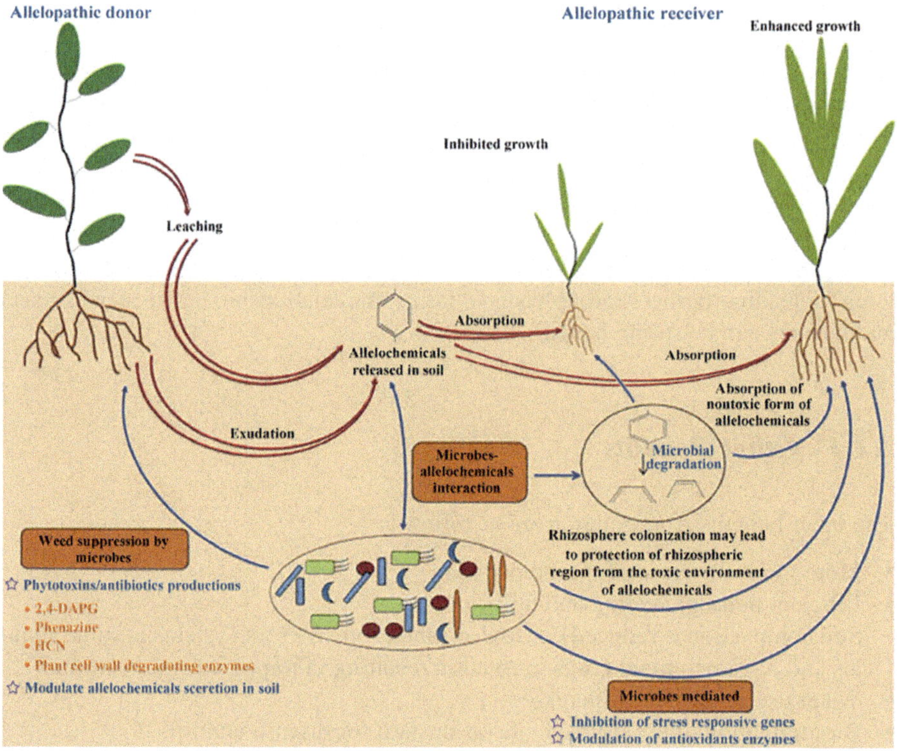

Fig. 7.6 Allelopathic effect for weed suppression

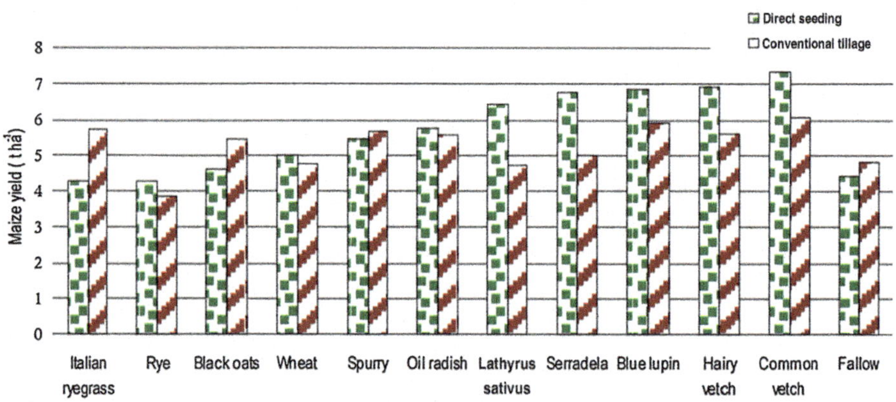

Fig. 7.7 Effect of integration of crop rotation with cover crops and direct seeding on maize yield (Calegari 1998)

7.7 Conclusions

Even though the crop rotations play a major role in most successful crop production enterprises, they are particularly essential for crop production by conservation tillage. Crop rotation can be considered as an integral part of a successful reduced-tillage cropping program. The benefits of no-till crop production can be spread to soil types or circumstances where success is difficult, through carefully planned crop rotations. The farm profitability associated with crop production can be considerably enhanced by integrating the timeliness and reduced-labor benefits of no-till with the yield advantages and reduced inputs with a good crop rotation. Some amount of initial investments in machinery or herbicides would be necessary to make these crop rotations work efficiently.

References

Calegari A (1998) Espécies para cobertura do solo. In: Darolt MR (ed) Plantio Direto. Pequena Propriedade Sustentável. IAPAR Circular 101, Londrina, pp 65–94

Calegari A, Mondardo A, Bulisani EA, Wildner LP, da Costa MBB, Alcântara PB, Miyasaka S, Amado TJC (1993) *Adubação verde no sul do Brasil. AS-PTA, Rio de Janeiro.* 2nd edn, 346 pp

Clark MS, Horwath WR, Shennan C, Scow KM (1998) Changes in soil chemical properties resulting from organic and low-input farming practices. Agron J 90:662–671

Cook RL, Ellis BG (1987) Soil management: a world view of conservation and production. Wiley, New York

Flint ML, Roberts PA (1988) Using crop diversity to manage pest problems: some California examples. Am J Altern Agric 3:163–167

Haque I, Jutzi S (1984) Nitrogen fixation by forage legumes in Sub-Saharan Africa: potential and limitations. ILCA Bull 20:2–13

Reis EM (1983) Selective medium for isolating *Cochliobolus sativus* from soil. Plant Dis 67:68–70

Unger PW, McCalla TM (1980) Conservation tillage systems. Adv Agron 33:2–53

Chapter 8
Agro-forestry

Abstract The integrated land use that combines elements of trees (woody perennials) with crops (herbaceous crops), pastures (grasses), and/or livestock in a sustainable production system on the same land by spatial arrangement/temporal sequence is called agro-forestry. Based on the components present, agro-forestry systems are classified into agrisilvicultural (trees with crops), silvopastoral (trees and pastures/livestock), and agrosilvopastoral (trees with crops, pastures, and animals) systems. The agro-forestry sustains wide range of products such as craft products, ecological services, fiber, fodder and forage, food, fuel, gardening materials, gums and resins, medicinal products, recreation, thatching and hedging materials, and timber. The benefits of agro-forestry include climate change mitigation and adaptation, enhancement of biodiversity, pest and disease control, protecting the environment and providing a number of ecosystem services, and regulation of soil, water, and air quality. Agro-forestry has a great potential impact in improving the local economy through diversification of local products and economies and rural skills, improved food and fuel security, improvements to the cultural and natural environment, increasing economic stability, and landscape diversification.

Keywords Agro-forestry • Trees • Crops • Pastures • Livestock • Biodiversity • Agrisilvicultural • Silvopastoral • Agrosilvopastoral • Products • Benefits • Ecological services • Pest control • Climate change mitigation

8.1 Introduction

In areas where farm sizes are small, soils are degraded, and agriculture outputs are low; conditions for mono-cropping agriculture are often harsh and unpredictable. Hence, shrubs and trees integrated into the agricultural system are vital assets of different products for farmers.

Agro-forestry is defined as an intentional combination of trees (woody perennials) with agricultural crops (herbaceous crops), pastures (grasses), and/or livestock on the same land by spatial arrangement/temporal sequence to create sustainable farming (intensive land-use) systems (Fig. 8.1). It helps to increase crop production, protection, and conservation and diversification of income benefits. In this system,

P.P. Reddy, *Sustainable Intensification of Crop Production*,
DOI 10.1007/978-981-10-2702-4_8

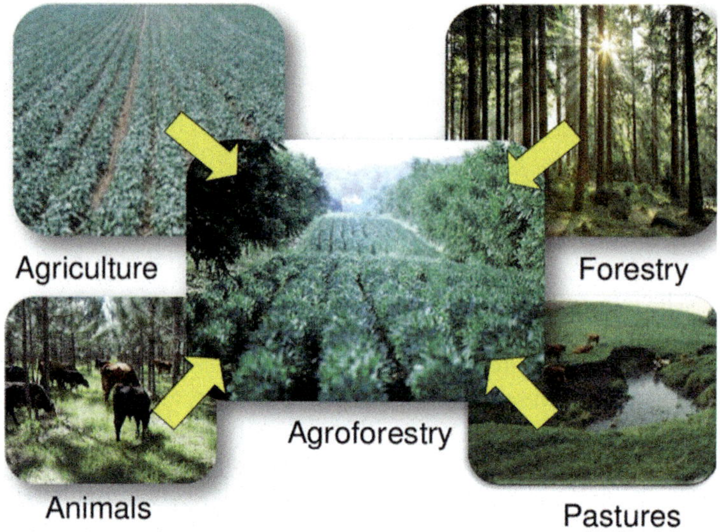

Fig. 8.1 Agro-forestry—combination of trees, crops, pasture, and livestock

forestry trees are planted with wide spacing and high-value crop plants are growing in between tree rows. The trees provide correct shade to crop plants, and they also give additional income (selling of tree products).

8.2 Objectives

The objectives of agro-forestry are as follows:

- Available farm resources to be appropriately used
- Ecological balance to be restored
- Physiological and biological resources to be optimized
- Productivity of food, fodder, and fuel to be increased
- Soil erosion from wind and water to be prevented
- Soil fertility and soil moisture to be enhanced

8.3 Different Types

The agro-forestry systems comprise of only three basic sets of components such as woody perennials (trees), herbaceous plants (crops and pasture), and animals (livestock). Hence, the agro-forestry systems are classified based on the nature of these three components into agrisilvicultural (crops including shrubs/vines and trees), silvopastoral (pasture/animals and trees), and agrosilvopastoral (crops, pasture/animals, and trees) systems (Nair 1989).

8.3.1 *Agrisilvicultural Systems (Crops, Including Shrub/Vine/ Tree Crops, and Trees)*

The simultaneous growing of trees and crops on the same piece of land is called agrisilvicultural system (Fig. 8.2). Agricultural crops are intercropped between the rows of trees and grown for 2 years under protective irrigation and for 4 years under rain-fed conditions. Later, fodder crops, shade-loving crops, and shallow-rooted crops can be grown economically.

The agrisilvicultural system is classified into three categories, namely, alley cropping, hedgerow cropping, and forest farming.

8.3.1.1 Alley Cropping

Simultaneously growing of an annual/perennial crop in the alley ways between rows of a long-term tree crop is called alley cropping. As the longer-term tree crop matures, the crop generates annual income (Fig. 8.3). Preferred tree species include oak, walnut, ash, and pecan which provide timber of high value. Potential specialty high-value crops include fruits, vegetables, flowers, herbs, nursery stock, etc., which boosts economic production (Table 8.1).

8.3.1.2 Hedgerows

Hedgerows are established for economic and environmental purposes to redirect the wind by planting single or multiple rows of trees, shrubs, or grasses (Fig. 8.4). Tree species used for hedgerows include *Eucalyptus*, *Cassia*, *Prosopis*, *Leucaena*, *Casuarina*, *Acacia*, and *Dalbergia*. The major benefits of hedgerows include lowering of wind velocity, protection for plants, provision of shelter, reduction of soil erosion, increase in carbon storage, improvement in wildlife habitat, and reduction of airborne chemical drift.

Fig. 8.2 Agrisilvicultural system

Fig. 8.3 Agrisilvicultural system—alley cropping (*left*, walnut and pecan; *right*, potatoes and hazel coppice)

Table 8.1 Effect of alley cropping of maize with *Leucaena leucocephala* on grain yield

Agro-forestry system	Grain yield (kg ha)
Alley cropping (maize + *Leucaena leucocephala*)	4707
Control	3740
Difference	967
Relative %	25.8

Kang (1993)

Fig. 8.4 Hedgerows

8.3.1.3 Forest Farming

Shade-tolerant, high-value specialty crops such as medicinal plants (ginseng, goldenseal, pacific yew, black cohosh, witch hazel, and mayapple), food products (mushrooms, papaya, chestnuts, pecans, and persimmon), and decoratives (palms, ferns, forsythia, pine straw, bittersweet, corkscrew willow, and red-twig dogwood) are grown under the protection of forest trees which also provide high-value timber (Fig. 8.5). The benefits of forest farming include income diversification, improved diversity and value of existing forests, and cash flow increase from non-timber products.

8.3.2 Silvopastoral Systems (Trees + Pasture and/or Animals)

Production of forage, timber, and livestock is combined on the same land (Fig. 8.6). Trees protect livestock from temperature stresses by providing shelter (shade). Longer-term returns are provided by trees, while annual income is generated by

Fig. 8.5 Forest farming (mushrooms, ginseng, and decorative palms)

Fig. 8.6 Silvopastoral system

livestock. Leaves and branches of trees can be pruned and used as livestock feed. Some of the fodder-producing trees include *Acacia* spp., *Albizia* spp., *Sesbania* spp., *Dalbergia* spp., *Cassia siamea*, *Erythrina indica*, *Gliricidia sepium*, *Leucaena leucocephala*, and *Ficus* spp. The types of forage crops include cool season (grasses and legumes) and warm season (Bahia grass, Bermuda grass, switch grass, big bluestem, little bluestem, Indian grass, and legumes) crops. The common livestock species used include cattle, goats, sheep, horses, and poultry.

The multiple benefits of silvopasture include lower animal stress (increased milk yields, increased weight gain, higher conception rates, and less medical bills), improved plant vigor (improved nutrient cycling, healthier trees, faster tree growth, lengthened forage-growing season, and higher forage quality), improved wildlife habitat (suitable cover, sufficient food supplies, adequate habitat, and well-distributed water), reduced wildfire risk (reduced winds, stand density, fuel loading, and ladder fuels), and short- and long-term cash flow.

The silvopastoral system is classified into three categories, viz., protein bank, live fence of fodder trees and hedges, and trees and shrubs on pasture.

8.3.2.1 Protein Bank

In order to supply feeds (cut and carry fodder production) during the fodder deficit period in winter, several multipurpose trees (trees rich in proteins) are grown in or around farmlands and range lands. Some of the trees rich in proteins include *Acacia nilotica, Albizia lebbeck, Azadirachta indica, Leucaena leucocephala, Gliricidia sepium,* and *Sesbania grandiflora.*

8.3.2.2 Live Fence of Fodder Trees and Hedges

To protect the property from stray animals or other biotic influences, several fodder trees are grown at hedges as live fence. Some of the trees grown for live fence include *Gliricidia sepium, Sesbania grandiflora, Erythrina* sp., and *Acacia* sp.

8.3.2.3 Trees and Shrubs on Pasture

In view of supplementing forage production, different tree and shrub species are grown irregularly or arranged according to some systematic pattern. *Acacia nilotica, Acacia leucophloea, Tamarindus indica,* and *Azadirachta indica* are some of the trees and shrubs grown for pasture.

8.3.3 Agrosilvopastoral Systems (Trees + Crops + Pasture/ Animals)

The skillful management of trees, crops, and pasture/livestock on a same piece of land is called agrosilvopastoral systems (Fig. 8.7).

This system is classified into two categories, namely, home gardens and woody hedgerows.

8.3.3.1 Home Gardens

The primary objective of most home gardens is food production. Growing of various species of trees, bushes, vegetables, and other herbaceous plants in dense and in random or spatial and temporal arrangements in homesteads is called home garden. The maintenance of livestock can also be supported by home gardens. The family labor can intensively manage the whole tree–crop–animal units which are highly productive, sustainable, and very practicable.

8.3.3.2 Woody Hedgerows

For browse, mulch, green manure, and soil conservation, the different woody hedges, particularly fast-growing and coppicing fodder shrubs and trees, are planted. Some of the woody hedge shrubs and trees include *Erythrina* sp., *Leucaena leucocephala,* and *Sesbania grandiflora.*

8.3.4 Other Agro-forestry Systems

Some of the other agro-forestry systems include apiculture with trees, aquaculture with trees, and mixed woodlots.

Fig. 8.7 Agrosilvopastoral system

Fig. 8.8 Apiculture with trees

8.3.4.1 Apiculture with Trees

The boundary of agricultural fields is planted with different honey (nectar)-producing trees frequently visited by honeybees (Fig. 8.8).

8.3.4.2 Aquaculture with Trees

The fish production and bund stabilization around fishponds are the main objectives of this system. Around the fishponds, different trees and shrubs preferred by fish are planted so that leaves are used to feed fish.

8.3.4.3 Mixed Woodlots

For different purposes like firewood, timber, fodder, soil conservation, and soil reclamation, some location-specific multipurpose trees are grown mixed or separately planted on the less fertile portion of the farm which is set aside for trees. Tree species commonly established as woodlots include *Grevillea robusta, Markhamia lutea, Casuarina equisetifolia, Melia volkensii, Prunus africana, Gmelina arborea,* and *Terminalia brownii.*

8.4 Products

The wide range of products produced from agro-forestry systems are as follows:

- Craft products (basketry, floral arrangements, natural dyes)
- Ecological services (air and water quality, biodiversity conservation, carbon sequestration, and soil enrichment)
- Fiber (bark, cork, pulp for paper, rubber, and woodchip mulch)
- Fodder and forage (grasses and legumes)
- Food (animal products, arable crops, fruit, leaves, mushrooms, nuts, oils, and vegetables)
- Fuel (charcoal, fuel wood, willow or hazel coppice)
- Gardening materials (bean poles, fencing, hurdles, pea sticks)
- Gums and resins
- Medicinal products [ginseng, goldenseal, witch hazel (Rao et al. 2004)]
- Recreation (agritourism, hunting, sport)
- Thatching and hedging materials (binders, spars, and stakes)
- Timber (construction and furniture making)

These agro-forestry products improve the local economy by enhancing:

- Cultural and natural environment improvements
- Diverse range of local products and economies
- Diversification of agricultural landscape
- Food and energy security
- Rural skills
- Stability of economy

8.5 Benefits

The various agronomical, environmental, and socioeconomic benefits of agro-forestry are as follows.

8.5.1 Agronomical Benefits

The agronomical benefits of agro-forestry include soil management (reduce soil erosion and improve soil health and quality), water management, reducing inputs, and pest management.

8.5.1.1 Soil Management

The soil management practices are formulated in order to improve soil fertility, quality, and health and to reduce soil erosion and degradation.

1. *Soil erosion*: The tree roots, besides enhancing structural stability of the soil, can also increase water infiltration and improve water storage by enhancing the number of soil pores. The excess surface water flow is rapidly discharged through macropore channel and permits air and moisture to enter into the soil, thereby reducing runoff and soil erosion. The physical barriers such as leaf litter, pruning debris, and tree roots and trunks also reduce surface water flow and sediment (Udawatta et al. 2008).
2. *Soil fertility*: The trees in agro-forestry have deep taproot systems which access nutrients from lower soil horizons and return them to the upper soil layers through leaf fall, thereby increasing soil nutrient pools and turnover and reducing reliance on external nutrient inputs. The 6-year-old poplar trees add 7 kg N ha^{-1} yr^{-1} to the adjacent alley crop through leaf litter (Thevathasan and Gordon 2004). Similarly, by intercepting rainfall through leaf flow (rainwater falling through tree canopies) and stem flow (rainwater falling down branches and stems) from hybrid poplar and silver maple trees, it can add 10.99 and 15.22 kg N ha^{-1} yr^{-1}, respectively, to alley crops (Thevathasan and Gordon 2004).

   In tropical agro-forestry systems, the green mulch from leguminous trees such as black locust (*Robinia*), mesquites (*Prosopis*), alder (*Alnus*), and oleaster (*Eleagnus*) increases soil fertility of adjacent alley crops (Yobterik et al. 1994). The tree legume red alder (*Alnus rubra*) is able to fix about 32–58 % of the total nitrogen in alley-cropped maize (Jose et al. 2004). In maize intercropped with a nitrogen-fixing tree (*Gliricidia sepium*), a 3- to 4-fold yield increase was noticed in maize. Likewise, maize intercropped with a legume tree (*Faidherbia albida*) can sustain a maize yield of 4000 kg/ha with no supplementary fertilizers. The *Faidherbia albida* legume tree supplies 300 kg of complete fertilizer and 250 kg of lime to maize alley crop (Fig. 8.9) (Garrity 2010).
3. *Microbial diversity*: The root exudates, and litter quality and quantity can increase microbial diversity [including arbuscular mycorrhizal fungi (AMF)], increased enzyme activity, and greater stability in agro-forestry alley cropping systems (Lacombe et al. 2009). The tree roots are generally associated with AMF that increase uptake of plant nutrient (especially phosphorus) and plant growth, soil aggregation and stability, and litter decomposition rates and increase alley crop yields while reducing the need for chemical fertilizer inputs (Rillig et al. 2002; Schädler et al. 2010).

8.5.1.2 Water

By planting tree strips adjacent to water courses, nonpoint source water pollution from crops and grazed pastures can be reduced through reducing surface runoff from fields and bank erosion, filtering surface runoff, groundwater runoff, and stream water (Udawatta et al. 2010). The reduced runoff and increased infiltration can be achieved through riparian (riverbank) buffers (Fig. 8.10) and other agroforestry systems (Bharati et al. 2002).

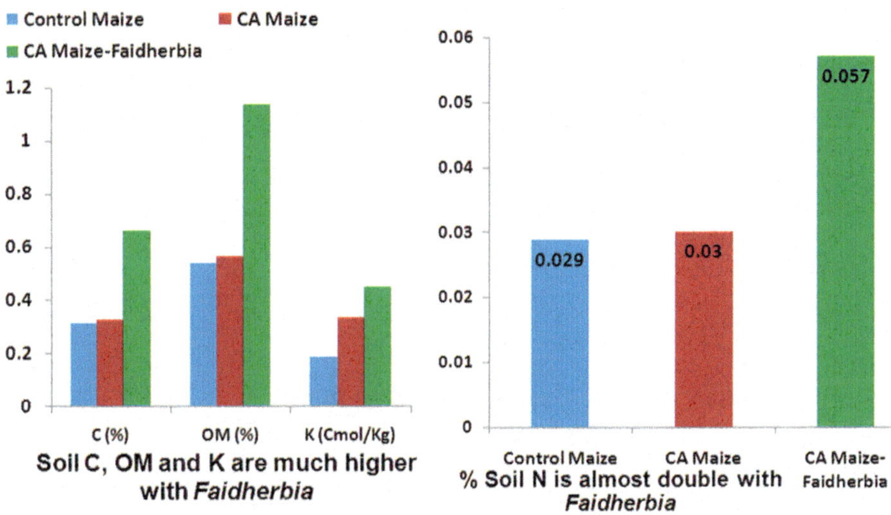

Fig. 8.9 Effect of fertilizer legume tree, *Faidherbia* sp., on soil fertility of maize

Fig. 8.10 Riparian (riverbank) buffers

8.5.1.3 Reducing Inputs

The aim of agro-forestry systems is to increase "eco-efficiency" and thereby reduce the need for inputs through minimum losses and maximum internal cycling of nutrients. The attributes for eco-efficient farming include increasing functional biodiversity conservation to support ecological processes and efficient use of resources with maximum inclusion of renewable resources (BCPC 2004).

8.5.1.4 Pest Management

The agro-forestry systems normally have reduced pest problems (Stamps and Linit 1998) due to the following reasons (Vandermeer 1989):

- Greater niche diversity and complexity make it more difficult for pests to find their host plants.
- Plant species may act as trap or repellent crops.
- Higher plant diversity increases natural enemy population and pressure on pests.
- Provision of alternative food sources (e.g., flowers supply pollen, nectar) for adult parasitoids.
- Provision of sites for mating, oviposition, and overwintering (Stamps and Linit 1998).

8.5.1.5 Climate Change Mitigation and Adaptation

The agro-forestry can be used as a potential tool to overcome the problems of climate change through adaptation and mitigation (Verchot et al. 2007; Lal 2004). Proper forest management helps to reduce atmospheric carbon and increase carbon sequestration through (Montagnini and Nair 2004):

- Conservation of soil carbon and biomass in forests, through agricultural systems, improved efficiency of wood processing, reduced logging effect, fire protection, and more effective use of burning in forests
- Reforestation, afforestation, restoration of degraded lands, and increasing growth rates by improved silvicultural techniques
- Substitution of carbon by increased use of biofuels, increased use of harvesting waste as feedstock for biofuel, and enhanced conversion of forest biomass into durable wood products to replace energy-intensive materials

In semiarid, subhumid, humid, and temperate regions, the average carbon sequestration in agro-forestry systems is estimated at 9, 21, 50, and 63 Mg C ha^{-1}, respectively (Montagnini and Nair 2004). Based on a worldwide estimate of 1023 million ha of agro-forestry, it is estimated that the carbon sequestration through agro-forestry accounts for 1.9 Pg of carbon over 50 years (Nair et al. 2009).

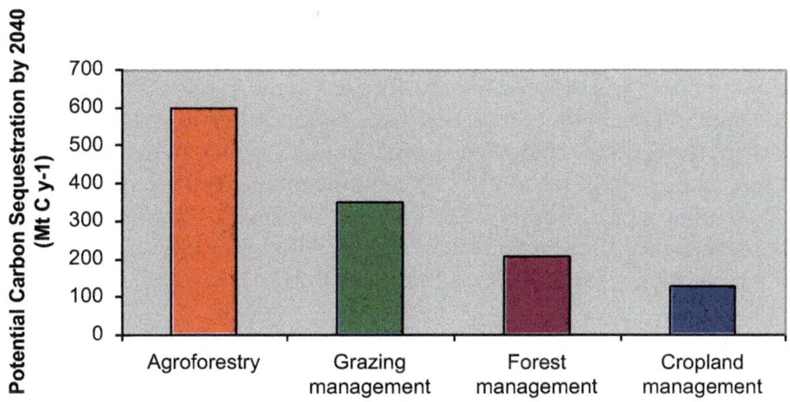

Fig. 8.11 Effect of four land-use systems on carbon sequestration (IPCC 2000)

Carbon sequestration potential is highest among four land-use systems ranging from 1.5 to 3.5 Mg C ha^{-1} yr^{-1} (Fig. 8.11).

8.5.2 Environmental Benefits

The integration of trees into farming systems (agro-forestry) helps to protect the environment and provide a number of ecosystem services by converting the waste products into valuable resources to enhance biomass production (Mirck et al. 2005).

8.5.2.1 Rehabilitation of Polluted Soils

The tree crops such as willows can absorb heavy metals from soil into their biomass and convert them to nontoxic breakdown compounds. The willows through evapotranspiration also control water dynamics including contaminated groundwater flow and water penetration into soils (Volk et al. 2006).

8.5.2.2 Recycling of Nutrients

In agro-forestry, the nutrients leached out of the crop rooting zone are captured by trees with the deep root system, thus reducing pollution. The nutrient use efficiencies are also increased in agro-forestry by recycling nutrients as leaf litter and root decomposition (Jose et al. 2004).

8.5.2.3 Reducing Pollution

The pollution runoff is reduced in agro-forestry to the extent of 70–90 % for suspended solids, 60–98 % for phosphorus, and 70–95 % for nitrogen through planting buffer strips (Borin et al. 2010). The switch-grass in woody buffer decreased 97 % of the sediment, 94 % of the total N, 85 % of the nitrate N, 91 % of the total P, and 80 % of the phosphate P in the runoff (Lee et al. 2003). The *Eucalyptus*-pasture system reduced the saline groundwater table by 2.0 m over a 7-year period as compared to pasture-only sites (Bari and Schofield 1991).

8.5.2.4 Enhanced Biodiversity

The diverse components chosen for agro-forestry such as the wild plants and animals occurring on the farmland enhance the associated biodiversity. The ways by which agro-forestry helps to maintain the biodiversity are as follows (Jose 2009):

- Alleviates resource use pressure
- Conserves biological diversity by providing ecosystem services like soil erosion control and water recharge
- Preserves germplasm of certain species
- Provides habitat for plant and animal species
- Provides habitats connectivity through corridors

8.5.2.5 Prevention of Greenhouse Gas Emissions

There is a potential for mitigating greenhouse gases such as CO_2 and N_2O and increasing the CH_4 sink strength by tropical agro-forestry systems (Mutuo et al. 2005). Since the recycled nitrogen from leaf litter provides a quantifiable contribution to adjacent crops, there is a reduced need for applications of inorganic N additions and thus reduce N_2O emissions. The leaching of nitrates in agro-forestry systems is reduced by 50 % due to denitrification, thus reducing NOx emissions (Thevathasan and Gordon 2004).

8.5.3 Socioeconomic Benefits

Improving the livelihoods of rural poor small holders is the main objective of implementing agro-forestry systems in the tropics. Some of the socioeconomic advantages of integrating trees into the tree-based intercropping system include diversification of local products and economies, improved food and fuel security, increased economic stability, diversification of rural skills, landscape diversification, and improvements to the cultural and biological environment (Garrity 2004).

8.5.3.1 Economics

The financial benefits in agro-forestry systems are due to enhanced diversity and productivity that are affected by price fluctuations of timber, livestock, and crops. It also provides annual and periodic revenues from multiple outputs that reduce the risks associated with monoculture (Benjamin et al. 2000). Silvoarable (Benjamin et al. 2000) and silvopastoral (Benavides et al. 2009) systems gave increased profitability than monoculture systems.

The ecosystem services provided by agro-forestry systems such as soil formation, soil carbon accumulation, nitrogen regulation, hydrological flow into groundwater reserves, biomass production, food and forage production, pest management, and wild pollinators generated a total income of US \$1074 ha^{-1} (Porter et al. 2009).

8.5.3.2 Diversification of Local Products

The diversified range of agro-forestry products should be produced locally to reduce transportation costs. This stimulated local economy by establishing major links and relationships between local community businesses and the end user, so that the money paid to obtain these products is spent locally (Volk et al. 2006). The reduced input costs and increased "eco-efficiency" of the farming system can be obtained by using tree products on the farm for fence posts, fodder, or bioenergy.

8.5.3.3 Employment of Rural Skills

The tree-based cropping systems are intensive and require high manual labor input for their management which generates skill-based (coppicing and hedge laying) employment in rural areas (Yamada and Gholz 2002).

8.5.3.4 Reduced Reliance on Fossil Fuels

The fossil fuel consumption can be reduced by the production of renewable energy through coppice systems, and the use of a by-product of timber production can reduce the use of fossil fuels for heating and cooking. The use of agrochemicals such as inorganic fertilizers and chemical pesticides can be substituted by internal cycling of nutrients and enhanced pest and disease control through agro-forestry systems. The need for long-distance transportation of goods can be prevented by producing diversified range of agro-forestry products locally.

8.5.3.5 Aesthetics

The agricultural landscapes can be made attractive for their visual appeal by integration of trees with crops as in the case of grazed orchards, parkland, and wood pastures (McAdam et al. 2008).

8.6 Limitations

- Adoption under land tenure system is difficult.
- Allelopathic effect may reduce the yield.
- Labor requirement is increased.
- Skilled management is needed.

8.7 Conclusions

Agro-forestry (integration of trees with agriculture) has the potential to increase productivity, protect the environment, and deliver ecosystem services including soil, water, and air quality regulation, biodiversity support and cultural services such as carbon sequestration, biodiversity conservation, soil enrichment, and air and water quality. Research, policy, and dissemination of information are essential to promote agro-forestry into the mainstream.

There is a need for research on species interactions, economic as well as environmental benefits, watershed impacts on nitrate leaching, biodiversity enhancement, homegrown timber and fuel to reduce imports and increase renewable energy production, and climate change mitigation and adaptation.

The agro-forestry systems can be adopted widely by giving incentives to establish new agro-forestry systems and eliminating constraints through supportive policies of the government (Current et al. 2009). In order to promote agro-forestry as a mainstream land-use system, it is necessary to raise awareness among policy makers.

Transfer of technology through extension projects and outreach support is very important to bring awareness among farmers by demonstrating the practical management skills and economic viability for adoption of agro-forestry practices on a wider scale (Graves et al. 2008; Current et al. 2009). Research, extension, and policy are the three key areas of activity required for promoting agro-forestry into the mainstream.

References

Bari MA, Schofield NJ (1991) Effects of agroforestry-pasture associations on groundwater level and salinity. Agrofor Syst 16:13–31

BCPC (2004) *Enhancing the eco-efficiency of agriculture*. British Crop Protection Council, Alton, Hampshire

Benavides R, Douglas GB, Osoro K (2009) Silvopastoralism in New Zealand: review of effects of evergreen and deciduous trees on pasture dynamics. Agrofor Syst 76:327–350

Benjamin TJ et al (2000) Defining competition vectors in a temperate alley cropping system in the Midwestern USA 4. The economic return of ecological knowledge. Agrofor Syst 48:79–93

Bharati L et al (2002) Soil-water infiltration under crops, pasture and established riparian buffer in Midwest USA. Agrofor Syst 56:249–257

Borin M, Passoni M, Thiene M, Tempesta T (2010) Multiple functions of buffer strips in farming areas. Europ J Agron 32:103–111

Current DA et al (2009) Moving agroforestry into the mainstream. Agrofor Syst 75:1–3

Garrity DP (2004) Agroforestry and the achievement of the Millennium Development Goals. Agrofor Syst 61:5–17

Garrity D (2010) Creating an evergreen agriculture in Africa: for food security and environmental resilience. World Agroforestry Centre, Nairobi

Graves AR et al (2008) Farmer perceptions of silvoarable systems in seven European countries, Chapter 4. In: Rigueiro-Rodríguez A et al (eds) Agroforestry in Europe: current status and future prospects. Springer, Dordrecht

IPCC (2000) Summary for policymakers on land use, land-use change, and forestry. IPCC Special Report, Intergovernmental Panel on Climate Change. ISBN: 92-9169-114-3

Jose S (2009) Agroforestry for ecosystem services and environmental benefits: an overview. Agrofor Syst 76:1–10

Jose S, Gillespie AR, Pallardy SG (2004) Interspecific interactions in temperate agroforestry. Agrofor Syst 61:237–255

Kang BT (1993) Alley cropping: past achievements and future directions. Agrofor Syst 23(2):141–155

Lacombe S et al (2009) Do tree-based intercropping systems increase the diversity and stability of soil microbial communities? Agric Ecosyst Environ 131:25–31

Lal R (2004) Soil carbon sequestration impacts on global climate change and food security. Science 304:1623–1627

Lee KH, Isenhart TM, Schultz RC (2003) Sediment and nutrient removal in an established multi-species riparian buffer. J Soil Water Conserv 58:1–8

McAdam J et al (2008) Chapter 2: Classifications and functions of agroforestry systems in Europe. In: Rigueiro-Rodríguez A et al (eds) Agroforestry in Europe: current status and future prospects. Springer, Dordrecht

Mirck J et al (2005) Development of short-rotation willow coppice systems for environmental purposes in Sweden. Biomass Bioenergy 28(2):219–228

Montagnini F, Nair PKR (2004) Carbon sequestration: an underexploited environmental benefit of agroforestry systems. Agrofor Syst 61:281–295

Mutuo P et al (2005) Potential of agroforestry for carbon sequestration and mitigation of greenhouse gas emissions from soils in the tropics. Nutr Cycl Agroecosyst 71:43–54

Nair PKR (ed) (1989) Agroforestry systems in the tropics. Kluwer, Dordrecht

Nair PKR, Kumar BM, Nair VD (2009) Agroforestry as a strategy for carbon sequestration. J Plant Nutr Soil Sci 172(1):10–23

Porter J et al (2009) The value of producing food, energy and ecosystem services within an agro-ecosystem. Ambio 38(4):186–193

Rao MR, Palada MC, Becker BN (2004) Medicinal and aromatic plants in agroforestry systems. Agrofor Syst 61:107–122

Rillig MC, Wright SF, Eviner VT (2002) The role of arbuscular mycorrhizal fungi and glomalin in soil aggregation: comparing effects of five plant species. Plant Soil 238:325–333

Schädler M, Brandl R, Kempel A (2010) "Afterlife" effects of mycorrhisation on the decomposition of plant residues. Soil Biol Biochem 42:521–523

Stamps WT, Linit MJ (1998) Plant diversity and arthropod communities: implications for temperate agroforestry. Agrofor Syst 39:73–89

Thevathasan NV, Gordon AM (2004) Ecology of tree intercropping systems in the North temperate region: experiences from southern Ontario, Canada. Agrofor Syst 61:257–268

Udawatta RP et al (2008) Agroforestry and grass buffer effects on pore characteristics measured by high-resolution X-ray computed tomography. Soil Sci Soc Am J 72(2):295–304

Udawatta RP, Garrett HE, Kallenbach RL (2010) Agroforestry and grass buffer effects on water quality in grazed pastures. Agrofor Syst 79(1):81–87

Vandermeer J (1989) The ecology of intercropping. Cambridge University Press, Cambridge

Verchot L et al (2007) Climate change: linking adaptation and mitigation through agroforestry. Mitig Adapt Strateg Glob Chang 12:901–918

Volk TA et al (2006) The development of short-rotation willow in the Northeastern United States for bioenergy and bioproducts, agroforestry and phytoremediation. Biomass Bioenergy 30:715–727

Yamada M, Gholz HL (2002) An evaluation of agroforestry systems as a rural development option for the Brazilian Amazon. Agrofor Syst 55:81–87

Yobterik AC, Timmer VR, Gordon AM (1994) Screening agroforestry tree mulches for corn growth: a combined soil test, pot trial and plant analysis approach. Agrofor Syst 25:153–166

Part V
Selection of Suitable Cultivars

Chapter 9
High-Yielding and Adapted Varieties

Abstract The technology that is aimed at increasing plant productivity, product quality, nutritional value and health, resistance to biotic and abiotic stresses, and reduction of risks associated with climate change (e.g., yield losses due to pests, diseases, or droughts) is by the introduction of new cultivated species and improved crop varieties which also improve the livelihoods of farmers. The biotechnological approaches such as genetic engineering and molecular biology can play a major role in increasing the productivity of global agriculture, helping farmers to meet the food, feed, and other demands of a rapidly rising world population, while saving water and forest lands. The strengthening of production, distribution, and use of improved staple crops seeds for enhanced crop productivity, food security, and food self-sufficiency are some of the potentials of improved seeds for food security.

Keywords Biotic stresses • Abiotic stresses • Productivity • Nutritional quality • Genetically modified (GM) crops • Climate change

9.1 Introduction

It is projected that the world population is expected to reach over 9 billion by 2050. A significant population of over 800 million people is still suffering from hunger and malnutrition. In order to meet the food demand for the growing population, the food production has to be increased by 70 % globally and doubled in developing world by 2050. Hence, there is an urgent need and increased efforts to accelerate food production and productivity.

Food production and productivity can be enhanced by two ways:

- Adopting improved cultural methods such as application of different kinds and amounts of nutrients, increasing plant density, and pest and weed control that are responsible for about 50 % of yield gains.
- Genetic improvements of crop varieties (very high-yielding varieties/F1 hybrids of crops with high nutritive values, well adapted, resistant to diseases, pests, and all sorts of environmental stresses, and production hazards) are responsible for another 50 % of the yield gains.

P.P. Reddy, *Sustainable Intensification of Crop Production*,
DOI 10.1007/978-981-10-2702-4_9

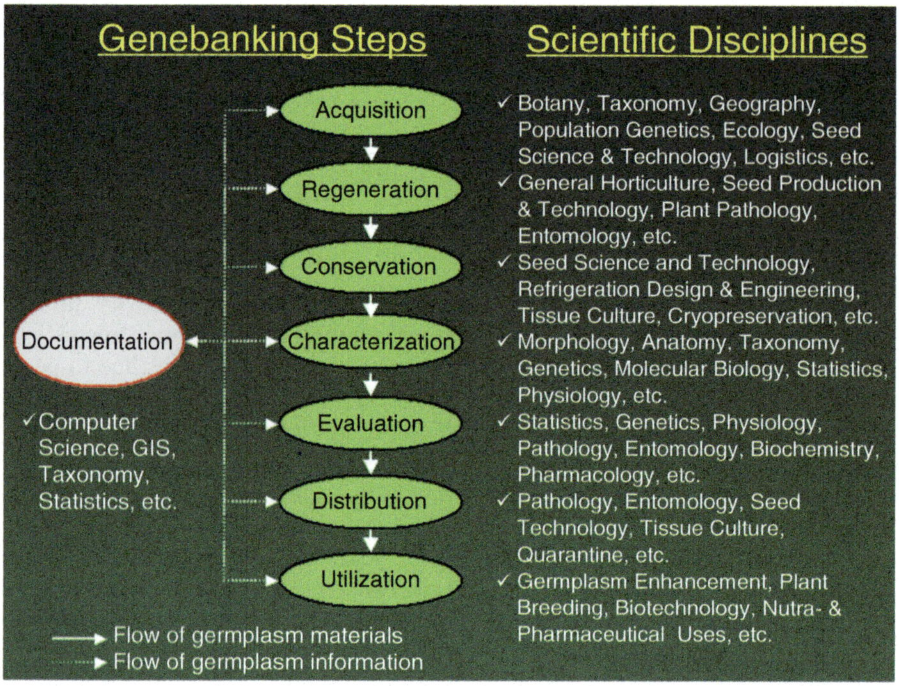

Fig. 9.1 Exploration, conservation, evaluation, cataloguing, distribution, and utilization of germplasm

The development of improved, fertilizer-responsive high-yielding varieties of wheat [developed by the International Maize and Wheat Improvement Center (CIMMYT), Mexico] and rice [developed by the International Rice Research Institute (IRRI), Philippines] during the early 1960s and their widespread adoptions by farmers, first in Asia and then in Latin America, marked the beginning of what is known as the green revolution. Since then, the development of new varieties through well-focused crop breeding programs have been extended to many other food and feed crops in developing countries.

9.2 Plant Genetic Resources

Immensely rich genetically variable forms of crop plants and their wild relatives are found in the primary and secondary centers of diversity. Crop genetic resources are collected from centers of diversity, gene sanctuaries, farmers' fields, gene banks, seed companies, and markets. The genetic resources for each crop have to be systematically explored, conserved, evaluated, catalogued, distributed, and utilized in order to meet the future needs of plant breeding (Fig. 9.1). The sum total of genes in a crop species is called germplasm. The genetic resources can be collected within the country (indigenous) or collected from foreign countries. This is the base material for launching a crop improvement program.

9.2.1 Conservation

The protection of genetic diversity of crop plants from genetic erosion can be achieved by two methods—in situ and ex situ conservation.

9.2.1.1 In Situ Conservation of Genetic Resources

The conservation of genetic resources in their natural habitats itself (which are also called national parks, gene sanctuaries, natural or biosphere reservoirs, protection of endangered species or areas) is called in situ conservation. In this method, land races, weedy relatives, wild relatives of plants, modern cultivars, obsolete cultivars, breeding stocks, and mutants are conserved.

The conservation of germplasm by this method is costly, since several areas of diversity have to be conserved for one species, and several problems are posed to manage such areas.

9.2.1.2 Ex Situ Conservation of Genetic Resources

This is the practical technique for conserving genetic resources under controlled experimental conditions (gene bank) outside their natural habitats. This method of germplasm conservation is cheap, since the entire genetic diversity of a crop species can be preserved at one place, and it is easy to handle, relatively safe, and requires minimum space. It can be conserved in the form of seed, pollen or meristem culture. The germplasm is conserved under short term (3–5 years), medium term (10–15 years), or long term (50–100 years). The international crop institutes which maintain the gene banks of different crops are presented in Table 9.1.

Based on the duration and importance of conservation, ex situ collections are classified into four groups—base collections, active collections, working collections, and core collections.

(i) *Working collections*: The working collections are short-term germplasm collections spread over 3–5 years. The material is maintained at 8–10 % moisture content and 5–10 °C temperature. These collections are made by breeders which are used for different breeding purposes.
(ii) *Active collections*: The active collections are medium-term germplasm collection spread over 10–15 years. The seeds are stored at moisture level of 8 % and at zero degree temperature. Under medium-term storage, cultured materials are also preserved.

Table 9.1 The international crop institutes which maintain the gene banks of different crops

Crops maintained	Location of gene bank
Rice	International Rice Research Institute (IRRI), Philippines
Maize, wheat, triticale, barley	International Maize and Wheat Improvement Center (CIMMYT), Mexico
Cassava, beans, rice, and maize	International Center for Tropical Agriculture (CIAT), Colombia
Cowpea, soybean, lima bean, cassava, sweet potato, yam, rice, maize	The International Institute of Tropical Agriculture (IITA), Nigeria
Potato	International Potato Center (CIP), Peru
Sorghum, pearl millet, pigeon pea, chickpea, groundnut	International Crops Research Institute for the Semi-Arid Tropics (ICRISAT), India
Durum wheat, barley, beans	International Center for Agriculture Research in the Dry Areas (ICARDA), Syria
Sugarcane	United States Department of Agriculture (USDA), Florida, USA

(iii) *Base collections*: The base collections are long-term germplasm collection spread over 20 years. The seeds are stored at zero degree temperature and low moisture levels (3–6 %), while other plant parts (including meristem cultures) are stored under cryopreservation.

(iv) *Core collections*: The core collections are a subset of the entire genetic diversity of a species with all the useful characters so that identification of useful entry becomes easily accessible to the breeders.

Ex situ conservation may be either in vivo or in vitro.

(v) *In vivo conservation*: Ex situ conservation of germplasm under field conditions (field gene banks) is called in vivo conservation.

(vi) *In vitro conservation*: The conservation of tissues, cells, genomes, genes, and different plant parts under controlled environmental conditions (at very low temperature of−196 °C in liquid nitrogen) is called in vitro conservation or in vitro gene banks.

9.2.2 Evaluation

The evaluation includes screening of germplasm with respect to economic, physiological, biochemical, genetical, morphological, entomological, and pathological attributes. It also involves identification of useful traits such as higher yields, good nutritional qualities, early maturity, dwarfness, and resistance to biotic and abiotic stresses. Evaluation is also useful to classify the germplasm into different groups and to find out the importance of individual collections.

9.2.3 Documentation

The compilation of the latest information about different activities of gene bank on characterization, conservation, distribution, and utilization of plant genetic resources is called documentation.

9.2.4 Distribution

The diverse plant genetic resources are made available at free of cost by the gene bank to the plant breeders and other users.

9.2.5 Utilization

The germplasm can be utilized as a variety, as a parent material in the hybridization, and as a variant in the gene pool. It can be utilized in the crop improvement programs for developing cultivars resistant to biotic and abiotic stresses, wider adaptability, and to improve the agronomic characters.

9.3 Crop Improvement

The crop improvement depends on finding good crop varieties that can give a good yield. Improved modern varieties of crops can be developed by using the following methods:

- Selection
  - Mass selection
  - Pure line selection
- Hybridization
  - Crosses with wild relatives
  - Hybrids
- Mutation
  - Irradiation
  - Chemicals
- Tissue culture
  - Embryo rescue
  - Somaclonal variation

9.3.1 Selection

The plant populations have been undergoing continuous speciation through natural selection under wild conditions. The crop plants of today were domesticated from their wild relatives or introduced from other continents or geographical areas. The natural selection is nothing but the differences (genetically based) in survival or reproduction which results in genetic change in a population.

In self-pollinating species of crop plants, two methods of selection are generally followed:

- Mass selection
- Pure line selection

9.3.1.1 Mass Selection

A crop strain is improved by selection of a large number of superior plants through visual assessment, pooling of their seeds, and developing a new variety. The easily observable characters include size of seed or grain, the panicle nature, height of plant, resistance to biotic or abiotic stresses, etc. The selected plant population raised shall be more superior and uniform in comparison with the original population. The selected population, however, exhibits some variations with respect to quantitative characters, which shall be likely subjected to progeny test. The performance assessment technique of a crop population by analyzing its progeny is called the progeny test.

In annual crops, the mass selection procedure involves identification of superior plants from a population and bulking their seeds, preliminary yield trials, multilocation trials, and seed multiplication and distribution, which shall take about seven to eight generations.

A huge number of visually similar superior plants are selected and their seeds are bulked during the first season. In a preliminary yield trial, the bulk seeds are grown during the second season in order to determine the improvement in characteristics of the superior plants over the variety from which the original selection is made. The multilocation trials are carried out from the third season onward by subjecting the superior seeds to the procedure of the variety release. Then the seeds of the superior variety are multiplied and distributed to the farmers.

The benefits of varieties developed through mass selection include:

- The procedure of mass selection is simple and less sophisticated.
- Since the superior varieties are selected from a large number of plants, they are more widely adopted compared to pure line selection.
- Considerable degree of genetic variability is retained through mass selection.

The limitations of mass selection procedure include:

- The level of superiority is limited.

- They are nonuniform as compared to pure line selection.
- The genetic stability of a variety cannot be assessed without the progeny test.

9.3.1.2 Pure Line Selection

In a self-pollinated crop, pure line selections are made in the first season based on phenotypical characteristics from a large number of superior plants and harvested individually from a single homozygous self-pollinated plant. Hence, all the individuals within a pure line selection have the same genotype, and variations, if any, are due to the environmental influences. Hence, the technique of pure line selection is also called individual plant selection.

The individual progenies are raised and evaluated and the best progeny is selected, the undesirable progenies are rejected during the second season, and the seeds of each line are collected separately. During the third season, preliminary yield trial of the remaining progenies is carried out, inferior progenies are rejected, and the seeds of each line are collected separately. From the fourth to seventh season, multilocation replicated yield trials are carried out to select the superior progeny and reject inferior progeny lines. The most adaptable and superior progeny is released as a new improved variety after multilocation trials during the eighth season. The seeds are multiplied and distributed to farmers.

The benefits of developing crop varieties from pure line selection are as follows:

- Extremely uniform in their behavior (flowering, maturity, etc.).
- In seed certification programs, offtypes can be easily identified.
- Maximum possible improvements over the original variety are incorporated.

The limitations of pure line selection include:

- The varieties are short lived.
- Technique is very tedious and takes a long time to release the variety.
- Very narrow range of adaptations to environmental fluctuations is possible.

9.3.2 Hybridization

Crop variety improvement is done by hybridization through plant breeding. The crossing between genetically different plants to obtain crop cultivars having useful characteristics like high yields, good quality, and disease resistance is called hybridization. Hybridization can be carried out between different varieties of the same species (intervarietal), between different species of the same genus (interspecific), or between different genera (intergeneric).

The crop improvement through application of genetics principles is called plant breeding. This is carried out by crossing the plants with useful characteristics and selecting progeny with superior performance and/or improved combinations of traits.

The crop improvement can also be undertaken through directed manipulation of the genotype at the DNA sequence level and introduction of new genes by genetic manipulation.

The steps for crop improvement to develop improved cultivars through plant breeding involve identification of genetic variability (good agronomic characteristics) within each crop, incorporation of useful agronomic characters to produce improved cultivars, their field evaluation for further improvement, release of superior cultivars, mass production of seeds, and distribution to the farmers.

The crossing between genetically different plants is undertaken to obtain crop cultivars having the following useful characteristics:

- Higher yield
- Improved quality
- Biotic and abiotic resistance
- Change in maturity duration traits
- Wider adaptability
- Desirable agronomic traits

9.3.2.1 Higher Yield

High-yielding crop varieties/hybrids have to be developed in order to increase the productivity per unit area. Genetically enhanced cultivars of crops have been developed in rice, maize, and wheat that have an increased growth rate and an increased percentage of usable plant parts. The practical ways to increase yield potential are as follows:

(i) *Plant architecture*: The plant architecture varies from crop to crop (e.g., more upright leaves in wheat, corn, and rice; smaller tassels in maize; larger panicles/spikes and larger stems in wheat and rice).
(ii) *Harvest index*: The harvest index (yield) can be increased in wheat, corn, rice, and soybeans by increasing biomass and thereby increasing the number of grains/kernels per unit area.
(iii) *High-density planting*: The soybean yield can be enhanced by increasing the plant population from 33 to 50–100 plant m^{-2}.
(iv) *Increased efficiency in utilizing soil nutrients*: The yield of new soybean cultivars (post-1976) became increasingly greater than that of the old cultivars (pre-1976) due to increased efficiency in utilizing soil nutrients by new cultivars (Fig. 9.2) (Specht et al. 1999).

The successes of plant breeding in enhancing the yields include:

- The increase in grain yield to the extent of 75 kg ha^{-1} yr^{-1} for maize can be attributed to breeding (2.5 tons/ha increase in yield every 30 years). In the USA, maize yields have increased from 60 % to 120 % (Cooper et al. 2004).
- Two- to three-fold increases in yields have been achieved in wheat and rice (green revolution varieties) (Knight 2003).

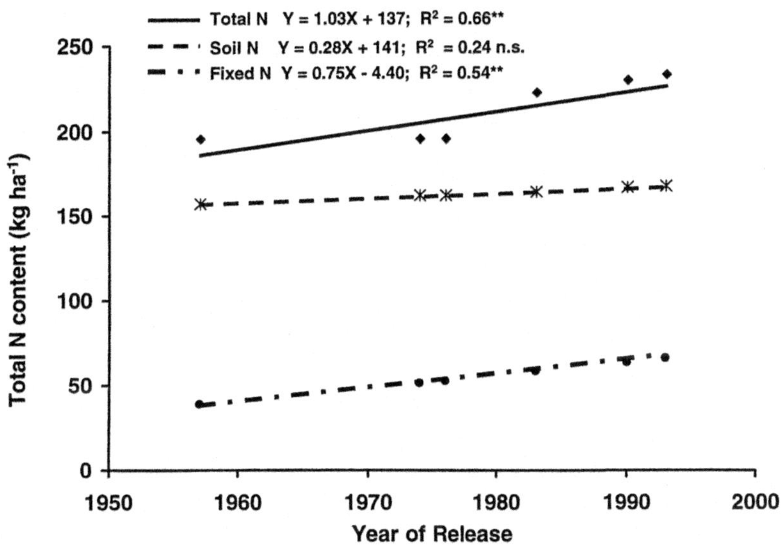

Fig. 9.2 Effect of year of release of soybean cultivars on total N content (Specht et al. 1999)

9.3.2.2 Improved Quality

There is a need to improve quality considerations of crop products which vary from crop to crop (e.g., fruits and vegetables, keeping quality; pulses, protein quality; oilseeds, oil quality; wheat, baking quality).

9.3.2.3 Biotic and Abiotic Resistance

The major factors for reduction in crop yields are abiotic and biotic stresses. For a successful crop production to meet with future food demands, there is a need to increase the tolerance/resistance levels to biotic (insects, mites, diseases, nematodes, weeds, etc.) and abiotic (drought, cold, heat, salinity, etc.) stresses, although each plant has some degree of innate tolerance to such stresses.

(i) *Biotic stresses*: Resistance to biotic stresses such as insect and mite pests; fungal, bacterial, viral, and nematode diseases; and weeds can be achieved through conventional breeding. However, the durable resistance is the need of the hour. The long-term theory and genetics for improved durable resistance is through molecular biology. Genetically modified crops (transgenics) can be developed by introducing one (or a few) foreign "good" gene/s into the best accepted cultivars.

(ii) *Abiotic stresses*: None of the cultivars are perfect since there is no completely stress-free environment. Increased tolerance to abiotic stresses such as drought, cold, heat, salinity, etc. is commonly associated with higher yield as well as with more stable performance.

9.3.2.4 Change in Maturity Duration Traits

The duration of the crop (from sowing to harvesting) plays a major role in multiple cropping systems. The shorter the duration of the crop, the more economical is the variety. Hence, the farmer can grow multiple crops in a year using short-duration crops. The harvesting process is made easy with uniform maturity and reduces losses during harvesting.

9.3.2.5 Wider Adaptability

The wider adaptability is essential to grow crops under different climatic conditions.

9.3.2.6 Desirable Agronomic Traits

The desired agronomic traits vary from crop to crop (e.g., cereal crops, should be dwarf, for higher productivity; fodder crops, more tillering, increased height, and profuse branching; rice, vigorous seedlings, drought tolerant, deep rooting system, erect leaves, input responsive, lodging resistant, rapid biomass development, and high harvest index). Hence, varietal development with desirable agronomic traits that are tailor-made will ensure higher yield.

9.4 Biotechnological Approaches

Modern biotechnology provides a new tool for breeding plants with much greater precision. A crop variety can be improved by inserting one or few genes to achieve desired traits. This produces genetically modified (GM) crops.

9.4.1 GM Crops and Conventionally Derived Varieties/Hybrids

In conventional breeding, in order to obtain one desirable disease resistance gene, 20,000–100,000 potentially negative genes are introduced through wide hybridization. For decades, induced mutagenesis (through irradiation chemicals) has been used to create genetic variants.

Genetic engineering has opened new opportunities which require new safeguards. GM crops are evolved by introducing one (or a few) foreign "desired" gene/s into the best accepted cultivar. Bridging the species barriers is the new thing with genetic engineering. This method can use recombinant DNA technology to move one or few foreign genes from one organism to any another organism.

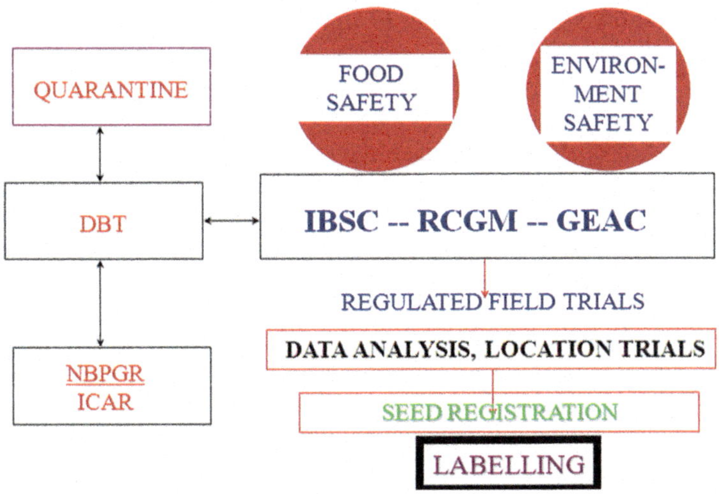

Fig. 9.3 System of release of transgenics and biosafety procedures

Even though the biotechnology may not be able to increase yield spectacularly in near future, but it is essential in developing countries over the long term to help yield gains to keep pace with global food needs. It can help plant breeders to develop insect- and disease-resistant cultivars. The biggest contribution of biotechnology to crop improvement is to enhance in depth knowledge about gene action and how to modify it to suit farmers' needs. It raises hopes of high-profit margins.

The systems of release of transgenics and biosafety procedures to be followed are presented in Fig. 9.3.

9.5 Systems of Varietal Development and Release

The systems of variety development and release under private and public sectors are presented in Figs. 9.4 and 9.5.

9.6 Seed Production and Distribution

9.6.1 Production

The production and productivity of crops can be increased to the extent of 50 % by using good quality seeds. The ancient Hindu manuscript aptly states that "Good quality seed on good land yields abundant produce." The seed is the most vital yet cheapest input which gives higher returns. Hence, for the successful crop

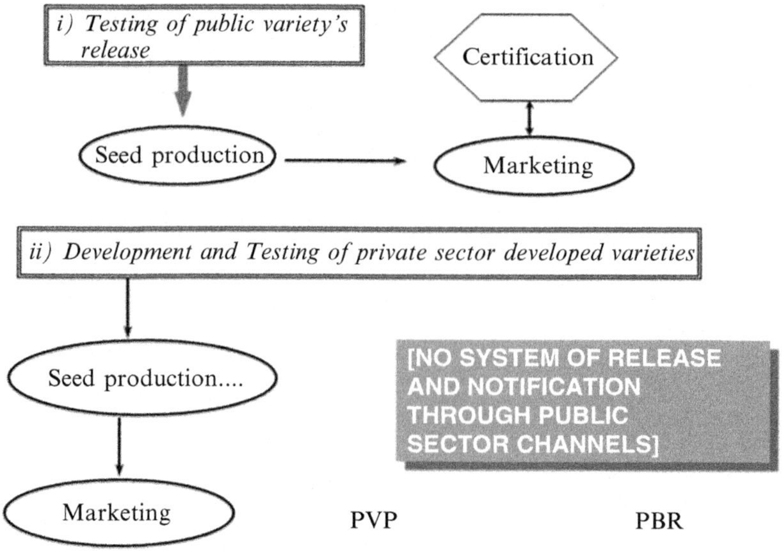

Fig. 9.4 System of variety development and release under private sector

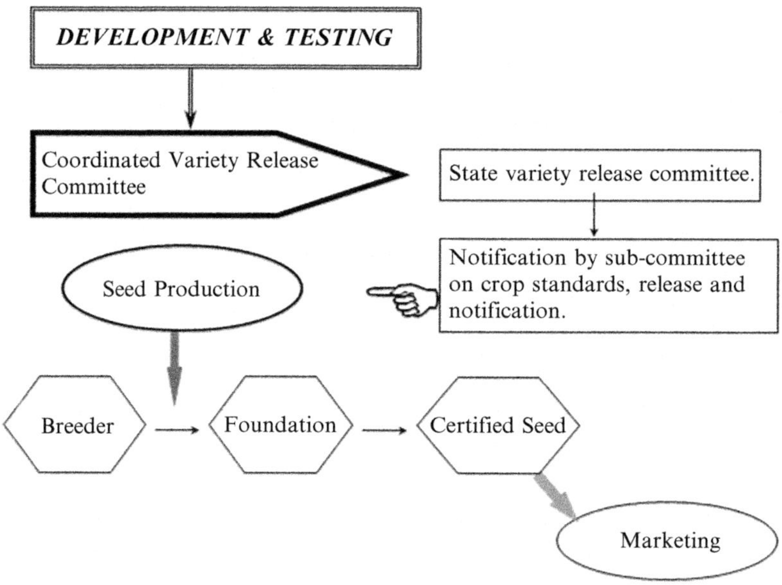

Fig. 9.5 System of variety development and release under public sector

production, the use of good quality seed is indispensible. Therefore, it should be ensured that good quality crop seeds are made available to the farmers.

The characteristics of good quality seed include:

- High yielding
- Superior quality
- Genetical purity
- Uniformity in size and other characters
- Higher germination percentage
- Cheap and easy availability
- Vigorous plant
- Biotic and abiotic stress tolerance
- Higher adaptability
- Low moisture content
- Free from seed-borne pests and diseases

The production of good quality seeds involve:

- Providing proper isolation distance
- Discarding offtypes
- Proper monitoring at critical crop growth stages
- Management of pests and diseases
- Selection of healthy fruits and seeds
- Proper seed extraction
- Correct drying, processing, packaging, and storage of seeds

9.6.2 Distribution

Distribution of seeds is a vital part of meeting the farmers' needs and requirements of quality seeds. This can be accomplished in one step, if seeds are supplied to farmers directly, or in multiple steps involving intermediate wholesalers and retailers (Table 9.2). The government, public sector agencies, cooperatives, and the private sector, or by a combination of all of these, are the channels through which seeds are supplied to the farmers.

9.7 Conclusions

The better performing and adapted cultivars have been developed using highly diverse and variable genotypes which have increased food production significantly. Multidisciplinary approach involving geneticists, plant breeders, pathologists, entomologists, weed scientists, biochemists, etc. is the need of the hour in developing varieties resistant to biotic and abiotic stresses. The critical area for the development

Table 9.2 Different channels for marketing seeds

Direct	Single level	Multilevel
Seed producer ↓ Intermediate storage possible ↓ Own sales force Own sales outlet Direct dispatch ↓ Farmer	Seed producer ↓ Intermediate storage possible ↓ Retail outlet ↓ Farmer	Seed producer ↓ National distributor wholesaler ↓ Regional distributor ↓ Possible sub-distributor ↓ Retail outlet ↓ Farmer

of climate-resilient agriculture would be in situ and ex situ conservation and sustainable use of genetic resources for food security. Biotechnological approaches need to be employed wherever conventional breeding is difficult. There is a need for developing crop varieties adapted to less favored areas and production systems, suited to a range of agro-ecosystems and farming practices, and resilient to climate change, which can produce food with higher nutritional value and desirable organoleptic properties and help improve the provision of ecosystem services. The crop varieties that are better adapted to ecologically based production practices can facilitate in achieving "evergreen revolution" to increase food production and productivity to meet the demands of growing world population.

References

Cooper M, Smith OS, Graham G, Arthur L, Feng L, Podlich DW (2004) Genomics, genetics, and plant breeding: a private sector perspective. Crop Sci 44:1907–1913

Knight J (2003) Crop improvement: a dying breed. Nature 421:568–570

Specht JE, Hume DJ, Kumudini SV (1999) Soybean yield potential – a genetic and physiological perspective. Crop Sci 39:1560–1570

Chapter 10
Seed Quality, Treatment, Rate, Depth, and Metering

Abstract In order to establish good plant stands, seed quality and seed rates are very crucial. Seed quality, treatment, rate, depth, and metering are important to enhance crop yields. This chapter discusses the enhancement of seed quality (through seed hydration, seed hardening, seed scarification, seed coating, seed pelleting, and integrated seed enhancement), seed treatment (to facilitate ease of sowing, pest management, and nutrient availability), seed rate (for stand establishment, to decrease the cost of seed, for weed control, and increase crop yields and higher benefits), and seed depth and metering (for precise placement of seeds and fertilizers for consistency in emergence of seedlings). The effect of drilling depth on seedling emergence of no-tilled wheat and lupine are also discussed.

Keywords Seed quality • Rate • Depth • Placement • Treatment • Seed metering • Seed delivery • Seedling emergence

10.1 Introduction

Seed is a ripe, fertilized ovule and a unit of reproduction of flowering plants. Seeds of superior quality are the most important inputs in agriculture. They are prerequisite for better crop production and productivity. Seed purity, viability, and vigor are the most important characters that determine the quality of seeds. Seed serves as a carrier of new technologies and act as a basic tool for secured food supply through increased crop yields.

10.2 Quality

The seed quality is dependent on the following three main characteristics:

- Varietal purity characteristics—true to its kind or variety
- Seed lot characteristics—contain a minimum of impurities
- Seed viability characteristics—have high germination and establishment rates in the field

P.P. Reddy, *Sustainable Intensification of Crop Production*,
DOI 10.1007/978-981-10-2702-4_10

10.2.1 Varietal Purity Characteristics

The genetic purity of the cultivar can be described by its physical, chemical, and crop characteristics. The physical characteristics of each variety such as specific length and width, shape, size, color, and aroma are tested by determining the 1000 grain weight and measuring color using the naked eye or a colorimeter. Several chemical tests such as alkali digestion color, gel consistency, amylase content, and brown rice protein can be used to differentiate between varieties. Normally, a crop variety always has the same plant growth habit, plant height, plant color, and time to maturity. However, slight variations are likely to occur particularly under unfavorable growth conditions.

10.2.2 Lot Characteristics

The seed lot characteristics include the seed size; damaged, deformed, or diseased seed; and level of impurities. The quality seed characteristics are fullness and plumpness which means that the seeds have enough food resources to provide healthy and vigorous seedling growth. The good quality seeds should be free from mechanical injury (damaged, deformed), seed-borne diseases, insects, and other extraneous matter which are likely to affect seed germination and vigor of seedlings. They should also be free from impurities such as seeds of other crops, weed seeds, and inert material (dirt, stones, or twigs).

10.2.3 Viability Characteristics

The main seed viability characteristics include stored moisture level, germination potential, and its vigor. For long-term seed storage, the moisture level should be preferable below 12–13 %. The germination percentage of seed should also be high and result in high establishment rates in the field. The good seed vigor is responsible for early and uniform crop stands that give good resistance for the growing seedlings against various biotic and abiotic stresses.

10.2.4 Enhancement of Quality

The seed quality can be enhanced by seed hydration/hardening (prehydration, priming), seed coating (pelleting, film coating), and integrated seed enhancement (Fig. 10.1) (Taylor et al. 1998).

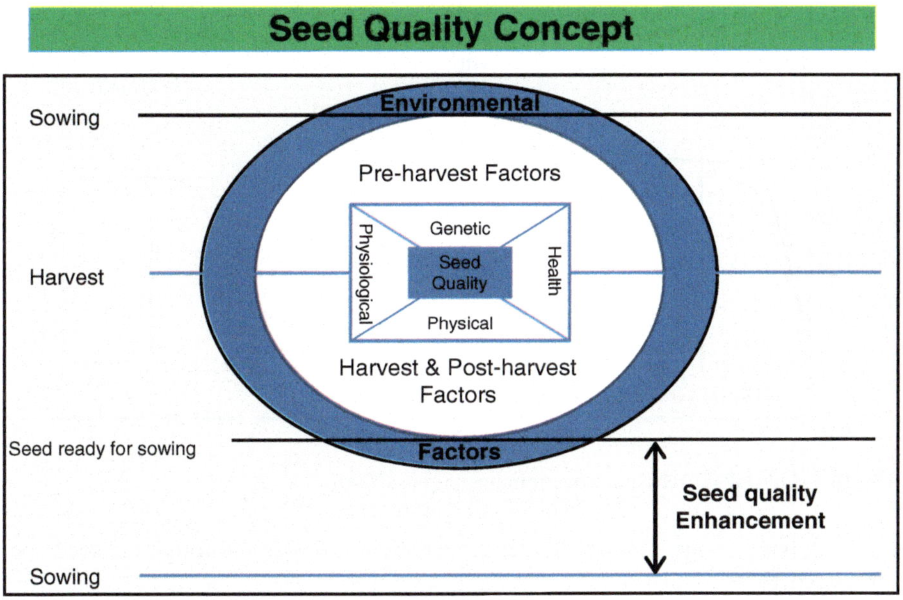

Fig. 10.1 Enhancement of seed quality (Taylor et al. 1998)

10.2.4.1 Hydration

1. *Prehydration*

In order to induce early germination, better seedling growth, root growth, and enhancing the yield potential of the crop variety, the pre-seed hydration process is carried out by soaking the seeds in water or dilute solution of growth-regulating compounds. The different types of hydration include prehydration (seed fortification, seed infusion) and priming (osmo-conditioning, halo-priming, bio-priming, hydro-priming, solid matrix priming, or matrix conditioning).

The seed fortification involves prehydration where seeds are soaked either in water or dilute solution of bioactive chemicals such as growth regulators, micronutrients, seed protectants, and vitamins. The impregnation of seeds with bioactive chemicals through organic solvents instead of water is called seed infusion, which helps to avoid the damage caused to the seed coat due to soaking in water, particularly the pulses (Halmer 2006).

The benefits of seed prehydration include promotion of early germination and good crop stand, faster water imbibitions, and swelling of seeds. Limited supply of O_2 to seed, toxicity of chemicals, and difficulty in handling large quantity of seed are some of the limitations of the above technique.

2. *Priming*

The seeds are soaked in osmotic solution before sowing which allows the seeds for controlled absorption of water that permits pre-germination metabolism to

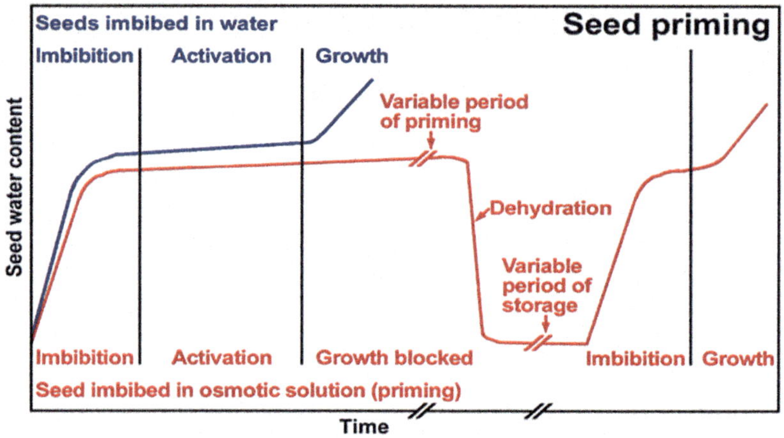

Fig. 10.2 Physiological basis of seed priming (Agrawal 2002)

proceed, but prevents actual emergence of radical protrusion through the seed coat (Bradford 1986).

The hydro-priming (drum priming) is a cheap and useful technique that is carried out by seed incubation in warm water for a limited period. The use of salts of chlorides, sulfates, nitrates, etc., in water is called halo-priming. In osmo-priming (osmo-conditioning), seeds are incubated in well-aerated solutions with a low water potential and then washed and dried (Halmer 2006). In order to confer slow absorption, the seed incubation in a solid, insoluble matrix (calcinated clay, vermiculite, peat moss, sand, Microcel) with a limited amount of water is called matrix priming (solid matrix conditioning) (McDonald 2000). The bio-priming (seed conditioning) is a biological seed treatment that refers to combination of seed hydration (physiological aspect of disease control) (Fig. 10.2) and inoculation (biological aspect of disease control) of seed with the help of beneficial organisms [fungi (*Trichoderma viride*, *T. harzianum*, *Pupureocelium lilacinum*) and bacteria (*Bacillus subtilis*, *Pseudomonas fluorescens*)] to protect seed (Halmer 2006).

The factors affecting priming include duration of priming, drying method, priming temperature and light, O_2 availability, and osmotic potential of the solution (Halmer 2006). The benefits of priming are prevention of absorption injury, controlling water absorption and salt priming, and supplying seeds with nitrogen and other nutrients for protein synthesis. The limitations of priming are limited O_2 supply to seed, chemicals toxicity, and difficulty in handling huge seed quantity.

10.2.4.2 Hardening

Seed hardening is a process of treating the seeds with certain chemicals (cotton with Cycocel or succinic acid, maize with KCl, sorghum and pearl millet with KH_2PO_4, groundnut with $CaCl_2$) or other materials (pulses with ash) to enhance drought

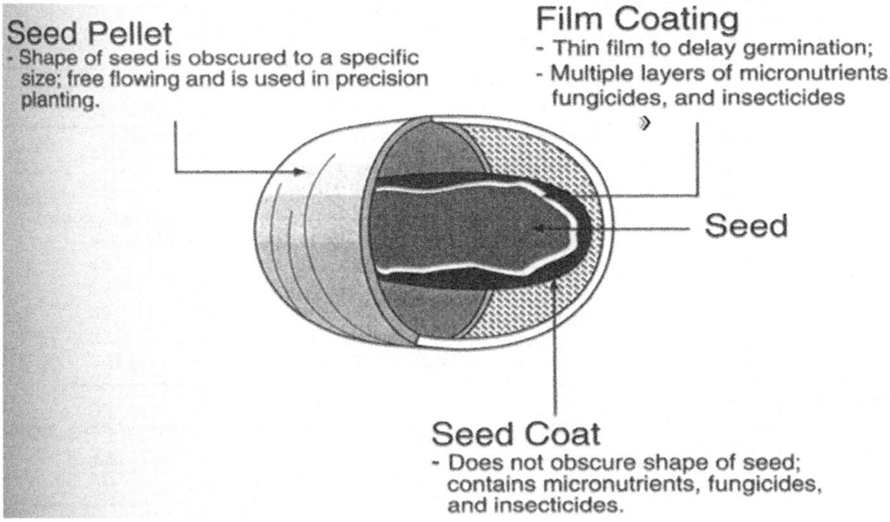

Fig. 10.3 Types of seed coating

tolerance in dry areas. The benefits of seed hardening include seed viability and water absorption enhancement, improvement of root development, and increases in relative leaf water content.

10.2.4.3 Scarification

Scarification is the process of breaking seed dormancy carried out at the time of sowing/few days before sowing by scratching, mechanically altering, or softening the seed coats to make them permeable to water and gases. The dormancy of seeds due to physiological reasons is also broken by seed treatment with cytokinins, ethylene, and gibberellic acid (GA_3).

10.2.4.4 Coating

Placement of chemicals on the seed coat to regulate and improve germination and to enhance seed placement and performance without altering the shape is called seed coating (Copeland and McDonald 2001). There are two types of seed coating, namely, seed coatings and film coatings. The application of seed coating with micronutrients, microbiological treatments, or fungicide directly placed on the seed without altering its shape is called seed coating. Seed coating with the precise amount of active ingredients in the form of thin film along with the liquid material directly onto the seed surface without obscuring its shape is called film coating (Fig. 10.3) (Copeland and McDonald 2001).

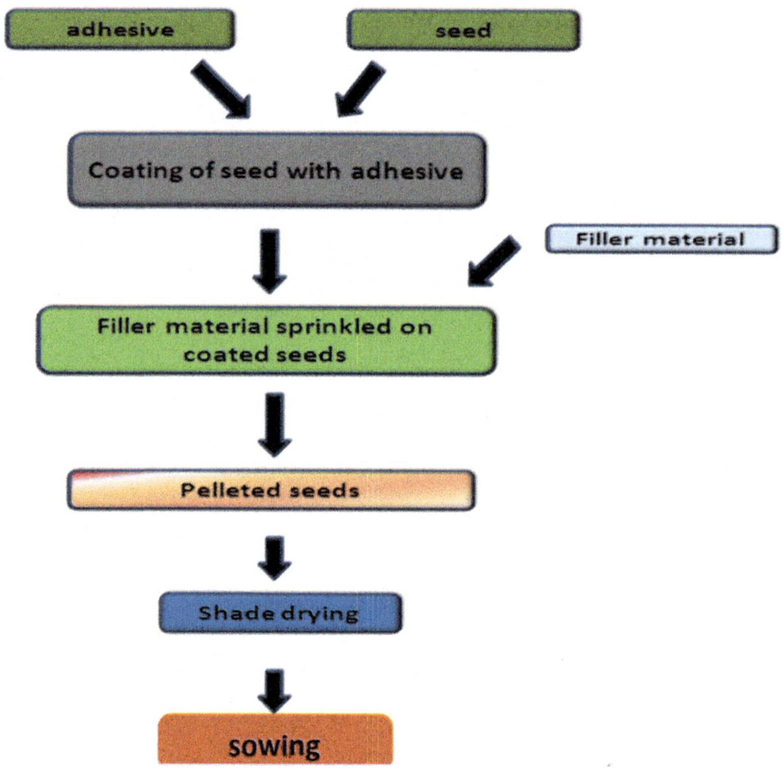

Fig. 10.4 The process of seed pelleting

The benefits of seed coating include direct application of micronutrients, insecticides, and fungicides on seeds, ease of seed flow in automatic seeding, dust-free handling, appearance improvement, reduction in chemical wastage by application of accurate and even dose of chemicals, temperature switch action, and regulation of water intake. The limitations of seed coating are deterioration of the whole seed lots by improper coating and improper dilution of coating material and high seed cost for seed coating.

10.2.4.5 Pelleting

The covering of seeds with a small quantity of adhesive (gelatin, gum arabic, polyvinyl alcohol, methyl cellulose, maida/starch gruel) and sprinkling of the filler inert material (limestone, vermiculite, clay, calcium carbonate, tamarind leaf powder) to facilitate precision planting is called seed pelleting (Fig. 10.4). It facilitates seed soil interface (Halmer 2006). The inert material contains water-holding media which provide small amount of nutrients to younger seedlings (Halmer 2006). In addition to adhesive and inert material, growth regulators and fungicides can also be included

Table 10.1 The different types of seed pelleting and the materials used

Type	Material used
Inoculant pelleting	Biofertilizers, viz., *rhizobia*, *phosphate-solubilizing bacteria* (PSB), *Azospirillum*, *Azotobacter*, *arbuscular mycorrhizal fungi* (AMF)
Protective coating	Biocontrol agent like *Rhizobacteria bataticola*, *Bacillus* spp., *Streptomyces* spp., pesticides, fungicides
Herbicide coating	Filler antidote or absorbent coating, herbicide antidote like 1.8 naphthalic anhydride (NA)
Nutrient coating	Coating with micro- and macronutrients, e.g., $ZnSo_4$, $FeSo_4$, Borax
Hydrophilic coating	Starch graft polymers, magnesium carbonate
Oxygen supplier coating	Peroxides of zinc and calcium

Halmer (2006)

for seed pelleting in order to encourage plant growth and control of seed-borne fungal diseases. The different types of seed pelleting and the materials used are presented in Table 10.1.

The benefits of seed pelleting include precision placement, increase in size and shape, supply of nutrients, moisture absorption, singling of seeds to prevent clogging, and protection from birds/animals. The limitations of seed pelleting are high cost of seeds, increased seed weight, and removal of pellet, if proper machines are not used.

10.2.4.6 Integrated Enhancement

The seed quality enhancement can also be achieved by integrating two or more methods (Table 10.2).

The benefits of enhancement of seed quality include improved nursery management, seed rate reduction, application of nutrients, growth regulators, beneficial microbes, reduced time of emergence and early emergence under stress conditions, uniformity in field stand, facilitating seedling to dominate weeds in competition for nutrition, direct seeding of conventionally transplanted vegetable seeds, minimum exposure to toxicant, and high turnover.

10.3 Treatment

The seed treatment is an age-old practice in most of the crops followed throughout the world. Before sowing, seeds are treated with physical, chemical, and biological agents for different purposes such as convenience in sowing, enhancing germination, inoculating biofertilizer culture, bird repellence, and control of pests and diseases which provide protection to seeds and seedlings and improve the establishment of healthy crops.

Table 10.2 The seed quality enhancement by integrated methods

Integrated methods	Benefits	References
Priming + pre-germination	Early emergence	Pill (1986)
	Increased shoot weight	
Seed coating + pelleting	Ease of sowing of flat seeds	Halmer (1988)
	Addition of bioactive chemicals, nutrients, and microbes	
Priming + pelleting	Complements individual effects	Valdes et al. (1985) and Bennett (1988)
Osmoticum/solid matrix carrier + growth regulators/nutrients/pesticides	Additive effect	Khan (1992) and Osborn and Schroth (1989)
Polyethylene glycol + gibberellic acid	Improved germination and prevented induction of secondary dormancy	Khan (1992)

10.3.1 Ease of Sowing

In certain crops, difficulties are encountered in sowing of seeds such as cotton seeds which cling together due to the presence of fuzz and lint. In order to overcome this problem, the seeds are rubbed together with a paste made of earth or wet dung or both and then dried. The seeds can also be treated for 2 minutes with concentrated sulfuric acid and then washed repeatedly with water and dried properly before sowing. The extra acid can also be removed by washing the seeds in running tap water for 10–20 min before sowing. To ensure uniformity in sowing, the coriander seeds are split into two locules, and the garlic compound bulbs are separated into individual cloves before sowing, which is also an economical practice. The seeds of small crops (pearl millet, finger millet, tobacco, sesame, etc.) are mixed with soil or sand in different ratios (5:1 to 20:1) to achieve even sowing.

10.3.2 Pest Management

For the control of seed-borne target diseases and insect pests to enhance the establishment of healthy and vigorous seedlings, the seeds are treated with hot water maintained at about 45 °C for 5–15 min, pesticides (thiram, captan, mancozeb, carbendazim, captfol against diseases and chlorpyrifos or monochrotophos or phosalone against insect pests), and biological control agents (*Bacillus subtilis*, *Pseudomonas fluorescens*, *Trichoderma viride* against diseases and *Beauveria bassiana*, *Nomurea rileyi*, *Metarhizium anisopliae* against insect pests). Based on the specific active ingredient, the new active ingredients, formulations of pesticides,

Table 10.3 Yield potential of maize under different plant population

Plant population/ acre	Mean yield (lb./acre)
25,000	6686
28,000	6690
31,000	6226
34,000	6369
Variable rate	6362

special wetting agents, colorants, and sometimes bird repellents give long-term, broad-spectrum control of diseases and pests. The stored grain pests are controlled by seed fumigation with acrylonitrile (16–20 g/m^2), carbon disulfide (24–32 g/m^2), carbon tetrachloride (300–400 g/m^2), ethyl bromide (24–30 g/m^2), methyl bromide (32 g/m^2), and phosphine.

10.3.3 Nutrient Availability

The treatment of the legume seeds with the culture of *Rhizobium* spp. (the symbiotic bacteria) reduces nitrogen requirement (by fixing atmospheric nitrogen) and enhances crop yields. The free-living microorganisms such as blue-green algae (BGA), *Azolla*, *Azotobacter chrococcum* (used in rice, cotton, and sugarcane crops), and *Azospirillum* (used in cereal crops) can also fix the atmospheric nitrogen. The seed treatment with phosphobacteria (*Azophos*) increases the availability of phosphorus to crops.

10.4 Rate

The seed rate is defined as the amount of seed sown per hectare [shown as kilos per hectare (kg/ha)] to ensure normal density of sprouts and a maximum yield. It depends on the cultivation purpose (grain, silage), requirements of plant for feeding space, soil fertility, size of the seed, germination of the seed, growing habit of the crop, climatic conditions, row spacing, type of farming, time of sowing, variety, and other factors. The ultimate crop stand desired is governed by the seed rate.

The variable seed rate is justified to match potential yield with plant population (Table 10.3), to adjust for stand establishment, and to decrease the cost of seed.

The seed rate and row spacing affect the weed population. By reducing row spacing, or increasing seed rate, or both increases crop competition with weeds (rye grass) in wheat (Fig. 10.5). The weed seed set can be reduced by increasing the competition with crops. This is a very cost-effective method of weed control in wheat.

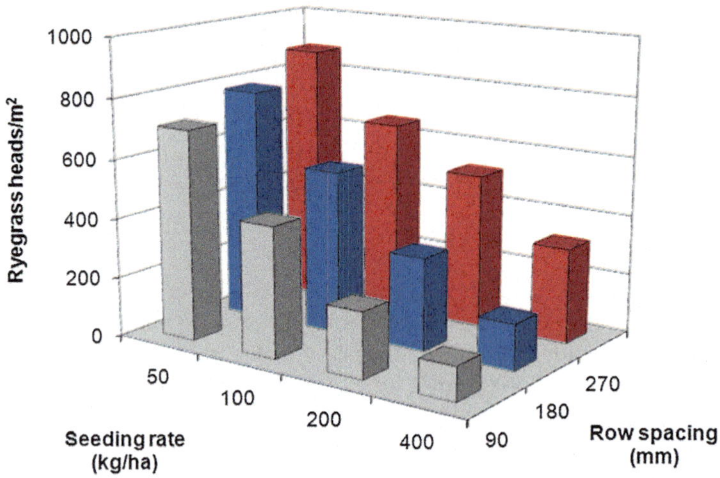

Fig. 10.5 Effect of seed rate and row spacing on weed (rye grass) population in wheat

Table 10.4 Effect of row spacing on wheat yield and net benefit

Treatment	Row spacing	
	19 cm	25 cm
Reduction in row spacing (cm)	11	5
Wheat yield benefit (kg/ha)	88	40
Benefit ($/ha)	23.76	10.80
Cost ($/ha)	10.56	2.01
Net benefit ($/ha)	13.20	8.79

In wheat, narrow row spacing resulted in more crop yield and higher benefits (Table 10.4), mainly because of lesser weeds. At narrow row spacing, percentage emergence of seeds sown is increased. By using modern farm machinery, the narrow row spacing can be achieved.

10.5 Depth and Metering

The seed depth is highly critical for consistency in emergence of seedlings. Hence, modern direct seeders should offer more consistent depth across the frame to conform to specific field conditions under no-till conditions. The depth is controlled by adjusting each individual seed opener up or down on each shank and then clamped tightly in place. The fertilizers can be placed below or side of the seeds at about 5–15 cm depth. The direct seeder should be able to place the seed and fertilizer precisely at proper depth with minimal soil disturbance to optimize farming operations.

Table 10.5 Effects of drilling depth on seedling emergence in no-tilled wheat and lupine

	Seedling emergence* (plants/m^2 in parenthesis)	
Drilling seed depth	Wheat	Lupine
20 mm	79 % (209) a	93 % (66) a
30 mm	80 % (210) a	87 % (62) b
50 mm	73 % (192) a	60 % (43) c
70 mm	61 % (160) b	24 % (17) d

Unlike letters in a column denote significant differences at $P < 0.05$
*% seedling emergence = % of the estimated number of seeds sown from the known weights of seeds sown

The hydraulic shanks (seeding tool uses a rigid frame for strength and support) provide consistent depth control for enhanced seed placement, resulting in more even seedling emergence. In order to achieve a more uniform depth across the entire width of the frame, the openers are designed to allow independent downforce. The banding of fertilizer should be at 15 cm depth and 11 cm away from the seed.

Hadfield (1993) measured the variations in germination and emergence of wheat (*Triticum aestivum*) and lupine (*Lupinus angustifolius*) drilled in inverted-T-shaped no-tillage slots at various depths. The lupine was more sensitive to depth of sowing than wheat at depth range of 20–50 mm. The seedling emergence with lupine decreased by 17 % for each 10 mm increase in drilling depth between 20 mm and 70 mm. In comparison, wheat emergence declined by 4 % for each 10 mm increase in depth between 20 and 70 mm. However, at depths higher than 50 mm, emergence of both crops was seriously affected. The subsurface mortality of seedlings that had already been germinated was the main cause for reduction in seedling emergence in both crops (Table 10.5).

The mechanism to deliver seeds and fertilizers from their respective hoppers to the seed and fertilizer placing mechanism at a selected rate is the function of seed and fertilizer metering and delivery device. It is a simple, low-cost, and trouble-free device.

10.6 Conclusions

In order to establish good plant stands, the seed quality, rate, depth, and treatment are very important. The information on seed purity, germination, vigor, and the presence of seed-borne diseases and pests can be obtained from seed testing laboratories. The suitability of seed for planting can be determined by the above characteristics. Seed germination percentage gives us an idea about the number of seeds that are expected to germinate, while the vigor indicates whether the seedlings would thrive under stressful conditions. The seed-borne pests and diseases can be managed by treatment with physical (hot water treatment) or chemical methods (using pesticides).

References

Agrawal AA (2002) Herbivory and maternal effects: mechanisms and consequences of transgenerational induced plant resistance. Ecology 83:3408–3415

Bennett MA (1988) Evaluation of seed coating and priming treatments for stand establishment of processing tomatoes. Proceedings of international conference on stand establishment in horticultural crops. American society of horticultural sciences, Lancaster, PA

Bradford KJ (1986) Manipulation of seed water relations via osmotic priming to improve germination under stress conditions. Hortic Sci 21:1105–1112

Copeland LO, McDonald MB (2001) Principles of seed science and technology, 4th edn. Kluwer Academic Publishers, Norwell, 488 pp

Hadfield AB (1993) Seeding depth effects on the performance of wheat and lupin seedlings under no-tillage. Thesis, Massey University Library, Auckland, New Zealand, 115 pp

Halmer P (1988) Technical and commercial aspects of seed pelleting and film coating, In: Applications to seed and soil. Monograph No. 39, pp 191–204. British Crop Protection Council, Thornton Heath

Halmer P (2006) Seed technology and seed enhancement. In: XXVII International Horticultural Congress-IHC2006: international symposium on seed enhancement and seedling production 771:17–26

Khan AA (1992) Preplant physiological seed conditioning. Hort Rev 13:131–181

McDonald MB (2000) Seed priming. In: Black M, Bewley JD (eds) Seed technology and its biological basis. CRC Press, Boca Raton, pp 287–325

Osburn RM, Schroth MN (1989) Effect of osmopriming sugar beet seed on germination rate and incidence of *Pythium ultimum* damping-off. Plant Dis Rep 73:21–24

Pill WG (1986) Parsley emergence and seedling growth from raw, osmo-conditioned and pregerminated seeds. Hortic Sci 21:1134–1136

Taylor AG, Allen PS, Bennett MA, Bradford KJ, Burris JS, Misra MK (1998) Seed enhancements. Seed Sci Res 8:245–256

Valdes VM, Bradford KJ, Mayberry KS (1985) Alleviation of thermo-dormancy in coated lettuce *Lactuca sativa* cultivar 'Empire' by seed priming. Hortic Sci 20:1112–1114

Part VI
Balanced Plant Nutrition

Chapter 11
Soil Organic Matter

Abstract Soil organic matter (SOM) benefits the soil and the health of plants. The SOM is built and maintained by retaining crop residues, growing cover/green manure crops, crop rotation, and addition of compost and farmyard manure. Addition of organic matter to soil helps in water-holding capacity, soil quality, nutrient cycling, drainage improvement, and reduction in soil compaction and crusting. SOM is also responsible for healthy plants (which are resistant to insect pests and diseases and tolerant to drought) with increased yields.

Keywords Soil organic matter • Crop residues • Green manures • Pastures • Drainage • Water-holding capacity • Nutrient recycling

11.1 Introduction

Soil organic matter (SOM) is any material that is produced originally by living organisms (plant or animal) which is returned to the soil that undergoes the decomposition process and releases plant nutrients. The plant needs at least 16 chemical elements for their growth and reproduction. These nutrients have to come from air, water (C, H, and O_2), and soil (N, P, K, Ca, Mg, S, Fe, Mn, Zn, Cu, B, Mo, Cl). The plants depend for some of these elements on SOM (N, P, S, and micronutrients are released during its decomposition by the microorganisms). Organic matter contains organic carbon which is the key to soil health and structure.

11.2 Components

The soil organic matter contains three main components, namely, biomass (living organisms), detritus (dead tissues and waste), and humus (nonliving and non-tissue). The organic components of the soil are as follows (Fig. 11.1):

- Crop and animal residues
- Soil macroorganisms
- Soil microorganisms

P.P. Reddy, *Sustainable Intensification of Crop Production*,
DOI 10.1007/978-981-10-2702-4_11

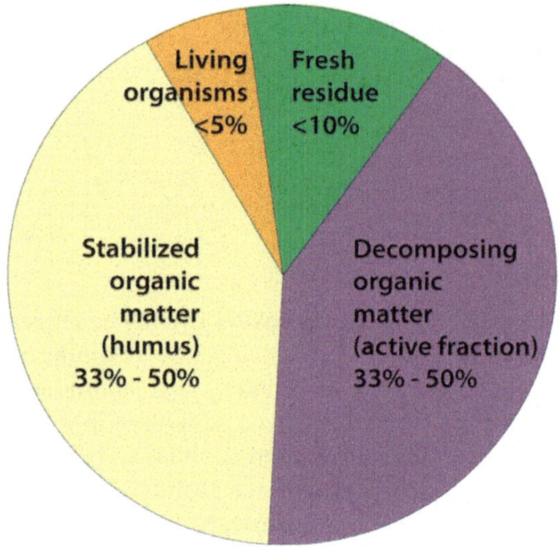

Fig. 11.1 Components of soil organic matter

- Stable organic matter (humus)
- Actively decomposing organic matter

11.2.1 Crop and Animal Residues

Residues depend on how much fresh plant (leaves, stems, roots of higher plants) and animal (the bodies of soil animals such as worms and insects) material is added to the soil. The majority of soil organic matter is contained in the nonliving component that includes plant, animal, and microbial debris and soil humus.

11.2.2 Soil Macroorganisms

Macroorganisms include small animals and predators (ants, insects, mites, beetles, and spiders), herbivores, and detritivores (millipedes, woodlice, earthworms, slugs, snails, rabbits, mice, springtails, ants, beetles, grubs). Earthworms, insects, and rodents are the most visible components of the living soil team. They work in tandem with soil microorganisms and fungi to contribute to aeration and nutrient cycling as part of a soil factory team effort. Earthworms and ants (Fig. 11.2) are the major mixers of soil that cause openings in soil which allow air and water movement. Earthworms are active when conditions are favorable such as around 10°C soil temperature, pH 6–7, and enough soil moisture. They are responsible for refining their main food source (dead plant remains) and other soil organic matter.

Fig. 11.2 Earthworms and ants

Fig. 11.3 Springtails and mites

Herbivores and detritivores such as slugs, snails, and arthropods [such as springtails (Fig. 11.3), millipedes, nematodes, and mites (Fig. 11.3)] play a major role in the decomposition of soil organic matter. They mix, mechanically alter, and add organic matter to the soil.

11.2.3 Soil Microorganisms

Microorganisms include detritivores, predators or parasites (nematodes, protozoa, and rotifers), various bacteria, actinomycetes, fungi (mushrooms, yeasts, and molds), and algae. Soil microorganisms and fauna make up a relatively small portion (1–8 %) of the total soil organic matter. They function as an important catalyst for transformations of soil organic matter to N and other nutrients.

Bacteria (Fig. 11.4) which are present in large numbers (10^8–10^{10}/g of soil) play a major role in the decomposition of soil organic matter, minerals, and other substances. They are single-celled organisms which exist in colonies or clumps. They are mostly heterotrophic and get their energy from the carbon in soil organic matter.

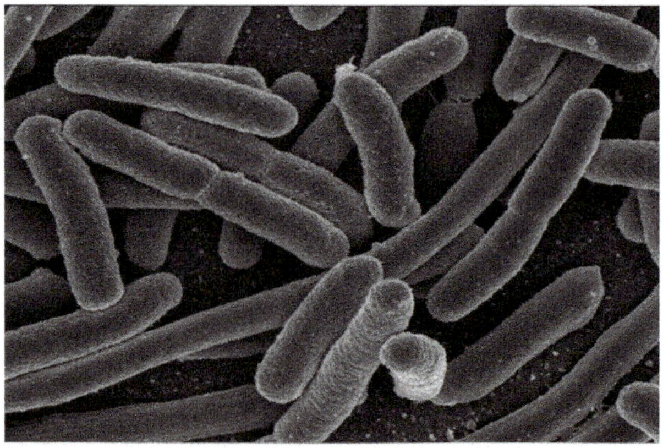

Fig. 11.4 Soil bacteria which are present in greater number of organisms present than all others combined

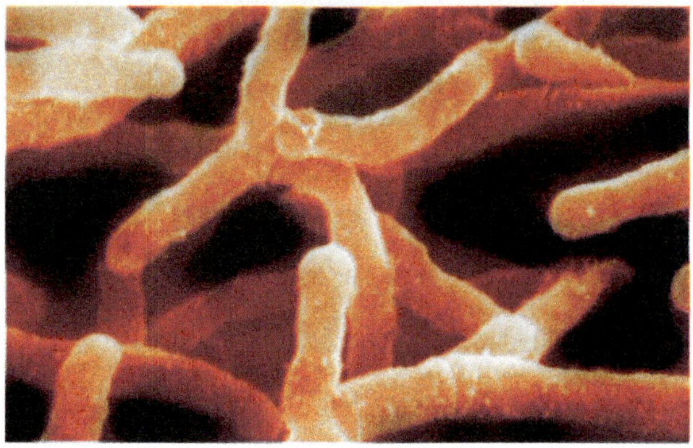

Fig. 11.5 Soil actinomycetes

Actinomycetes (Fig. 11.5) are single-celled organisms slightly larger than bacteria (fine, thread like), which decompose complex organic materials in later stages in conversion to humus. They are capable of producing antibiotics which are useful for managing crop diseases.

Fungi (Fig. 11.6) are dominant flora when pH is less than 5. These are heterotrophic microorganisms which can decompose highly resistant and complex organic compounds.

11.2.4 Stable Organic Matter (Humus)

Soil humus (Fig. 11.7) is the fully decomposed and stable organic matter, which is most reactive and important component of soil organic matter. The major constituents of humus include cellulose (2–10 %), hemicelluloses (0–2 %), lignin (35–50 %), proteins (28–35 %), fats, waxes, etc. (1–8 %).

Proteins in humus are nitrogenous compounds which store and slowly release N into the soil. Humus contains concentrations of P, S, and high cation exchange capacity (CEC) and is capable of absorbing large volumes of water. It has low C/N

Fig. 11.6 Fungi (mushroom)

Fig. 11.7 Humus which is fully decomposed and stable organic matter

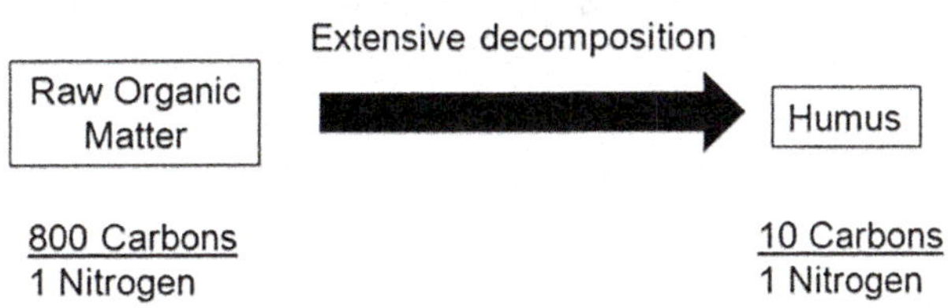

Fig. 11.8 C/N ratio of humus

ratio of 10:1 (Fig. 11.8). It also increases soil fertility because of high CEC and anion retention.

11.2.5 Actively Decomposing Organic Matter

The actively decomposing organic matter fraction is used and transformed by living plants, animals, and microbes.

11.3 Factors Affecting Soil Organic Matter Level

Normally the topsoil contains between 1 and 6 % of organic matter, and the subsoil contains less than 2 % OM (Fig. 11.9).

The factors affecting SOM include vegetation, climate, tillage, drainage, and soil texture.

11.3.1 Soil Texture

The OM decomposition process speeds up in soils with more oxygen (soil aeration). Soil organic matter is generally higher in fine-textured soils because soil humus forms stable complexes with clay particles. Coarse-textured soils have faster gas exchange, which leads to more CO_2 loss.

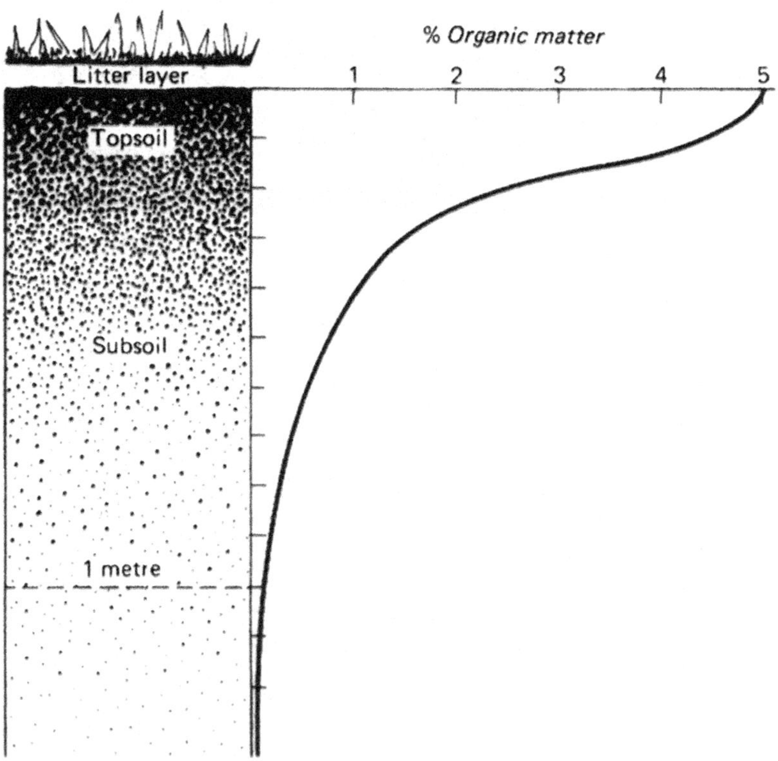

Fig. 11.9 The levels of soil organic matter at different soil depths

11.3.2 Climate

In warm and humid climates, the SOM decomposes more quickly than cool dry climates.

11.3.3 Vegetation

More organic material is added to the soil in prairie soils than in forest soils because of vegetation. Soils under grass generally have a relatively high percentage of organic matter on their surface. Organic matter levels are generally higher in a topsoil supporting hay, pasture, or forest than in a topsoil used for cultivated crops. Soils which develop under perennial trees often have a low organic matter percentage in the surface mineral soil, but do contain a surface litter layer.

11.3.4 Tillage

Soils which are tilled frequently are generally low in organic matter. Tilling the soil (plowing) increases the aeration of the soil, which increases the organic matter decomposition rate.

11.3.5 Drainage

Soil organic matter is generally higher in poorly drained soils because of less oxidation, which slows down the overall biological decomposition process.

11.4 Benefits

The benefits of SOM include (Fig. 11.10):

- Prevents soil erosion (13–77 %) and runoff (1–68 %).
- Supply nutrients to crop plants. Each percent of SOM releases 9–13.5 kg of N, 2–3 kg of P_2O_5, and 0.9–1.35 kg S per year.
- Food source for microorganisms. Both dead crop residue (58 % C) and live crop biomass (46 % C) are food sources for soil organisms.
- Encourages soil aggregation process. SOM helps the soil to clump and form soil aggregates which improve soil structure and water infiltration in soil.
- Serves as excellent soil conditioner.
- Increases cation exchange capacity (CEC).

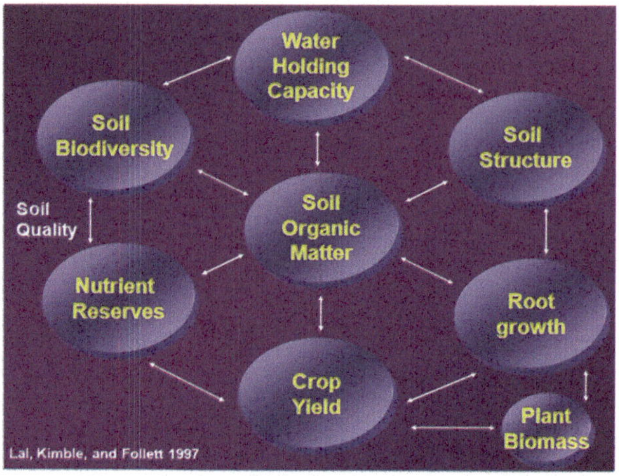

Fig. 11.10 The benefits of SOM (Lal et al. 1997)

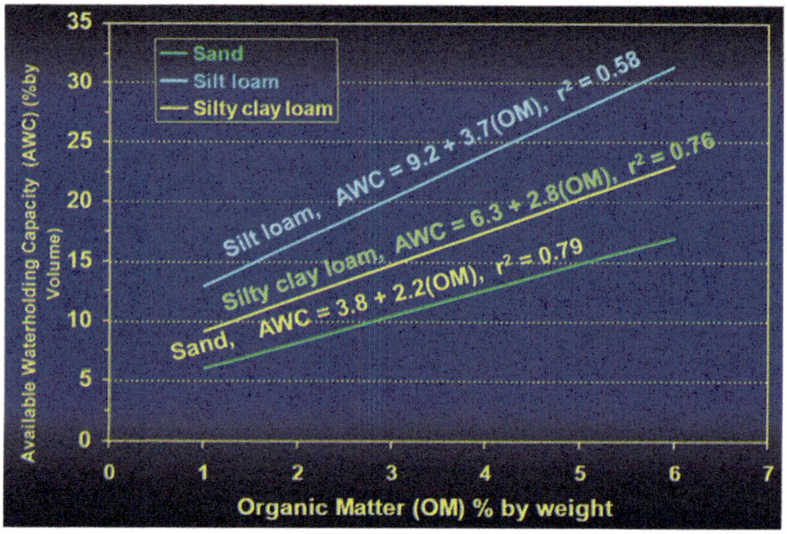

Fig. 11.11 SOM increases available water-holding capacity (Hudson 1994)

- Increases water-holding capacity of the soil (Fig. 11.11) (Hudson 1994). Soil organic matter acts like a sponge for water retention and release to plants (Fig. 11.12). Soil high in carbon is rich in spongy organic matter. SOM has the ability to absorb and hold water up to 90 % of its weight. Most of the water absorbed by SOM is released for plant growth.
- Reduces nitrogen (N) leaching potential and replaces limited resource phosphorus (P) fertilizer. Efficient use of manure nutrients means not purchasing fertilizer unless it is needed.
- Increases crop productivity.
- At low pH, reduces aluminum toxicity.
- Increases soil organic carbon sequestration (Fig. 11.13).

Fig. 11.12 SOM "sponge" for water

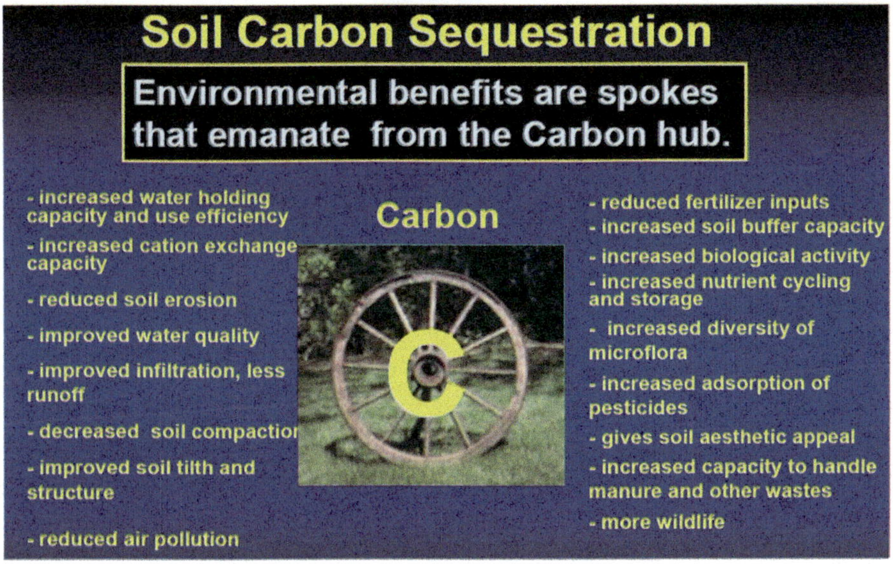

Fig. 11.13 Soil carbon sequestration

11.5 Decomposition

The decomposition of SOM is biological breakdown and biochemical transformation of complex organic molecules of dead litter into simpler organic and inorganic molecules (Fig. 11.14) (Juma 1998). It is important for the release of nutrients held up in dead organic matter and returning it back to the soil.

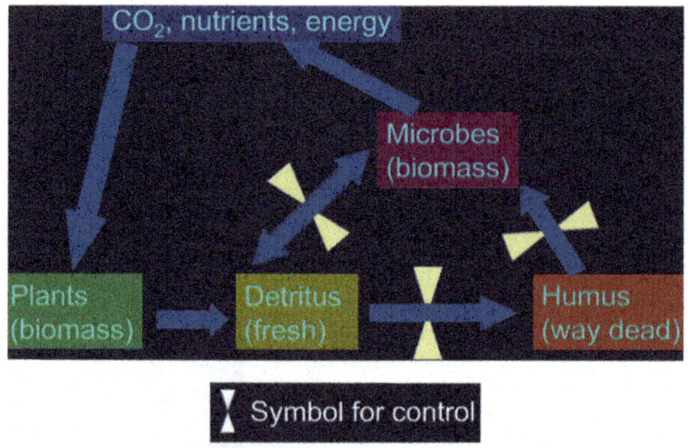

Fig. 11.14 Soil organic matter decomposition

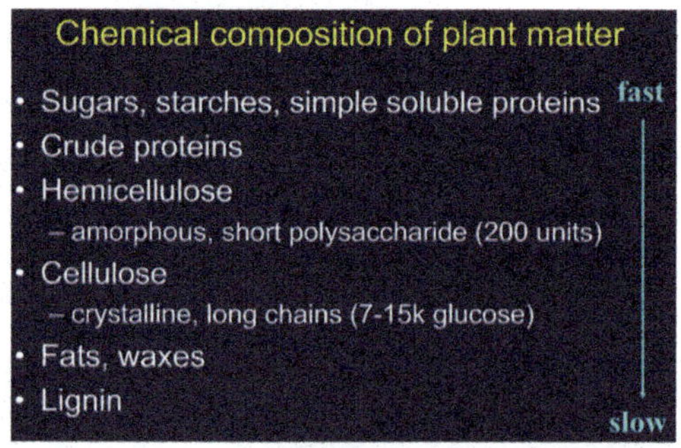

Fig. 11.15 The rate of decomposition of SOM

The rate of organic matter decomposition depends on physical interactions, chemical recalcitrance, environmental controls, microbial communities, and soil aeration.

11.5.1 Physical Interactions

Physical nature of the litter material, sorption, and disconnection are some of the physical interactions which affect the decomposition of organic matter. Small material (increased surface area) decomposes faster than bigger material (decreased surface area).

11.5.2 Chemical Recalcitrance

The rate of decomposition of SOM depends on the chemical composition of the plant matter such as chemical nature of the litter material—microbial and plant inputs—and their transformation in soils (Fig. 11.15).

The cellulose generally accounts for the largest proportion (15–50 %) of fresh organic material which decays rapidly in the presence of nitrogen. The rate of decomposition of hemicelluloses (glucose, galactose, mannose, and xylose) (accounts for 5–35 % of fresh OM) is moderate to slow. Lignin decomposes slowly; hence the nutrients bound in lignin forms are not available for plant growth immediately. Lignin (accounts for 15–35 % of fresh OM) is soluble in alkali solutions, but it is insoluble in hot water and neutral organic solvents. Polysaccharides (chitin, starch, pectins, and inulin) decompose rapidly in soils and serve as an immediate source of energy (carbon) for microorganisms. Crude proteins which account for 1–10 % of fresh OM decay rapidly.

The C/N ratio of the litter is indirectly correlated with the rate of SOM decomposition (Fig. 11.16).

The C/N ratio also affects the release of N to plants. At a low C/N ratio (below 24: 1), sufficient nitrogen is supplied to meet the needs of decomposing microorganisms through the decomposition of the organic materials, resulting in net release and buildup of inorganic nitrogen in the soil (mineralization). At a high C/N ratio (above 24:1), microorganisms need more nitrogen from soil (in the form of nitrate or ammonium) to decompose the carbon in the organic materials, resulting in immobilization of nitrogen (unavailable to plants) until these microorganisms die and the nitrogen is released (Fig. 11.17).

11.5.3 Environmental Controls

Water, oxygen, pH, and temperature of the soil environment are some of the environmental factors affecting decomposition of organic matter. Microbial activity responds exponentially to increased temperature until enzymes denature. Microbial activity requires optimum moisture. Lower moisture content leads to desiccation and slow diffusion; while high moisture content prevents lignin degradation and low availability of oxygen. The optimum range of pH for rapid decomposition of various organic wastes and crop residues is 6.5–8.5. Generally, the microbial activity is optimum near pH 7. Fungal activity is more in acid soils, while bacterial and actinomycetes activity is more in moderate soil pH (neutrality). Hence, soil fungi often become dominant in acid soils.

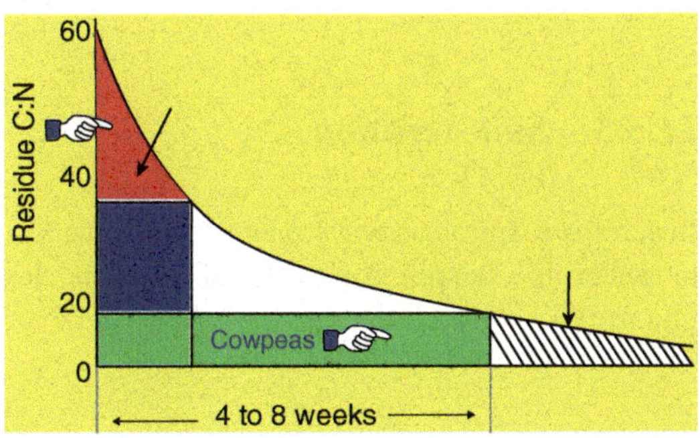

Fig. 11.16 The effect of C/N ratio on rate of SOM decomposition

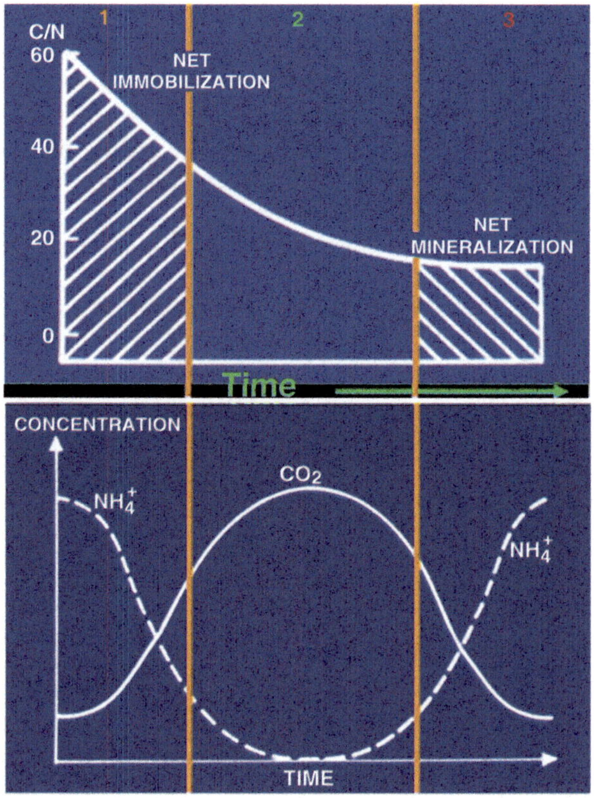

Fig. 11.17 Effect of C/N ratio on mineralization and immobilization of nitrogen

11.5.4 Microbial Communities

Microbial communities (kind and number of soil fauna—enzyme pathways) convert organic matter into inorganic compounds (CO_2, H_2O, NH_4^+, NO_3^+, etc.). Fungi break down the woody components of litter into inorganic molecules (lignin, cellulose), while bacteria and other microorganisms break down just about anything else. Mycorrhizal fungi decompose by breaking down proteins in organic matter into amino acids that are transferred to host plant.

11.5.5 Soil Aeration

In aerobic respiration with oxygen, more energy is released (38 ATP/glucose), while in anaerobic respiration without oxygen, less energy is released (two ATP/glucose).

After decomposition, nutrients released from the residues are available for use by the growing plants. Heat generated during decomposition kills many disease-causing pathogens.

11.6 Management

SOM is the primary nutrient and energy source for bacteria, fungi, insects, and other soil microorganisms. SOM can be increased in the soil by adopting the practices such as prevention of soil erosion, conservation tillage, winter cover crop, soil fertility management, elimination of summer fallow, forage-based rotation, use of improved crop varieties, organic amendments, and irrigation management (Fig. 11.18) (Table 11.1).

11.6.1 Prevention of Soil Erosion

Soil erosion can be reduced to the extent of 20–33 % by increasing the SOM (through soil aggregation and water infiltration) from 1 to 3 %.

11.6.2 Conservation Tillage

No-tillage or reduced tillage will reduce the rate of soil organic matter decomposition. Ten years of no-tillage farming practices with corn increased soil organic carbon in 30 cm surface area by 0.25 % (Fig. 11.19) (Blevins et al. 1983). Organic matter levels increased in the native prairie soils since no cultivation was ever employed for thousands of years. There was approximately 4.0 % soil organic matter in the surface 30 cm of soil, when the prairie soils of Oklahoma were first cultivated in the late 1800s. There were over 9000 kg of N/ha in that 4.0 % organic matter.

11.6.3 Soil Fertility Management

Proper fertilization encourages plant growth including root growth, which can help to build and maintain SOM.

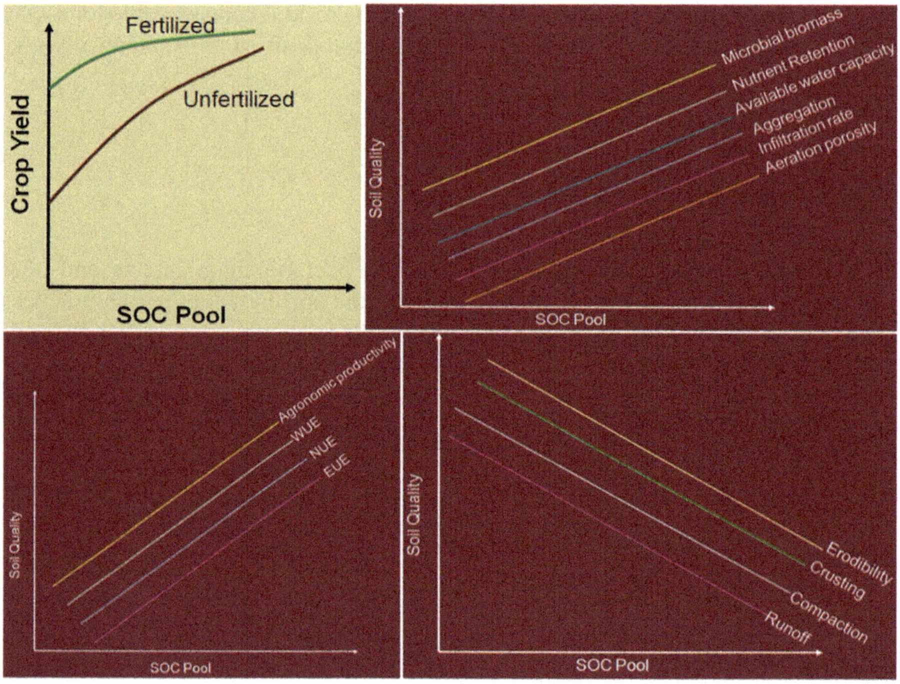

Fig. 11.18 Beneficial effects of soil organic pool on crop yield and soil quality

Table 11.1 Recommended practices to increase organic carbon in soil

Recommended practices	C sequestration potential (Mg C/ha/yr)
Conservation tillage	0.10–0.40
Soil fertility management	0.05–0.20
Winter cover crop	0.05–0.10
Elimination of summer fallow	0.05–0.20
Forage-based rotation	0.05–0.20
Use of improved varieties	0.05–0.10
Organic amendments	0.20–0.30
Irrigation management	0.05–0.10

Lal et al. (1998)

11.6.4 Crop Residue Management

Crop residues are retained as organic soil mulch. They are used as organic soil amendments for carbon sequestration and soil quality improvement to achieve global food security.

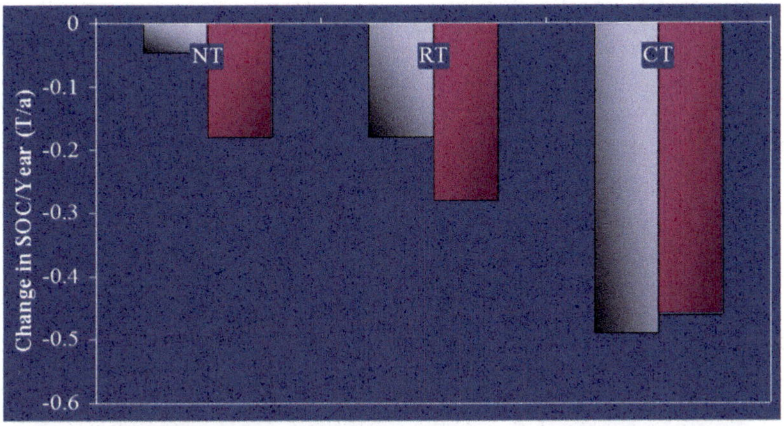

Fig. 11.19 Effect of no-tillage (NT) and reduced tillage (RT) on soil organic carbon compared to conventional tillage (CT) (Blevins et al. 1983)

11.6.5 Forage-Based Rotation

Forage-based crop rotations contribute organic matter to the soil and improve organic matter stability.

11.6.6 Cropping Systems

Crop rotation, winter cover/legume green manure cropping, and relay cropping are some of the cropping systems that can be used for enhancing the soil organic matter. Including leguminous green manure cover crops in the crop rotation cycle not only increase SOM but also N in soil. The duration of crop rotation and accumulation of SOC are directly correlated (Fig. 11.20). Cropping grasslands reincorporates organic matter into the system. Growing cover crops adds huge quantity of biomass which helps in building up and maintain SOM.

11.6.7 Elimination of Summer Fallow

There is a need for crop rotations that reduce fallow periods during summer to increase soil carbon levels.

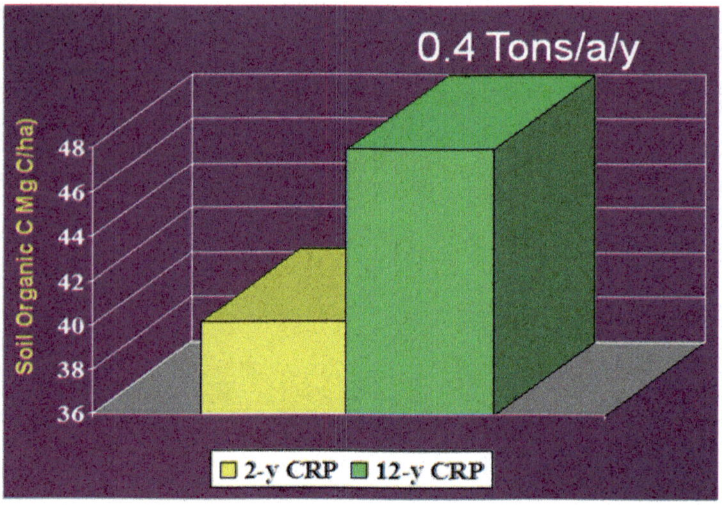

Fig. 11.20 The effect of duration of crop rotation on SOC (Baer et al. 2002)

11.6.8 Irrigation Management

Irrigation helps for growth of plants which shall increase the amount of roots and crop residues added to the soil every year. Water can be conserved by following subsurface drip irrigation and water harvesting.

11.6.9 Introduction of Earthworms

Introduction of earthworms facilitate stabilization and accumulation of soil organic matter in agricultural systems. Earthworms help with the decomposition of dead plants and animals, breaking them down into soil and thereby supply an abundance of organic fertilizer (castings).

The effects of various management practices on the soil organic matter and the resultant nutrient supplying power of the organic pools are well understood. The methods to increase soil organic matter are presented in Table 11.2.

11.7 Conclusions

The organic matter is an important component of soil which is associated with various important physical, chemical, and biological properties. Cool climate favors the retention of soil organic matter to a larger extent than in warmer climate. Enough evidence is available that the SOM levels and crop yields are directly correlated.

Table 11.2 Methods to improve SOM

Objective	Strategy	Practices
Rapid improvement of soil structure and water retention	Addition of stable organic matter with high carbon content	Composted manures
		Spent mushroom compost
		Composted municipal waste
		Heavily bedded manure
		Waste paper
Supply nutrients	Incorporate organic matter with high nutrient content	Manures (FYM, compost)
		Composted municipal waste
		Sewage sludge
	Change rotation to incorporate high nitrogen crops	Green manure
		Grass or legume lay
Long-term soil structure improvement	Change crop residue management	Incorporate crop residues
		Grass or legume lay
	Introduce reduced or minimum tillage	Reduced tillage
		Minimum tillage
	Incorporate stable organic matter with high carbon content	Heavily bedded FYM
		Paper pulp

Hence, farmers should adopt management practices such as crop residue management, including green manure crops and pastures in crop rotation to build SOM.

References

Baer SG, Kitchen DJ, Blair JM, Rice CW (2002) Changes in ecosystem structure and function along a chronosequence of restored grasslands. Ecol Appl 12(6):1688–1701

Blevins RL, Thomas GW, Smith MS, Frye WW, Cornelius PL (1983) Changes in soil properties after 10 years of continuous non-tilled and conventionally-tilled corn. Soil Tillage Res 3:135–136

Hudson BD (1994) Soil organic matter and available water capacity. J Soil Water Conserv 49(2):189–194

Juma NG (1998) The pedosphere and its dynamics: a systems approach to soil science, vol 1. Quality Color Press Inc, Edmonton, p 315

Lal R, Kimble JM, Follet RF, Cole CV (1998) The potential of U.S. cropland to sequester carbon and mitigate the greenhouse effect. Sleeping Bear Press, Chelsea

Lal R, Kimble JM, Follet RF, Stewart BA (eds) (1997) Soil processes and the carbon cycle. CRC Press, Boca Raton

Chapter 12
Application of Manures and Fertilizers

Abstract The fertilizers are applied for plant growth and development mostly in the solid form (90 %) as compared to liquid form (10 %). The solid fertilizers are applied through broadcasting (basal application and top dressing), band placement (hill and row placement), and pellet application. Starter solutions, foliar application, fertigation, injection into soil, and aerial spraying are some of the methods of liquid fertilizer application. Increasing fertilizer use efficiency through (controlled uptake long-term ammonium nutrition) CULTAN fertilization is also discussed.

Keywords Fertilizer application • Solid fertilizers • Liquid fertilizers • Fertigation • CULTAN fertilization

12.1 Introduction

The plants need 16 essential nutrients (nine macronutrients and seven micronutrients) for normal growth and reproduction. The macronutrients which are needed in large amounts include carbon (C), hydrogen (H), oxygen (O), nitrogen (N), phosphorus (P), potassium (K), calcium (Ca), magnesium (Mg), and sulfur (S). The micronutrients which are used in small quantities include boron, (B), chlorine (Cl), copper (Cu), iron (Fe), manganese (Mn), molybdenum (Mo), and zinc (Zn) (Fig. 12.1). These nutrients can be supplied to plants through organic manures or inorganic fertilizers.

12.2 Organic Manures Versus Inorganic Fertilizers

Manures and fertilizers contain one or more of the essential elements, e.g., nitrogen, phosphorus, potassium, calcium, magnesium, etc. They are applied to soils to encourage the growth of crops. Soil or plant analysis forms the basis for the amount of manures and fertilizers to be added. Manures are mainly organic (made from plant and animal wastes), while fertilizers are inorganic materials that are manufactured in factories. The difference between manures and fertilizers are presented in Table 12.1.

P.P. Reddy, *Sustainable Intensification of Crop Production*,
DOI 10.1007/978-981-10-2702-4_12

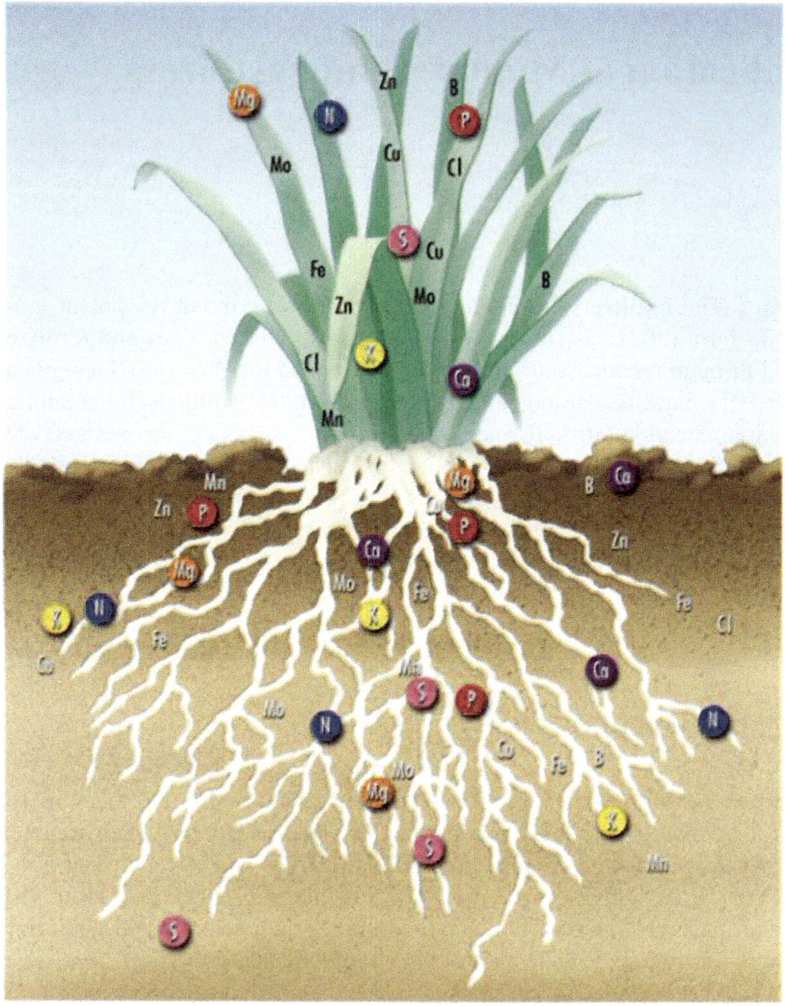

Fig. 12.1 Nutrient requirements for plant growth and development

Table 12.1 The difference between manures and fertilizers

Manures	Fertilizers
Organic material	Chemical substance
Natural material	Inorganic substance
Small amounts of plant nutrients	Large amounts of plant nutrients
Add humus to the soil	No addition of humus to the soil
Prepared in fields	Manufactured in factories

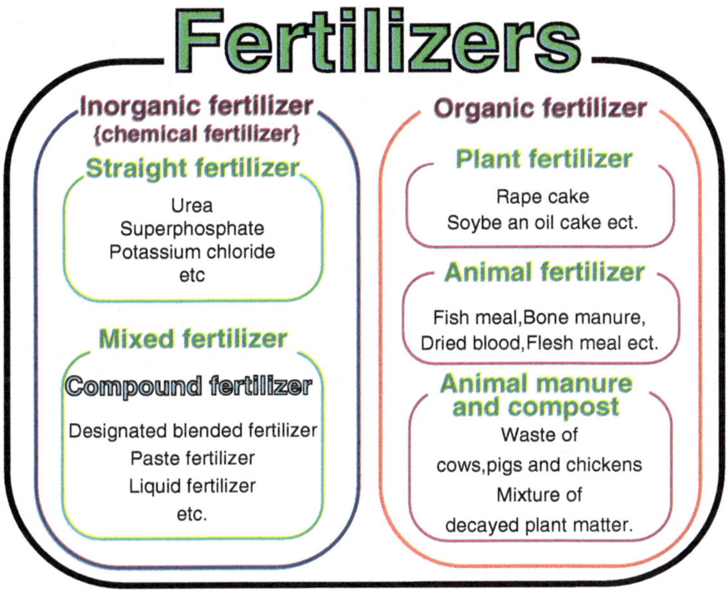

Fig. 12.2 Different types of manures and fertilizers

12.2.1 Organic Manures

The manures are waste organic materials of plants and animals (crop residues, farmyard manure, compost, oil cakes, animal slurry, sewage sludge, fish meal, bone meal, flesh meal, dried blood, etc.) (Fig. 12.2). The manuring is done to replenish the soil with nutrients resulting in stronger plants and better crops. The benefits of manuring include improvement in soil texture and moisture retention capacity of the soil, enhancement of nutrients and soil biota (beneficial microbes) in soil, and increase soil porosity to facilitate exchange of gases.

12.2.2 Inorganic Fertilizers

The fertilizers are chemical substances rich in nutrients such as N, P, and K either singly or in a combination. There are two types of fertilizers, namely, straight and compound fertilizers (Fig. 12.2). The straight fertilizers contain one nutrient element (urea, super phosphate, potassium chloride), while compound fertilizers have two or more nutrient elements in combination (blended fertilizers). The compound fertilizers are named based on their percentage content of N, P, and K (e.g., 10:10:20 contain 10 % N, 10 % P, and 20 % K). They are balanced fertilizers formulated based on the requirements of the particular crops.

Fig. 12.3 The manure shredder and spreader

The fertilizers facilitate farmers to get higher crop yields and better economic returns. The limitations include pollution of water bodies, soil salinization, and susceptibility to pests and diseases.

12.3 Manure Application

The farmyard manure (FYM) (bulky material), which is a mixture of urine, feces, and bedding material (crop residues), is a by-product of housing of animals. FYM is beneficial to the soil because it contains huge quantity of organic matter. The FYM is filled in the muck spreader utilizing a front loader and a tractor. The FYM is shredded and applied to soil from the side of the spreader (Fig. 12.3).

The animal slurry (bulky material) is mainly a mixture of feces and urine that contains lower amounts of minerals (nitrogen, 0.5 %; phosphorous, 0.15 %; and potassium, 0.60 %). The slurry spreader sucks slurry into the tank through a tractor-driven pump and applies it to the soil in a 6 m wide band (Fig. 12.4).

12.4 Fertilizer Application

The soils normally may not supply the desired nutrients for healthy crop growth. When the crop is harvested, they remove certain nutrients from the soil. Hence, the fertilizers are applied regularly to increase plant growth and yield and to increase

Fig. 12.4 The slurry spreader

farmers' income. Since the fertilizers are costly, only sufficient amount which gives maximum yield should be added (which is most profitable to the farmer). The basis of fertilizer application is soil and plant tissue analysis. The quantity of fertilizer application depends on several factors such as soil type (clay and organic matter content), soil moisture, crop type, row spacing, type of fertilizer, and spread of seed and fertilizer (type of opener). The nitrogenous fertilizers should be applied in split doses at the time of sowing, booting stage, and grain filling stages of crop growth. The potassic and phosphatic fertilizers have to be applied at the time of sowing in the root zone, because of their restricted mobility.

The placing of fertilizer materials to the side of and below seed or mixing fertilizer materials in the soil is the proper method of fertilizer application. The fertilizer should never be placed too close to seed or seedling. There are several methods of fertilizer application for both solid and liquid fertilizers.

12.4.1 Solid Fertilizers

The different ways of solid fertilizer application to the soil are as follows (Fig. 12.5).

12.4.1.1 Broadcasting

The spreading of fertilizers uniformly on the soil surface is called broadcast application. The two types of broadcasting include basal application and top dressing (depending upon the time of fertilizer application).

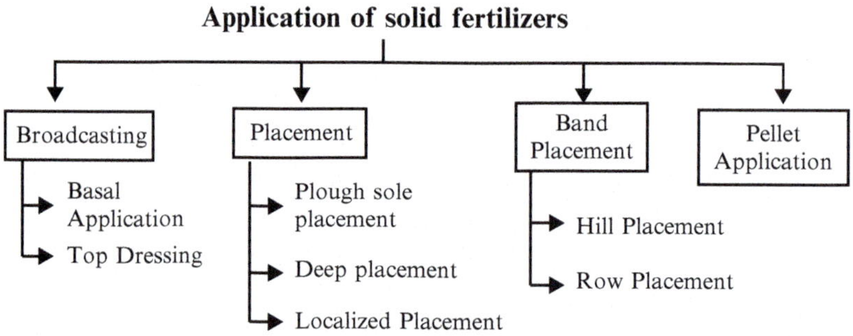

Fig. 12.5 Different methods of application of the solid fertilizers

Fig. 12.6 Broadcasting of fertilizers in rice (*left*) and wheat (*right*)

1. *Basal Application*

The application of larger quantities of fertilizers at the time of planting/sowing is called basal application.

2. *Top Dressing*

The broadcast application of second dose of fertilizers in the standing crop is called top dressing. The nitrogenous fertilizers (NO_3) are usually top dressed in rice (Fig. 12.6) and wheat.

The main limitations of broadcasting include stimulation of weed growth throughout the field, fixation of nutrients in the soil, and improper utilization of nutrients as they spread laterally over long distances.

12.4.1.2 Placement

The application of fertilizers in soil at a specific place with or without reference to the position of the seed is called placement. It is practiced when the quantity of fertilizers to be applied is small particularly for application of phosphatic and potassic fertilizers. There are three different types of placement, viz., plow sole placement, deep placement, and localized placement.

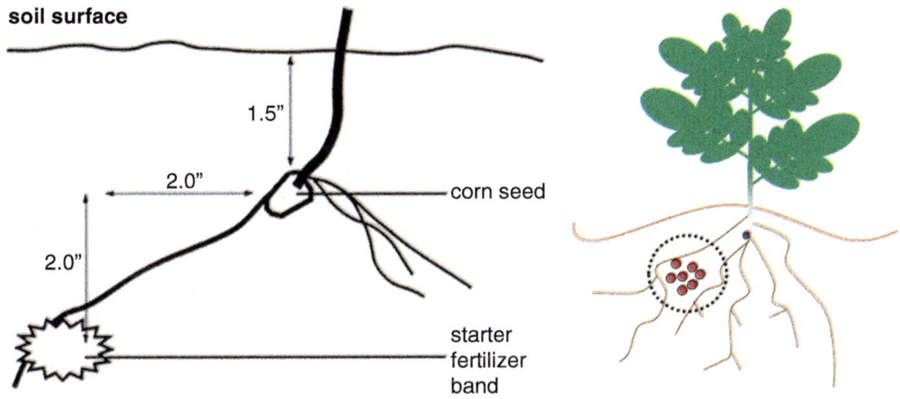

Fig. 12.7 Band placement of fertilizers

1. *Plow Sole Placement*

The placement of fertilizers at the bottom of the plow furrow in a continuous band during plowing is called plow sole placement.

2. *Deep Placement*

It is the placement of ammoniacal nitrogenous fertilizers (ammonium sulfate and urea) in the reduction zone of soil particularly in paddy fields, where ammoniacal nitrogen remains available to the crop. It prevents the loss of nutrients by runoff since the fertilizers are distributed in the root zone soil.

3. *Localized Placement*

The application of fertilizers into the soil close to the seed or plant so as to supply the nutrients to the roots of growing plants in adequate amounts is called the localized placement.

12.4.1.3 Banding

The placement of fertilizers 5 cm below and 5 cm to the side of the seed (to prevent injury to the seed) is called band placement (Fig. 12.7). It is commonly used in row crops.

There are two types of band placement, namely, hill placement and row placement.

1. *Hill Placement*

This method of fertilizer application is commonly practiced in orchard crops such as citrus, mango, guava, etc. The fertilizers are applied in bands close to the plants on one or both sides. Depending on the nature of the crop, the depth and length of the band vary.

2. *Row Placement*

In this method, the fertilizer is placed on one or both sides of the row in continuous bands. It is usually practiced in annual crops which are sown close together in rows such as potato, sugarcane, cereals, maize, etc.

12.4.1.4 Pellet Application

The small pellets of convenient size are prepared by mixing nitrogenous fertilizers with soil in the ratio of 1:10 which are placed at 2.5–5 cm deep between rows in the mud of rice crop. Neem-coated urea is prepared in this way.

12.4.1.5 Benefits

The benefits of application of solid fertilizers are as follows:

- Better use of phosphorus.
- Higher response of residual fertilizer.
- Improper use of fertilizers by weeds.
- Leaching of nitrogen is lower.
- Nutrient utilization for plant growth and development is higher.
- Reduction in fixation of nutrients.

12.4.2 Liquid Fertilizers

The different methods of application of liquid fertilizers are as follows:

- Starter solutions
- Foliar application
- Fertigation
- Injection into soil
- Aerial application

12.4.2.1 Starter Solutions

The application of liquid fertilizers (containing nitrogen, phosphorus, and potassium) in the ratio of 1:1:2 and 1:2:1 to seedlings at the time of transplanting is called starter solution. The benefit of starter solutions is to facilitate proper establishment and faster growth of transplanted seedlings. The limitations include requirement of additional labor and higher phosphate fixation.

Fig. 12.8 Foliar application of fertilizers

Table 12.2 Effect of soil vs. foliar application of NPK to banana

Treatment	Leaf nutrients (%)	Fruit yield (t/ha)
Soil appln. 300 g N, 100 g P_2O_5, and 300 g K_2O	6.38	58
¼ NPK through soil + ¾ NPK as 12 foliar sprays	7.12	68

12.4.2.2 Foliar Application

The spraying of the crop foliage with fertilizer solutions having one or more nutrients to facilitate their quick absorption is called foliar application (Fig. 12.8). Normally the micronutrients such as zinc, iron, boron, manganese, and copper in solution form are sprayed to the foliage to correct deficiencies. This method is the most efficient for application of micronutrients and offers the quickest response compared to any other fertilizing methods. Plants can absorb nutrients ten times more efficiently and faster through foliar application than through soil application (Table 12.2). Pesticides are also mixed along with micronutrients to manage crop pests.

12.4.2.3 Fertigation

The application of liquid fertilizers (normally nitrogenous fertilizers) through irrigation water is called fertigation or chemigation (Fig. 12.9). The benefits of fertigation are increase in fertilizer use efficiency (Table 12.3), saving in amount of fertilizers, yield increase (Table 12.4), and saving of labor cost.

Fig. 12.9 The fertigation system

Table 12.3 Effect of method of fertilizer application on efficiency of different nutrients

Element	Method of application	Fertilizer use efficiency (%)
N	Fertigation	90–95
	Surface flow	50–70
P	Fertigation	40–45
	Surface flow	10–20
K	Fertigation	75–80
	Surface flow	<50

Table 12.4 Effect of fertigation on saving fertilizer and increased yields

Crop	Saving in fertilizer (%)	Increase in yield (%)
Banana	20	11
Onion	40	16
Potato	40	30
Tomato	40	33
Onion	40	18
Broccoli	40	10

12.4.2.4 Injection into Soil

The injection of liquid fertilizers into the soil with or without the pressure is called injection into soil. The liquid fertilizers (anhydrous ammonia) can be applied in narrow furrows (at a depth of 12–15 cm) without pressure and covered immediately with soil to avoid loss of nutrients (ammonia).

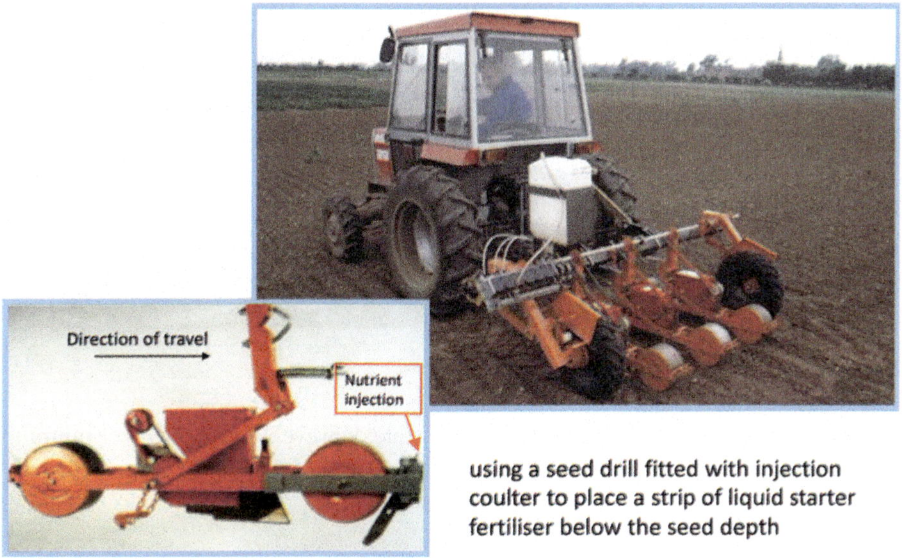

Fig. 12.10 Strip injection of liquid fertilizer

Fig. 12.11 Spot injection of liquid fertilizer

Liquid fertilizers can be applied through drip irrigation system or tractor drawn equipments [strip injection of liquid fertilizer (Fig. 12.10)/spot injection of liquid fertilizer (Fig. 12.11)].

The benefits of liquid fertilizer injection include increased fertilizer use efficiency throughout the crop cycle, protection of the environment (by preventing

Fig. 12.12 Aerial spraying

volatilization of ammonia and leaching of nitrates), prevention of fertilizer injury to plants, and an economical practice for the farmers.

12.4.2.5 Aerial Application

In difficult areas such as hilly areas, forest lands, grass lands or sugarcane fields, etc., where the ground spraying is impracticable, the spraying of fertilizer solutions can be achieved through aerial application (Fig. 12.12).

12.5 Fertilizer Use Efficiency

The goal of fertilizer use efficiency is to optimize the beneficial role of nitrogen, phosphorus, and potassium through recovery efficiency (percentage increase in uptake per unit of nutrient added) and agronomic efficiency (increase of crop yield per unit of nutrient added as kg grain/kg nutrient) in sustainable food production and to minimize negative effects of fertilizers on human health and the environment.

The scheduling of fertilizer should be based on plant tissue and soil test. The soil reaction should form the basis of selection of fertilizers, viz., basic fertilizers for acidic soil reactions and acidic fertilizers for alkaline soils. The fertilizers should be applied at 5 cm by the side or below the seed.

The nitrogenous fertilizers should be applied in three equal split doses, i.e., 1/3 as basal, 1/3 after 30 days of sowing, and the balance 1/3 about 50–60 days after sowing. In view of the restricted mobility of phosphatic and potassic fertilizers, they should be applied in the root zone as basal dose at the time of sowing/planting.

Table 12.5 Effect of balanced nutrition in increasing fertilizer use efficiency

	Treatment	
Crop	N	NPK
Wheat (11 years)	31	70
Corn (5 years)	35	66
Barley	28	51

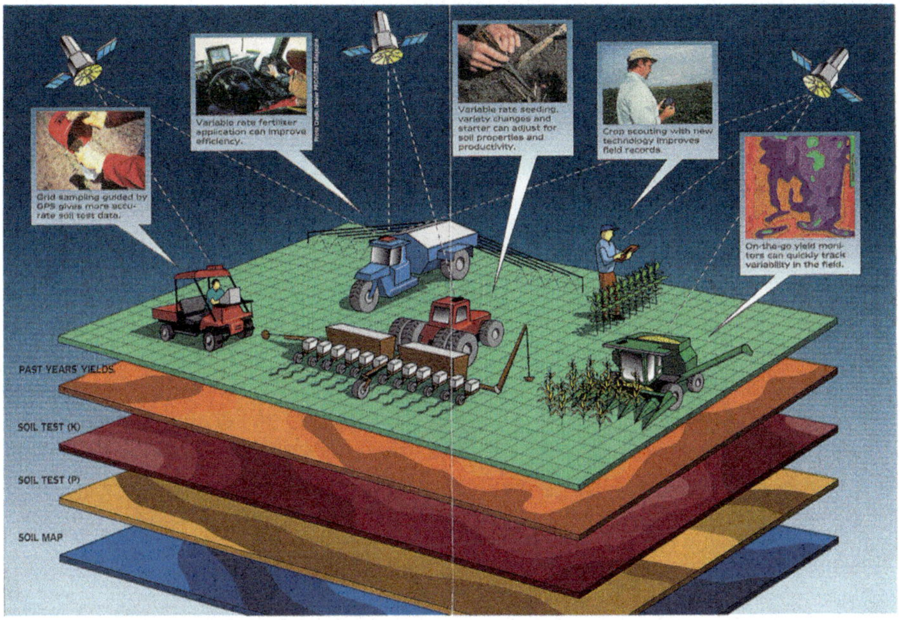

Fig. 12.13 Site-specific precision application of fertilizers using computer technology

12.5.1 Balanced Nutrition

The balanced nutrition instead of application of individual nutrients is very important in increasing the fertilizer use efficiency (Table 12.5).

12.5.2 Site-Specific Application/Variable Rate Technology

The use of computer technology to alter the rate of fertilizer application as the fertilizer applicator passes across the field is called site-specific precision application [also called variable rate technology (VRT) (Fig. 12.13)]. This method of fertilizer application minimizes the possibility of overfertilization and has potential to improve yield resulting in additional profit to the farmers.

Fig. 12.14 Deep placement of urea briquette by hand

Table 12.6 Effect of urea and neem-coated urea on mint oil yield

N (kg/ha)	Mint oil yield (kg/ha)		% increase
	Urea	Neem-coated urea	
120	130	170	31.6

12.5.3 Deep Placement

The surface application of urea by hand broadcasting in rice fields results in loss of ammonia by way of runoff, denitrification, and volatilization (atmospheric evaporation) leading to the extremely low (35 %) fertilizer use efficiency. A low-cost, simple, proven, and innovative fertilizer application technology which is responsible for about 18 % increases in yield and about one-third reduction in fertilizer use is achieved through deep placement of urea in the soil. The physical placement of the super granules below the soil surface is well suited for small-scale rice farmers (Fig. 12.14). The rice farmers who are adopting deep placement of urea in double cropping system (two rice crops per year) have obtained US$ 400 additional income per hectare compared to the conventional farmers.

12.5.3.1 Neem-Coated Urea

The prilled urea is coated with neem oil and formulated as super granules to facilitate deeper placement and to reduce nitrification loss. The benefits of neem-coated urea include enhanced nitrogen use efficiency, reduced losses of nitrogen due to denitrification and leaching, increased plant growth and grain yield (Table 12.6), slow release of nitrogen and longer availability, and regulation of nitrogen supply as per the requirement of crop growth and development.

12.5.3.2 Root Activity Studies Using Both Stable and Radioactive Isotopes

Another method to increase the fertilizer use efficiency is to apply nutrients in the active root zone for their effective utilization by the plant. The active root zone can be traced using both stable and radioactive isotopes. The active root zones of

Table 12.7 Active root zone of some important fruit crops and the placement of fertilizers

Fruit crop	Active root zone	Placement of fertilizer
Mango	82–85 % of active roots located within a radius of 300 cm with highest activity at 120 cm from tree trunk	1 m away from tree trunk under drip and mix thoroughly. Deep placement
Citrus	Within a radial distance of 120 cm at depth of 15–30 cm	Apply fertilizers in a circular band between 90 and 120 cm radial distance from trunk. Incorporate fertilizers superficially
Guava	At 100–220 cm distance from trunk at a depth of 30–60 cm	Apply fertilizers at 100 cm radial distance from trunk at a depth of 15–20 cm. Deep placement
Grape	In radius of 100–220 cm at a depth of 45–75 cm	Apply fertilizer within 15 days of pruning at radial distance of 60 cm from trunk. Deep placement
Papaya	100 % of active roots located within a radius of 100 cm at a depth of 0–45 cm from tree trunk	1 m away from tree trunk under drip and mix thoroughly
Litchi	Shallow on surface	Soil surface with light working
Pineapple	Highly restricted	9–10 cm away in 10–15 cm deep strip around plants
Ber	50–75 cm deep under drip	Under drip and mix in soil up to 30 cm
Sapota	Under canopy within a depth of 0–30 cm	Under canopy and mix in soil up to 15 cm depth
Date palm	Up to a radius of 1.7 m from the bole	Around tree up to a radius of 2.0 m

different fruit crops and the placement of fertilizers have been presented in Table 12.7 (Chundawat 1997).

The root activity studies have been carried out in selected fruit crops like grapes and mango. In grapes, it has been shown that the roots present between 15 and 25 cm depth and 45–60 cm lateral distance are more efficient in the absorption of P. Similarly, in mango, the highest zone of root activity has been found at a depth of 30–60 cm and at a lateral distance of 120–240 cm.

12.5.4 CULTAN Fluid Injection

The predominant nutrition of crops with ammonium instead of nitrate through subsurface fluid injection became known as "CULTAN" (controlled uptake long-term ammonium nutrition) fertilization in Germany (Sommer 2000). In view of the increasing negative environmental impacts of using nitrate form of N (e.g., NO_3, leaching, N_2O release), the CULTAN fertilization strategy has been adopted. The benefits of CULTAN fertilization are as follows:

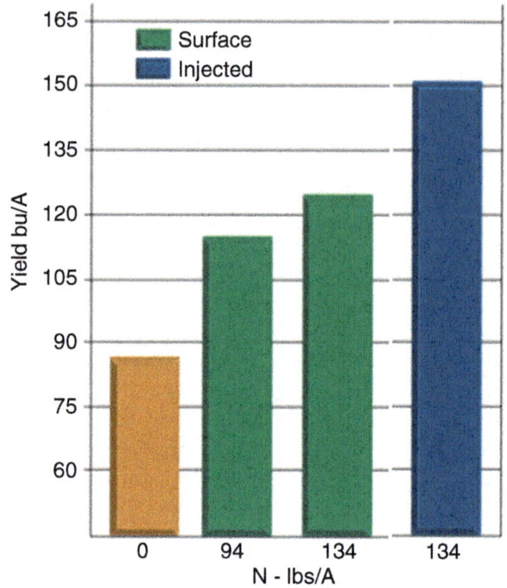

Fig. 12.15 Effect of surface application and CULTAN fertilization with urea ammonium nitrate on wheat yield

- Higher N use efficiency
- Reduced N fertilizer consumption (15–20 %)
- Increase in grain yield and quality (Fig. 12.15)
- Prevention of nitrifying bacteria (which transform nitrate N into more mobile nitrate form) from ammonium toxicity
- Supply of all nitrogen needed for vegetation at one dose
- Prevents nitrogen loss caused by washout and without risk of underground water contamination (Kücke and Scherer 2006)
- Development of dense plant root system
- Increased nitrogen uptake by plants compared to the conventional system (Sommer 2005)

12.6 Conclusions

The low level of nutrients supplied by organic manures alone (nitrogen, 0.5 %; phosphorous, 0.15 %; and potassium, 0.60 %) is not sufficient to meet the requirements of crop growth. The inorganic fertilizers supply large amounts of nutrients and thereby supplement the organic manures. Hence, the integration of organic manures and inorganic fertilizers is quite effective for the crop growth and development, since physical soil properties are improved by organic manures, while nutrient needs of crops are fulfilled by inorganic fertilizers. In order to realize maximum benefits from manure and fertilizer application, the optimum time, mode, and kind of fertilizers used are important.

References

Chundawat BS (1997) Nutrient management in fruit crops. Agrotech Publishing Academy, Udaipur
Kücke M, Scherer HW (2006) Injektionsdüngung in Deutschland. RKL, Rendsburg, pp 397–429
Sommer K (2000) CULTAN-cropping system: fundamentals, state of development and perspectives. In: Nitrogen in a sustainable ecosystem: from the cell to the plant. Backhuys Publishers, Leiden, pp 361–375
Sommer K (2005) CULTAN-Düngung. Verlag Th. Mann, Gelsenkirchen, p 218

Chapter 13
Integrated Nutrient Management

Abstract The enhancement of soil health (biologically based phenomenon), productive capacity, and function are the precursors for sustainable production intensification. The main principle for sustainable production is that farmers should fertilize their soils rather than their crops. In INM system, the soil biota and microorganisms, soil organic matter, and soil porosity are built up and maintained. Application of organic manures to the soil supports higher levels of biological activity. Enhanced productivity is unlikely to be attained without ensuring that plants have an adequate and balanced supply of nutrients, as long as agriculture remains as a soil-based enterprise. Hence, there is a need for utilizing organic, inorganic, and biological nutrient resources to meet crop needs for increased production and productivity.

Keywords Soil health • Organic manures • Inorganic fertilizers • Biofertilizers • Integrated nutrient management

13.1 Introduction

The inorganic fertilizers, organic manures, and biofertilizers are applied to soils to encourage plant growth and development. The inorganic fertilizers are one of the most expensive inputs in agriculture. Their excessive and indiscriminate use resulted in adverse effects such as salinization or acidification of soil, leaching of large amounts of nitrates, pollution of groundwater resources, deterioration of soil fertility, death of friendly microorganisms and beneficial insects, susceptibility of crops to diseases (with excess nitrogen), loss of nutrients through emission of greenhouse gases and fixation, and reduced nutrient use efficiency, leading to overall damage to the system.

The organic manures contain lower amounts of nutrients compared to inorganic fertilizers; hence huge quantities are required for soil application to obtain sufficient quantities of nutrients resulting in transportation and handling problems. The availability of organic manures in large amounts is also a problem. Since the decomposition of organic manures is a long process, the availability of nutrients is very slow. Therefore they may not release the nutrients when the plants need them. The fresh

P.P. Reddy, *Sustainable Intensification of Crop Production*,
DOI 10.1007/978-981-10-2702-4_13

manures contain weed seeds and when applied to soil spread the new weeds in the field.

The biofertilizers have short shelf life and difficult to store, have insufficient population of colony forming units, and are slow in action, and their growth and multiplication are affected by low soil moisture or high soil temperature and alkalinity or acidity in soil.

The use of organics alone may not yield enough due to its limited availability to fulfill the entire nutrient demands of crops which justifies for integrated use of fertilizers and manures.

In future, there is a need for judicious use of nutrients from all resources for plant growth and development. Hence, integrated nutrient management (INM) using inorganic fertilizers, organic manures, and biofertilizers is the need of the hour to maintain the overall balance and flow of soil nutrients.

13.2 Integrated Nutrient Management

FAO (2008) defined INM as follows:

- Balanced use of inorganic fertilizers combined with organic manures and biofertilizer sources of plant nutrients to enhance or maintain productivity of soil
- Enhancement of plant nutrient stock in the soil
- Preventing nutrient losses to the environment through improving nutrient use efficiency

The integrated nutrient management (INM) is a holistic approach to maintain optimum level of plant nutrient supply and soil fertility for attaining higher crop productivity by utilizing all possible resources (inorganic fertilizers, organic manures, and biofertilizers) of plant nutrients (in what form at what timing in best possible method) in an economically acceptable and eco-friendly way (Roy and Ange 1991). An integrated nutrient management and supply system should be adopted in order to encourage balanced nutrition and efficient use of plant nutrients. In INM, the main emphasis should be given to enhance balanced and proper use of chemical fertilizers to increase the crop productivity, while the use of organic manures and biofertilizers is considered supplementary and not substitutable. Based on the principle of Liebig's Law of Restitution, the fertility of soil can be sustained by replenishment of nutrients removed from the soil.

The native soil status, potential and targeted yields, type of rooting system and its mining ability, and nutrient management decide the nutrient requirement of crops. A 22 % increase in yield has been obtained in India under different crops by following INM practices rather than farmer's practice based on a 3-year data collected from 267 sites.

Enhancing the nutrient use efficiency (NUE) is an important aspect in INM through proper fertilizer placement close to the rhizosphere of the highest root activity. The highest root activity zones have been worked out in certain crops using

Table 13.1 Optimum leaf nutrient status of some important fruit crops

	Nutrients											
	%						ppm					
Fruit crop	N	P	K	Ca	Mg	S	B	Fe	Mn	Zn	Cu	References
Mango	1.23	0.66	0.54	1.71	0.91	0.12	–	171	66	25	12	Samra et al. (1978)
Banana	3.29	0.44	3.11	2.12	0.24	–	–	–	–	–	–	Ramaswamy (1971)
Citrus	2.50	0.14	0.90	4.20	0.43	0.25	–	90	112.5	62.5	10.5	Reuther et al. (1962)
Guava	1.70	0.50	1.30	2.10	0.46	–	10	–	–	77.5	75	Singh (1981)
Grape	1.75	0.56	1.66	0.94	–	0.20	–	–	125	92	–	Bhargava and Chadha (1993)
Litchi	1.47	1.03	–	–	–	–	–	–	–	–	–	Ghosh et al. (1986)
Sapota	1.66	0.08	0.80	0.83	0.48	0.07	–	100	39.3	15.7	6.7	Annapurna et al. (1988)
Papaya	1.66	0.50	5.21	1.81	0.67	0.38	–	–	–	–	–	Bhargava and Chadha (1993)
Pineapple	1.40	0.16	3.70	–	–	–	–	–	–	–	–	Chadha and Reddy (1993)
Pomegranate	2.50	0.19	1.47	–	–	–	96	199	196	55	–	Shende (1977) and Bhambal (1987)
Ber	2.50	0.25	1.82	1.21	0.45	–	–	–	–	–	–	Khanduja and Garg (1984) and Bhargava et al. (1990)

both stable and radioactive isotope techniques using labeled elements (Chundawat 1997). Soil and plant analysis (more particularly leaf and petiole analysis) should form the basis of crop fertilization in many countries. Hence, there is an immediate need for establishing plant tissue analysis laboratories at different identified centers both in public and private sector to work out leaf nutrient guides and offer means of assessing nutritional requirements of different crops. Leaf nutrient standards for a large number of crops have been developed and presented in Table 13.1.

The farmers need to be trained and educated with the help and effective coordination of State Agricultural Universities/ICAR Institutes/State Departments of Agriculture/Horticulture and NGOs for effective implementation INM practices.

13.2.1 Goals

The goals of INM are as follows:

- Using leaf and petiole analysis for fertilizer recommendations
- Supplementing inorganic fertilizers with organic manures and biofertilizers for plant nutrient needs
- Reducing the use of inorganic fertilizers

13.2.2 Needs

The several needs for adopting INM practices include:

- The fertilizers are the most costly inputs.
- The applied fertilizers are lost from soil–plant system due to runoff; leaching of nitrates and phosphates; denitrification, nitrification, and volatilization; and soil, water bodies, and atmosphere pollution.
- Only two-thirds of total requirement of plant nutrients should be met through the use of chemical fertilizers, and another one-third can be obtained from the organic manures and recycling of biological waste materials.
- The accumulation of nitrates and phosphates to the extent of more than 100–200 ppm can cause cancer in human beings and kill them.
- The inorganic fertilizers emit greenhouse gases such as nitrous oxide, methane, etc., which are responsible for destruction of stratospheric ozone.
- The presence of excess concentrations of heavy metals (Fe, As, Cd, Co, Cr, Hg, Ni, Pb, etc.) present in phosphatic fertilizers is harmful to the health of humans and animals.
- The fertilizers are not available at appropriate time.
- Increase in prices of fertilizers.

All these factors are responsible for adoption of integrated nutrient management practices which are very crucial (Fig. 13.1).

13.2.3 Benefits

- Enhanced and sustained physical, chemical, and biological functioning of soil
- Provided balanced crop nutrition
- Increased carbon sequestration
- Minimized antagonistic effects due to hidden deficiencies and nutrient imbalance
- Reduced water, soil, and ecosystem deterioration
- Reduced nutrient losses to surface and groundwater bodies and to atmosphere

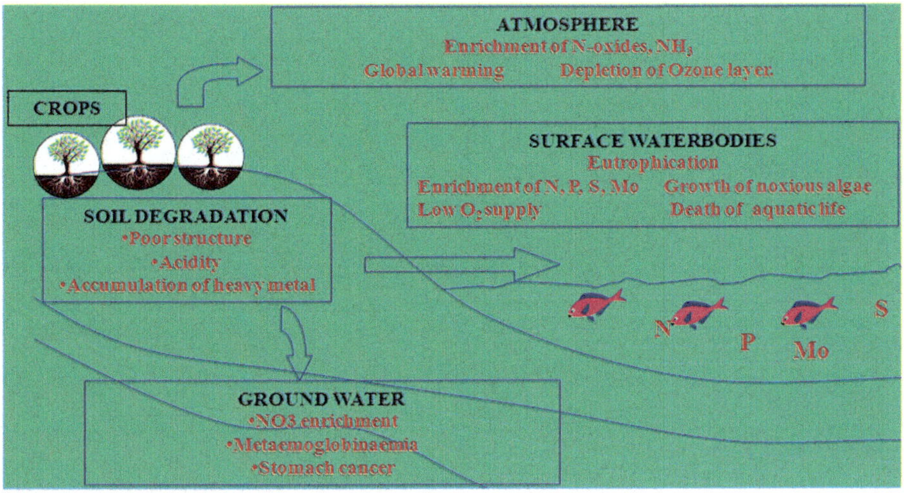

Fig. 13.1 Ecological constraints of inorganic fertilizers

Table 13.2 Commonly used nitrogen, phosphatic, and potassic fertilizers

Nitrogen fertilizers	Phosphorus fertilizers	Potassium fertilizers
Ammonium sulfate, calcium ammonium nitrate, urea	Single superphosphate	Potassium sulfate,
	Rock phosphate	Potassium nitrate,
		Muriate of potash (potassium chloride)

13.3 Components

Besides inorganic fertilizers as the major component, other components of INM include farmyard manure (FYM), composts, green manure, crop residues, crop rotation, and biofertilizers.

13.3.1 Inorganic Fertilizers

The commonly used nitrogen, phosphatic, and potassic fertilizers are presented in Table 13.2.

The nitrogen fertilizers are usually applied in split doses based on the requirement of crops at planting, earthing up/thinning, and flowering in order to get higher yield per unit N applied and greater B:C ratio. The entire amounts of phosphorus and potassium fertilizers are normally applied at the time of planting as basal application within the active root zone. The balanced application of N, P, and K fertilizers to the soil is done in the ratio of 2:1:1 or 3:1:1 ($N:P_2O_5:K_2O$). A number of

Table 13.3 Fertilizer recommendations for some tropical and subtropical fruit crops

	Nutritional recommendation (g/plant/year)		
Fruit crop	N	P_2O_5	K_2O
Mango	775	500	700
Citrus (acid lime)	900	250	500
Guava	360	180	180
Grape	270	450	900
Pomegranate	600–700	200–250	200–250
Ber	500	200	300
Sapota	1000	500	500
Amla	1500	750	1000
Date palm	1000	500	500

Table 13.4 Effect of green manure crops on biomass production and nitrogen accumulation

Green manure crop	Fresh biomass (t/ha)	N accumulation (kg/ha)
Dhaincha, *Sesbania aculeata*	26.30	185.00
Sesbania rostrata	24.90	219.00
Sun hemp, *Crotalaria juncea*	21.20	159.00
Pillipesara, *Phaseolus trilobus*	19.60	201.00
Senji, *Melilotus alba*	28.60	163.00

recommendations for the application of inorganic fertilizers are available for different crops for several agro-ecological regions (Table 13.3).

13.3.2 Farmyard Manure and Compost

The application of farmyard manure/compost to soil significantly increased beneficial microorganisms such as of actinomycetes, fungi, rhizobia, and bacteria (responsible for insect pest and disease control and plant growth), enhanced moisture retention capacity of soil, and improved soil fertility resulting in the realization of higher yields.

13.3.3 Green Manure

The application of green manure to the soil is responsible for addition of significant quantity of biomass and nitrogen accumulation (Table 13.4). Tillman et al. (2002) reported that intercropping of legume green manure cover crop in cotton increased predator population resulting in reduced insecticide use and higher returns. The in situ green manure crops are sun hemp (*Crotalaria juncea*), dhaincha (*Sesbania aculeata*), green gram (*Phaseolus* sp.), cowpea (*Vigna unguiculata*), khesari (*Lathyrus sativus*), and berseem (*Trifolium alexandrinum*), while green leaf manure crops include *Gliricidia maculata*, *Pongamia glabra*, *Calotropis gigantea*, *Tephrosia purpurea*, *Indigofera teysmannii*, *Sesbania speciosa*, *Ipomea carnea*, *Cassia tora*, and *Leucaena leucocephala*. About 20–30 million tons of green manure can be generated by growing green leaf manure crops all along the bunds and harvested at the right stage.

13.3.4 Crop Residues

The applications of crop residues play a major role in the recycling of nutrients. In the global harvest, more than one half of all dry matter is in the form of residues. The amounts of nutrients in residues are often several orders of magnitude higher than the quantities applied as fertilizers in most of the developing countries. The crop residues should be converted into bio-composts for their effective utilization by crop plants. Hence, proper management of crop residues for the maintenance of soil fertility cannot be overemphasized.

13.3.5 Biofertilizers

The biofertilizers, also called as microbial inoculants, are cultures of certain soil microorganisms multiplied artificially and packed in some carrier material for easy application in the field to increase the fertility of soil and productivity of crops. The soil microorganisms are capable of converting the non-available forms of nutrients to available form through biological processes.

The benefits of biofertilizers are eco-friendly, less expensive, sustainable, and long-lasting. They are also responsible for decomposition of organic material and recycling of nutrients, provide growth-promoting substances to crops, enhance health of soil and plants, control soilborne diseases, and help savings in inorganic fertilizers.

The different types of biofertilizers are as follows:

- Nitrogen-fixing microorganisms
- Phosphate-solubilizing bacteria (PSB)/phosphate-mobilizing bacteria
- Potash-mobilizing microorganisms
- Sulfur-mobilizing bacteria
- Arbuscular mycorrhizal fungi (AMF)

13.3.5.1 Nitrogen-Fixing Microorganisms

The nitrogen-fixing microorganisms include bacteria (*Rhizobium*, *Bradyrhizobium*, *Azospirillum*, *Azotobacter*, *Beijerinckia*), and blue-green algae (*Azolla*, *Anabaena*, *Nostoc*, *Plectonema*, *Aulosira*, *Oscillatoria*, *Tolypothrix*).

Table 13.5 The recommended biofertilizers, their method of application, and dose for different crops under field applications

Crop	Recommended biofertilizer	Method of application	Dose (kg/ha)
French bean, pea, cowpea, pigeon pea	*Rhizobium*	Seed treatment	2.5
Beet root	*Azotobacter*	Seed treatment	4.5–7.5
Ornamental plants	*Azotobacter*	Seedling treatment	3.75–5
Ginger, turmeric, garlic, rose, banana, gladiolus	*Azotobacter*	Buds/tuber/set treatment	7.5–10
Coffee, tea	*Azotobacter*	Soil treatment	7.5–10
Coconut, rubber	*Azotobacter*	Soil treatment	20–30 g/ plant
Potato	*Azospirillum*	Soil/tuber treatment	10
Radish, spinach, turnip, carrot, pointed gourd	*Azospirillum*	Seed treatment	10
Tomato, brinjal, chili, onion, cabbage, cauliflower	*Azospirillum*	Seedling treatment	10

The *Rhizobium* species fix atmospheric nitrogen in symbiotic association with certain legume plants forming nodules in roots. They are highly specific to particular legume crops. *Azotobacter* species are free living and fix atmospheric nitrogen nonsymbiotically in crop plants. They also facilitate plant growth and control soil-borne root pathogens. The inoculation of *Azotobacter* to cotton along with 40 kg N ha^{-1} is equivalent to addition of 60 kg N ha^{-1} (a saving of 20 kg N ha^{-1}). The *Azotobacter* is responsible for conversion of unavailable sources of atmospheric N and bound phosphates into available forms and thus helps to increase soil fertility and crop yields. *Azospirillum* species fix atmospheric nitrogen in non-leguminous plants. They also produce growth-promoting substances.

The recommended biofertilizers, their method of application, and dose for different crops are presented in Table 13.5.

The blue-green algae *Azolla* species are free living and fix atmospheric nitrogen in crop plants. The additional benefits of blue-green algae are enhanced soil aeration and moisture retention capacity, production of growth-promoting substances (including vitamin B12), and addition of biomass on decomposition after completion of life cycle.

13.3.5.2 Phosphate-Solubilizing/Phosphate-Mobilizing Microorganisms

The Indian soils are very poor in phosphorus (P) content. A small amount (25–30 %) of P applied to the soil is available for plant growth and the rest is fixed in soil and not available. A huge quantity (130 million tons) of P is in non-available form in the soil. The nutrient phosphorus, both inherent in the soil and applied as chemical fertilizers is mostly non-available to plants due to its low levels of solubility and mobility and its tendency to become fixed in soil. The phosphate-solubilizing bacteria (PSB) (*Bacillus megaterium, B. circulans, B. subtilis, Pseudomonas striata, P. flourescens*) and fungi (*Aspergillus awamori, A. flavus, A. niger, Penicillium digitatum*) help in dissolving the fixed P in soil and enhance phosphate uptake by plants. The soil application of PSB (*Bacillus megaterium* var *phosphaticum*) at 5 kg ha^{-1} along with vermicompost at 2 t ha^{-1} and rock phosphate is equivalent to supplying P through single super phosphate. This practice is very economical since the rock phosphate is cheaply available.

13.3.5.3 Potash-Mobilizing Microorganisms

The Indian soils are moderate in potash (K) content. The K is also fixed in soil. Potassium is an important factor in the assimilation of CO_2 by plants either directly or indirectly. The carbohydrates are more in plants well supplied with K compared to K-deficient plants. The K-mobilizing bacterium *Frateuria aurantia* (family: Pseudomonadaceae) (isolated from banana plant from Orissa soil) is capable of solubilizing 90 % of fixed form of K (from mineral source) within 22 days (Chandra et al. 1998). The application of *F. aurantia* at 2.5 kg/ha mixed with 200–500 kg FYM in furrows before sowing increased banana fruit yield by 20 %. A saving of 50–60 % toward the cost of chemical K fertilizer can be achieved by using K-mobilizing bacterium. The bacterium is capable of increasing the availability of K to the extent of 5–60 kg ha^{-1}.

13.3.5.4 Sulfur-Mobilizing Bacteria

The sulfur is associated with oxides of iron and aluminum and present in insoluble sulfate form at a depth of 30–35 cm in soil. The bacterium, *Acetobacter pasteurianus* (at 625 g ha^{-1}), is capable of converting this non-usable form to usable form in crops like vegetables, cabbage, turnip, onion, etc. (Chandra et al. 1998).

13.3.5.5 Arbuscular Mycorrhizal Fungi (AMF)

The arbuscular mycorrhizal fungi are obligate symbionts associated with plant roots and extend their extramatrical hyphae to far away areas from the roots and increase the surface area for absorption of phosphorus from soil. Besides P, AMF also absorbs zinc and copper to enhance plant growth and resist soilborne root pathogens. In order to colonize the seedling roots, the AMF is applied to nursery beds at 2 kg/m^2. The transplanting of colonized seedlings of chili, tomato, eggplant, potato, grapevine, apple, banana, tea, coffee, cocoa, and rubber facilitates establishment of AMF under field conditions. A saving of 25–50 kg P ha^{-1} can be achieved by the AMF in addition to increasing the yield to an extent of 10–12 %.

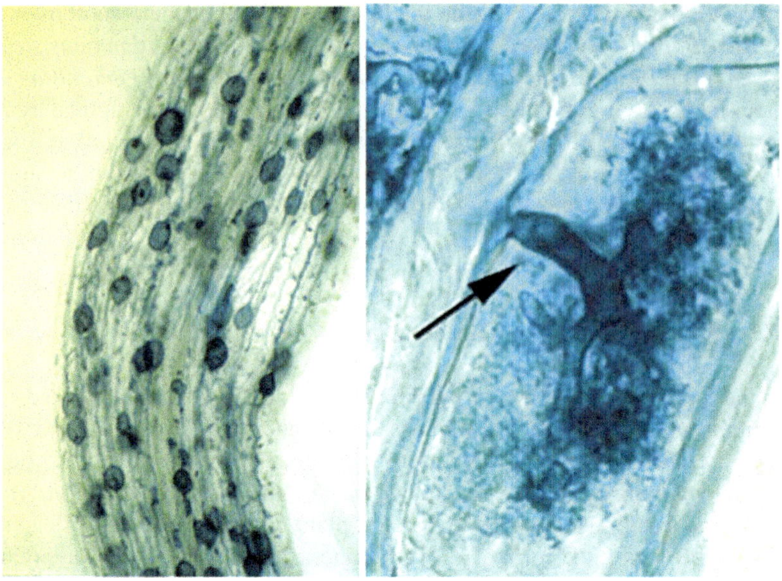

Fig. 13.2 Arbuscular mycorrhizal fungi. *Left*, vesicles in root; *right*, arbuscle in a root cortical cell

The most common AMF genera such as *Glomus*, *Gigaspora*, *Acaulospora*, and *Sclerocystis* (belong to class, Phycomycetes; order, Mucorales; family, Endogonaceae) produce arbuscules (highly branched fungal hyphae) and vesicles (bulbous swellings of hyphae) inside the root system (Fig. 13.2).

The benefits of AMF include savings in P fertilizers by 50 %, control of soil-borne root pathogens, increased moisture retention capacity, withstanding water stress and transplant shock, and enhanced plant growth.

13.3.6 Rotational Crops

The crop rotation is a very important component in sustaining nutrient supply. By growing legume crops in rotation improves fertility of soil by fixing atmospheric N to an extent of 135–488 kg ha^{-1}, utilizes residual nitrogen after harvest of the main crop, enhances soil properties, and controls insect pests, diseases, nematodes, and weeds.

13.4 Integration of INM Components

The integrated nutrient management system is characterized by the combined use of inorganic fertilizers along with organic manures (such as animal manures, crop residues, green manure, rotational crops, and composts) and biofertilizers (nitrogen-fixing microorganisms, phosphate-solubilizing/phosphate-mobilizing

Table 13.6 Effect of integration of organic manures and inorganic fertilizers on banana fruit yield

Treatment	Yield (MT/ha)
200 g N inorganic	68.90
25 % N – FYM + 75 % N – inorganic	77.03
25 % N – FYM + 25 % N – neem cake + 50 % N – inorganic	73.93
25 % N – FYM and green manure + 75 % N – inorganic	76.70

microorganisms, potash-mobilizing microorganisms, sulfur-mobilizing bacteria, and arbuscular mycorrhizal fungi). This can be achieved by adopting the following principles:

- Integrated and optimum usage (dose) of organic manures, inorganic fertilizers, and biofertilizers in a correct proportion
- To obtain full potential of all the resources
- To maintain soil fertility and plant nutrient supply at an optimum level for sustaining soil health and desired crop productivity

13.4.1 Organic Manures and Inorganic Fertilizers

Combined application of 25 % N in the form of FYM + 75 % N as inorganic fertilizer gave highest banana fruit yield followed by application of 25 % N in the form of FYM + green manure and 75 % N as inorganic fertilizer (Table 13.6). In a 2-year rotation, combined application of cotton crop residue, FYM, and vermicompost was on par with FYM + vermicompost + fertilizers in terms of yield (Basavaneppa and Biradar 2002). At Coimbatore (Tamil Nadu), an integration of organics, viz., FYM at 5 t ha^{-1} (15 days before planting), cotton residues at 2.5 t ha^{-1} (30 days before sowing), and sun hemp seeded at 15 kg ha^{-1} simultaneously in inter rows of cotton (as green manure) and buried at 45 DAP produced significantly higher seed cotton yield and lower pest population over the recommended dose of fertilizers (Praharaj et al. 2005).

13.4.2 Biofertilizers and Inorganic Fertilizers

Green manuring of dhaincha with *Azolla* inoculation at 3 tons ha^{-1} resulted in the grain yield similar to 90 kg N. Integration of *Azotobacter* (biofertilizer) with recommended dose of inorganic fertilizers significantly increased the fruit yield of guava (Table 13.7).

Table 13.7 Integration of organic manures and biofertilizers on fruit yield of guava (3 years)

Treatment/dose per tree	Yield (kg/tree)
350 g N + 200 g P_2O_5 + 350 g K_2O	8.09
20 kg FYM + 250 g *Azotobacter*	13.69

Table 13.8 Effect of biofertilizers and inorganic fertilizers on yield and TSS of Nagpur mandarin

Treatments	Yield/tree (kg)	TSS (°Brix)
RDF (600:200:400 g NPK/plant)	9.00	8.89
75 % RDF + biofertilizers (AMF + *Azospirillum lipoferum* + PSB + *Pseudomonas fluorescens*)	12.00	9.51

In Nagpur mandarin also, combined application of different biofertilizers (AMF + *Azospirillum lipoferum*, + PSB + *Pseudomonas fluorescens*) and recommended dose of inorganic fertilizers increased fruit yield and total soluble sugars in fruit (Table 13.8).

13.4.3 Biofertilizers, Organic Manures, and Inorganic Fertilizers

The application of biofertilizers, green manures, nitrification inhibitors, and split application of inorganic fertilizers (mostly N and K) enhances fertilizer use efficiency in cropping system. The fertilizer application should be balanced by integrating biofertilizers and organic and inorganic forms to correct imbalances.

13.4.4 Different Organic Manures

To achieve higher production efficiency with lower nutrient loss from soil (nutrient recharge), a set of organic materials/complexes are applied to the soil for better nutrient recycling in INM for sustainable farming. The soil application of *Gliricidia* green foliage loppings, green gram mulching, FYM, and sun hemp as green manure increased cotton yield by 15–32 % compared to control. Following this practice, there was a significant increase in available soil nutrient buildup (Blaise et al. 2003). The addition of organic manures to the soil is responsible for direct addition of N to the soil and facilitates higher multiplication of soil microbes which convert organic form of N to inorganic form resulting in restoring soil N pool. The microbial decomposition of organic manures also forms organic complexes with sesquioxides, and thereby, the P fixing capacity of the soil is reduced. In addition, the increased availability of K is facilitated by mineralization of organic manures which releases K to the soil and reduction in K fixation due to clay–organic matter interaction. Hence, INM using different organic manures has a potential role in nutrient cycling.

13.5 Future Thrusts

The following aspects need emphasis in future to attain sustainable nutrient management to increase productivity of crops without harming the environment:

- The inventory of the resources available in a region including the components of crop production such as managing crop with site-specific technology, farming/cropping system, tillage practices, water management, moisture retention practices, and abiotic and biotic stresses should form the basis for evaluation of micro-/secondary nutrients in INM practices.
- The maximum input use of efficient agro-technologies should form the integral components of the INM strategy.
- It is necessary that ecological compatibility, agronomic productivity, economic profitability, and social acceptability should be given due attention for conducting large-scale adaptive trials.
- The farmers should be made aware of unsustainable agricultural production practices such as the deteriorating soil health and environmental pollution due to nonuse of organics by extension agencies. They should also emphasize the importance of integration of organic manures, biofertilizers, and inorganic fertilizers in sustainable crop productivity.

13.6 Conclusions

A concerted effort should be made in using locally available INM components involving rational use of inorganic fertilizers, biofertilizers, and organic manures that will go a long way in providing a sustainable management of crop nutrition. In addition, micronutrients such as Mg, S, Zn, and B are likely to become critical in future due to multiple cropping, inadequate use of organic manures, and application of chemical fertilizers devoid of secondary/trace elements. Moreover, fertilizer management should be based on cropping systems rather than sole crop for higher nutrient use efficiency and economics. Therefore, integrated plant nutrient system has emerged as a necessity for sustainable agriculture. Developing a suitable nutrient management system that integrates the use of these three kinds of fertilizers may be a challenge to reach the goal of sustainable agriculture.

References

Annapurna K, Bojappa KM, Bhargava BS (1988) Leaf sampling guide for sapota (*Manilkara achras* M. Foseberg) cv. cricket ball. Crop Res 1:69–75

Basavaneppa MA, Biradar DP (2002) Integrated nutrient management practices on the production of cotton-maize-bengal gram sequence under irrigated ecosystem in Tungabhadra Project area. J Cot Res Dev 16:125–129

Bhambal SB (1987) Effect of foliar application of micronutrients on growth, yield, fruit quality and leaf nutrient status of pomegranate (Punica granatum *L.*) cv. Ganesh. M. Sc. (Ag.) thesis, Mahtma Phule Agri. Univ., Rahuri, Maharashtra

Bhargava BS, Chadha KL (1993) Leaf nutrient guide for fruit crops. In: Chadha KL (ed) Advances in horticulture, vol 2. Malhotra Publishing House, New Delhi, pp 973–1029

Bhargava BS, Raturi GB, Hiwale SS (1990) Leaf sampling technique in ber (*Zizyphus mauritiana* Lamk.) for nutritional diagnosis. Singap J Prim Ind 18:85–95

Blaise D, Singh JV, Venugopalan MV, Mayee CD (2003) Effect of continuous application of manures and fertilizers on productivity of cotton sorghum rotation. Acta Agron Hung 51(61):67

Chadha KL, Reddy BMC (1993) Pineapple nutrition. In: Chadha KL, Pareek OP (eds) Advances in horticulture, vol 2. Malhotra Publishing House, New Delhi, pp 899–905

Chandra K, Singh T, Srivastava H (1998) *Biofertilizer ready reckonor*. Agriculture Department, Govt. of Orissa, Bhubaneswar, 34 pp

Chundawat BS (1997) Nutrient management in fruit crops. Agrotech Publishing Academy, Udaipur

FAO (2008) Investing in sustainable agricultural intensification: the role of conservation agriculture – a framework for action. FAO Rome, August 2008 (available at www.fao.org/ag/ca/)

Ghosh B, Bose TK, Mitra SK (1986) Chemical induction of flowering and control of fruit drop in litchi (*Litchi chinensis* Sonn.). Proc 22nd Int Hort Cong, California, Abstr. p 1189

Khanduja SD, Garg YK (1984) Macro-nutrient element composition of leaves from jujube (*Ziziphus mauritiana* Lam.) tree in north India. Ind J Hort 41:22–24

Praharaj CS, Rajendran TP, Sankaranarayanan K (2005) Studies on the incorporation of various organics sources to soil on the performance of *hirsutum* cotton. Proc of the Intern Symp on Strategies for Sust Cot Prod.-A Global Vision, Crop Production 2:130–133. November 23–25, 2004, UAS, Dharwad

Ramaswamy N (1971) Studies on the effect of nitrogen on the growth and development of 'Robusta' (Musa cavendishi *L.*). M. Sc. (Ag.) thesis, Annamalai Univ., Annamalainagar

Reuther W, Batchelor HJ, Webber HJ (1962) Citrus industry, vol II. Univ. of California Press, Berkeley, California

Roy RN, Ange AL (1991) Integrated plant nutrient systems (IPNS) and sustainable agriculture. Proc FAI Annual Seminar, FAI, New Delhi, pp SV/1- 1/1-12

Samra JS, Thakur RS, Chadha KL (1978) Comparison of some mango cultivars in terms of their macronutrient status in fruiting and non-fruiting terminals. Ind J Hort 35:144–187

Shende DG (1977) Effect of N, P and K on growth, yield and quality of pomegranate (Punica granatum *L.*). M. Sc. (Ag.) thesis, Mahtma Phule Agri. Univ., Rahuri, Maharashtra

Singh NP (1981) Studies on guava nutrition with special reference to leaf analysis. National Symp Trop Subtrop Fruit Crops, Bangalore, p. 75

Tillman G, Schomberg H, Phatak S, Timper P, Olson D, Santen-e-van (2002) Enhancing sustainability in cotton with reduced chemical inputs, cover crops and conservation tillage. In: Proceedings 25th annual southern conservation tillage conference for sustainable agriculture, pp 366–368. Auburn, AL, USA

Part VII
Improved Water Management Practices

Chapter 14
Micro-Catchment Rainwater Harvesting

"The greatest potential increases in yield are in rain-fed areas where many of the world's poor live and where managing water is the key to such increases"

Molden (2007).

Abstract The potential practices that support sustainable agriculture include micro-catchment rainwater harvesting technique. The benefits of this technique are improved soil moisture, runoff and groundwater recharge, water infiltration, enriched soil nutrients, increased biomass production, and subsequent higher yields. The outcomes from adopting rainwater harvesting technique include higher food security, higher income, and delivery of positive impacts on other ecosystems.

Keywords Rainfall • Runoff • Rain-fed agriculture • Soil moisture • Water harvesting • Yield

14.1 Introduction

Out of the 1.5 billion hectares of cropland worldwide, about 80 % area depends on rainfall alone, contributing two-thirds of global food production (Rockström et al. 2007; Scheierling 2011). The majority of the rural poor affected by food insecurity can be found in semi-humid and semiarid areas that are dependent on rain-fed agriculture. In dry lands, which cover approximately 40 % of the global land area, yields of the major crops tend to be relatively low (Wani et al. 2009; Scheierling et al. 2013). The major constraints to crop production under rain-fed agriculture include poorly distributed rainfall, excessive nonproductive losses, and mid-season long dry spells along with poor soil fertility. About 25–40 % of the terrestrial precipitation accounts for the productive "green transpiration." Rainwater can be lost due to surface runoff (which is responsible for soil erosion), evaporation from the soil surface, and through deep percolation/drainage. The rainfall losses can be transformed into productive water: meaning soil water directly used by "green transpiration" for plant growth (Fig. 14.1). The runoff feeds water harvesting systems that

P.P. Reddy, *Sustainable Intensification of Crop Production*,
DOI 10.1007/978-981-10-2702-4_14

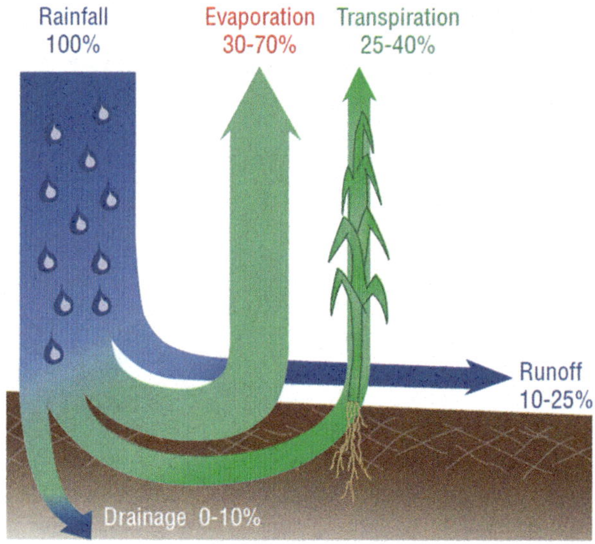

Fig. 14.1 The fate of rainwater for productive use (green transpiration) and losses (evaporation, runoff, and leaching)

store water directly in the soil profile. An extra 10–25 % of water runoff harvested and made available during critical periods of plant growth can double or triple yields (Liniger et al. 2011).

For the smallholder farmers, the only option to conserve water is in situ rainwater harvesting (RWH), since they do not have irrigation facilities. In situ rainwater harvesting is basically a prevention of runoff from a given cropping area by collecting, storing, conserving, and prolonging the time for infiltration of local surface water in arid and semiarid regions. It is an integral part of many farming systems worldwide which hold a significant potential for improving rainwater-use efficiency and sustaining rain-fed agriculture in the region. The micro-catchment techniques can increase the soil water content in the rooting zone by up to 30 % depending on rainfall patterns, appropriate application, and local soil characteristics. This is achieved mainly by the good agronomic practices, use of swales, contour farming, and box ridging. The evaporation of moisture is prevented through agronomic measures such as cover crops, mulching, etc. In situ rainwater harvesting technologies such as contour ridging, realigned ridges, and box ridging improve soil moisture condition for crop growth and development resulting in significant increase of grain yield by 20–25 %. The integration of rainwater harvesting and fertilizer use can enhance the crop yields by up to sixfold.

The RWH system generally comprises of the following three components (Fig. 14.2):

- A collection/catchment area which produces runoff because the surface is impermeable or infiltration is low
- A conveyance system through which the runoff is directed, e.g., by bunds, ditches, and channels

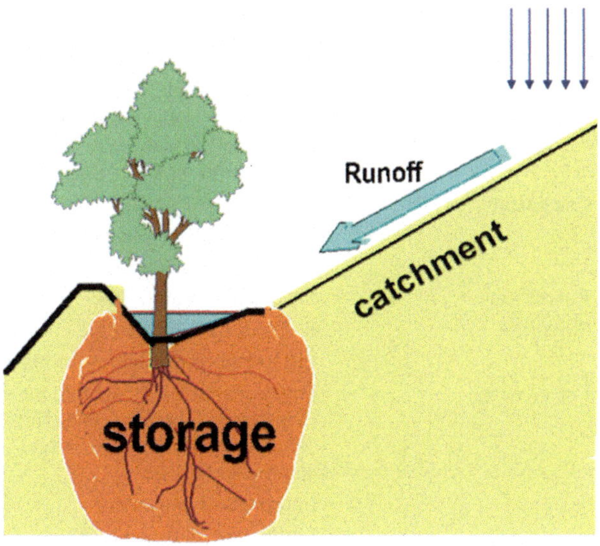

Fig. 14.2 Three components of micro-catchment of rainwater

- A storage system (target area) where water is accumulated or held for use—in the soil, in pits, or ponds

This method of rainwater harvesting is very useful for semiarid areas with common seasonal droughts for supplemental irrigation.

14.2 Micro-Catchments

The within-field systems of water harvesting are called micro-catchment systems consisting of small structures such as basins, pits, holes, and bunds formed for surface runoff water collection from within the cropped area. In micro-catchments, the farmer has control over both the catchment area (C) (less than 1000 m^2) and the cropped area (CA) (less than 100 m^2) which are relatively small with C:CA = 1:1–10:1. The nutrient management of micro-catchments is by the use of farmyard manure, compost, biofertilizers, and inorganic fertilizers.

The characteristics of micro-catchment water harvesting are presented in Table 14.1.

The most common micro-catchments of rainwater include (Critchley and Siegert 1991):

- Pits and basins
- Cross-slope barriers

Table 14.1 The characteristics of micro-catchment water harvesting

Group	Micro-catchment water harvesting
Strategy	Trap localized runoff within the field
Agroclimatic zone	Dry subhumid and semiarid climates
	Where rainfall is more reliable but scattered and/or poorly distributed within the season
Catchment	In field
	Catchment and application area distributed evenly over field
	System replicated many times with identical designs
Runoff water	Sheet and some rill flow
Storage	Soil moisture in root zone
	Pits, trenches, and bunds for planting
Use of water	For agricultural use, suitable for any crop, often perennial tree crop systems (orchards and afforestation), also for annual crops in cereal-based production systems (e.g., millet, sorghum, maize), and fodder bushes
	Improve soil moisture
	Limited groundwater recharge
Annual rainfall range	200 mm–800 mm
	Minor runoff events lost if not harvested, relatively frequent
	Poor rainfall distribution within the season
Terrain	Catchments on slopes
	Application areas on flatter land or depressions
General slope of catchment area	0–50 %
Runoff coefficient	High
Catchment surface	Natural, cleared, and often treated
Application area	Lowest point of each system
Soils	Soils only need to be relatively deep
	Systems can be applied on highly degraded soils to rehabilitate them
	Manure and fertilizers must be added
Landscape scale	Household level
	Local scale
Land/water use rights	Individual
Level of mechanization	None or little
Labor requirement	Relatively low labor requirement for establishment but high for maintenance
Level of technical know-how: establishment	Low
Level of technical know-how: maintenance	Low to medium
Investment	Low
Financial, material, and technical support required	Low to medium depending on system

(continued)

Table 14.1 (continued)

Group	Micro-catchment water harvesting
Examples of costs	Stone lines (Niger): 31 US$/ha
Examples of benefits	Half-moons and stone lines (Kenya, Burkina Faso): yield increase in annual crops 30–400 %
Benefit to cost	Short term: slightly positive
	Long term: positive
Climate change: resilience and adaptability	Brings considerable resilience to systems
	Can be adapted especially through manipulation of C: CA ratio (Catchment area: Cropped area).
	Vulnerable to long drought periods
Risk reduction	High
Main constraints	During heavy rainfall events, structures might be irreparably damaged
	Because of relatively small catchment area, these systems will always be vulnerable to prolonged droughts
	Depending on technology, crop planted has to be repeated each cropping season or annually
	Requires continuous maintenance
	Unprotected application area leads to reduced infiltration rates
	Possible on higher slopes, but costs increase quickly with a need for higher ridges and bunds
Technologies by group	*Pits and basins*
	Small planting pits, e.g., zaï/tassa
	Micro-basins, e.g., negarims, meskats, small semicircular bunds
	Cross-slope barriers
	Vegetative strips
	Contour bunds and ridges
	Tied ridges
	Stone lines and bunds
	Contour bench terraces (Fanya juu terraces)

14.2.1 Pits and Basins

The most common micro-catchments of rainwater through pits and basins include:

- Small planting pits, e.g., zaï/tassa
- Micro-basins, e.g., negarims, meskats, and small semicircular bunds

14.2.1.1 Small Planting Pits

The planting pits are also called *zai*, tassa, or chololo pits. In order to break the crusted soil surface, to store water and to build up soil fertility, and to increase yields (Mazvimavi et al. 2010), the planting pits of 15–20 cm in depth and 30 cm in diameter are dug. One or two seeds of either maize/millet or sorghum crop are

Fig. 14.3 Planting pits for water harvesting and conservation

planted in each pit and fertilized with FYM, compost, ash (to expel termites), and crop residues (Fig. 14.3) (Itabari and Wamuongo 2003). The pits are spaced 90 cm between rows and 60 cm apart within rows, with the rows running along the contour. The planting pits harvest rainwater and conserve moisture and fertility and yields have been tripled (Fig. 14.4) (Mazvimavi et al. 2010). The planting pits are used in areas which receive about 350–600 mm rainfall (Hatibu and Mahoo 1999).

14.2.1.2 Micro-Basins

Micro-basins are in the form of negarims, meskats, small semicircular bunds, eyebrow terraces, and mechanized Valerian basins. They collect surface runoff, increase water infiltration, and prevent soil erosion. The small pools are surrounded by soil ridges on all sides to collect the rainwater and surface runoff for supplying water for small bush or tree planting (Anschuetz et al. 2003). Depending on the planted crop, the micro-basins may vary in size from 10 to 100 m^2. The soil depth of at least 150–200 cm is required for micro-basin construction.

(i) *Negarims*: Negarim micro-catchments are diamond-shaped structures surrounded by small earth bunds, with an infiltration pit at the lowest corner (Fig. 14.5). They harness the surface runoff water for the production of fruit tree crops like papaya, citrus, banana, and mango. The requirements for negarims include about 300–700 mm rainfall annually, 1.5 % soil slope, and 2 m deep soils for runoff water storage. The benefits of negarims include effective under low rainfall (300 mm) conditions, enhanced establishment of fruit trees (60 %), yield increases, and increased income from farm.

(ii) Meskats: The meskat system of micro-catchment water harvesting is suitable for gentle slopes in semiarid to arid climate with about 300 mm annual rainfall. This system consists of catchment area of about 500 m^2 and cropped area of

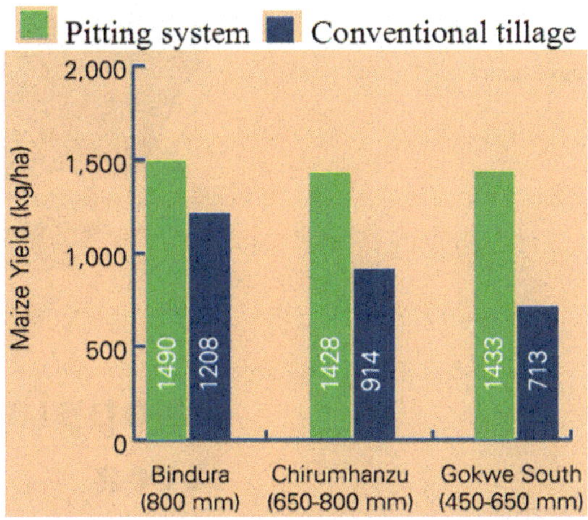

Fig. 14.4 Effect of planting pits and conventional tillage on maize yields (Mazvimavi et al. 2010)

Fig. 14.5 Negarims micro-catchments for tree crops (Critchley and Siegert 1991)

about 250 m^2 surrounded by small earth bunds (about 20 cm) and connected by spillways for discharging the excess water (Fig. 14.6) (Achouri 1994). The meskat system collects runoff water and responsible for groundwater recharge resulting in prevention of soil erosion and floods.

(iii) *Small Semicircular Bunds*: The earth embankments in semicircular shape with the tips of the bunds on the contour are called semicircular bunds (Fig. 14.7). This method is used to grow fodder, shrubs, trees, and sometimes crops in gently sloped area under arid or semiarid regions with an annual rainfall of 200–750 mm. The soil dug from the pit is used to construct a semicircular bund of 0.25 m height and 2–3 m radius on the lower side of the pit and they are mostly lined in staggered rows. The semicircular pits are filled with FYM/compost/crop residues to fertilize the soil and to conserve moisture.

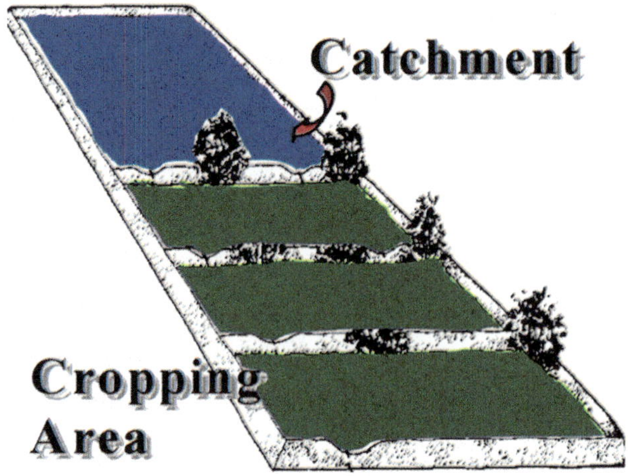

Fig. 14.6 Meskat system of micro-catchment water harvesting

Fig. 14.7 Semicircular bunds for micro-catchment water harvesting

14.2.2 Cross-Slope Barriers

The most common micro-catchments of rainwater through cross-slope barriers include:

- Vegetative strips
- Contour bunds and ridges
- Tied ridges
- Stone lines and bunds
- Contour bench terraces (e.g., Fanya juu)

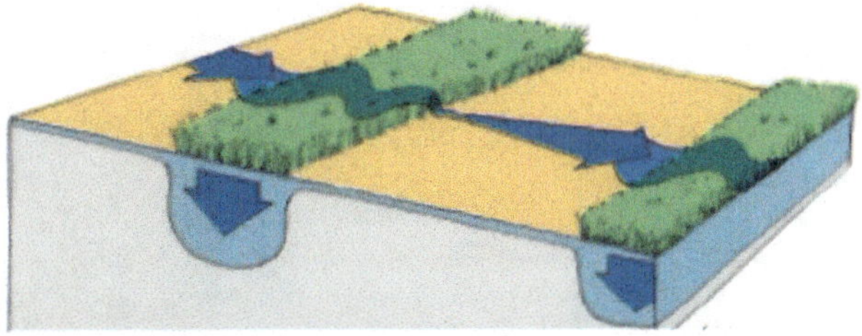

Fig. 14.8 Vegetative strips for micro-catchment water harvesting

Fig. 14.9 Contour bunds for micro-catchment water harvesting

14.2.2.1 Vegetative Strips

The strips of grasses planted (elephant grass, *Pennisetum purpureum*, and vetiver grass, *Vetiver zizanioides*) or a hedgerow of leguminous trees (*Gliricidia* and *Leucaena*) are called vegetative strips (Fig. 14.8). This method of micro-catchment restricts water movement in the slope and also collects sediment flowing from uphill. The soil is likely to build up behind the strips causing a terrace.

The vegetative strips can supply valuable fodder for livestock and mulch (by cutting strips) to cover the field, to conserve moisture, and to control weeds.

14.2.2.2 Contour Bunds and Ridges

To collect the surface runoff, to enhance water infiltration and soil moisture, and to reduce soil erosion, the contour bunds are constructed at regular intervals with earth or stones in sloping lands (Fig. 14.9). The contour bunds have significantly increased sorghum yields to the extent of 80 % in northwestern Somalia (Malesu et al. 2007; Ouessar et al. 2012).

Fig. 14.10 Tied ridges for micro-catchment water harvesting

14.2.2.3 Tied Ridges

The tied ridging increases moisture storage and reduces soil and water loss by conserving rainwater in situ where it falls within the root zone. The ridges are aligned along the contour at a grade of 0.4–1 %. In order to trap the runoff water, the cross ties are formed along the furrows on the ridges (Fig. 14.10). The planting is recommended only when the ridges are fully moist for better germination. The tied ridging is recommended in drier areas with about 500 mm rainfall.

14.2.2.4 Stone Lines and Bunds

The improved stone bunds are constructed aligned along the contour at an interval of 15–30 m (depending on labor and stone availability) to slow down and filter runoff, thereby increasing infiltration and capturing sediment (Fig. 14.11). The sediment and water harvested are used directly to improve the crop performance. The benefit of stone bunds is that there is no need for spillways or diversion ditches. The filtering effect of the semipermeable stone barrier all along its full length helps better distribution of runoff water compared to earth bunds. The stone bund has proved to be an effective method of micro-catchment which requires much less maintenance.

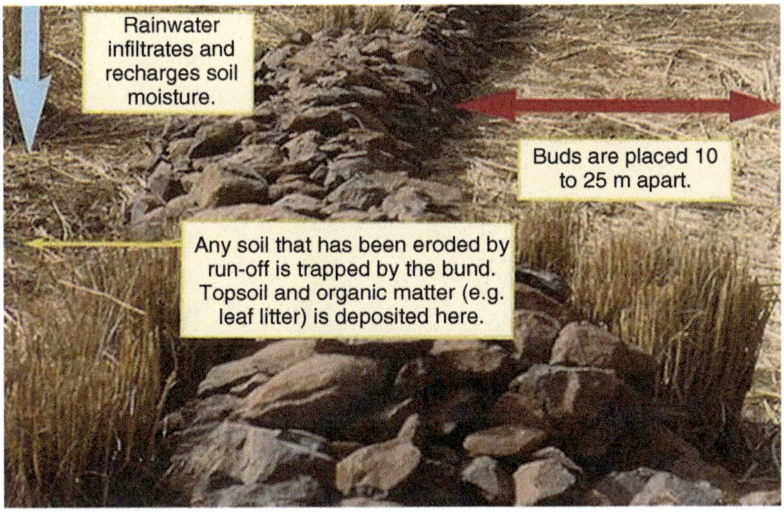

Fig. 14.11 Stone bunds

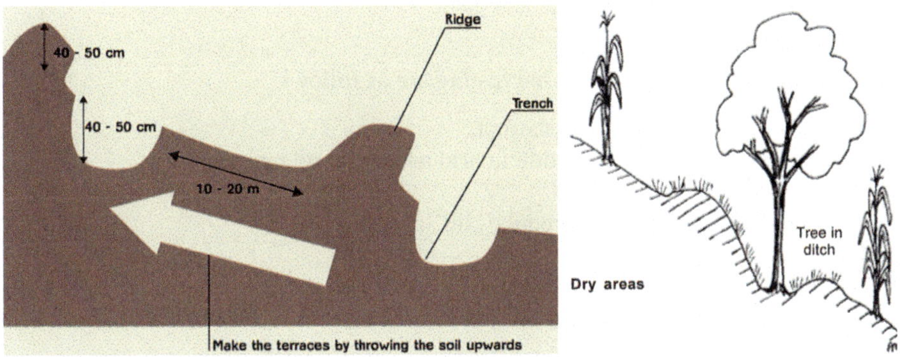

Fig. 14.12 Contour bench terraces (Fanya juu terraces)

14.2.2.5 Contour Bench Terraces (Fanya Juu Terraces)

The contour bench terraces (Fanya juu terraces) are constructed by excavating a drainage channel and placing the soil upslope to make a ridge (Fanya juu means "throw soil uphill" in Swahili) (Fig. 14.12). The channel is normally 0.5 m wide and 0.6 m along the contours and built at an interval of 10–20 m depending on the steepness of the field. In order to stabilize the ridges and to prevent soil erosion, the grass is planted on the ridges which can be used as mulch or to feed the farm animals. The Fanya juu terraces can be aligned along the contour and trees planted in the ditches in drier areas.

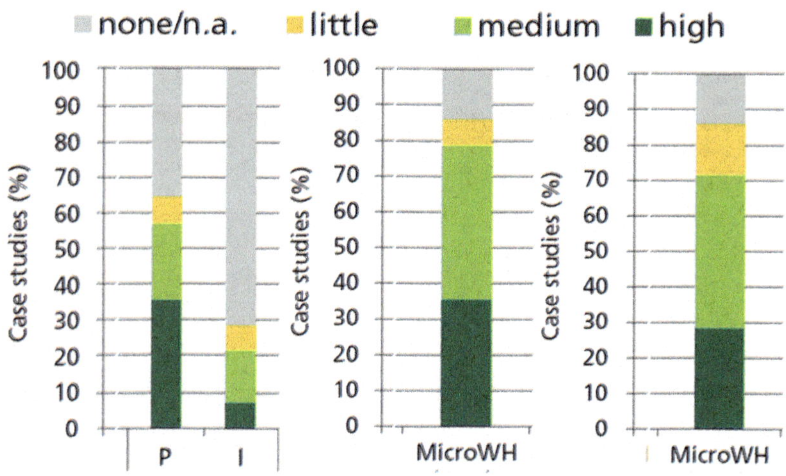

Fig. 14.13 Effect of micro-catchment water harvesting on enhanced production (P) and income (I) (*left*), increased water use efficiency (*center*), and reduction in soil erosion (*right*)

14.3 Benefits

The benefits of in situ rainwater harvesting are as follows:

- Simple and inexpensive to construct.
- Used for supplemental irrigation in arid and semiarid regions.
- Crop failure risk is decreased.
- Enhance leaching and often reduce soil salinity hazard.
- Less production risks.
- Soil erosion is reduced (Fig. 14.13).
- Availability of water is enhanced.
- Return land to productive use.
- Water use efficiency is increased (Fig. 14.13).
- Improves crop productivity and income (Fig. 14.13).
- Groundwater recharge is improved.
- Rainwater management is enhanced.
- Contribution to food security.
- Reduction of production cost.

14.4 Conclusions

The rainwater harvesting system increases the rainfall use efficiency and produces high-value crops (enables farmers to get higher returns from selling the product) that reduce poverty and food insecurity (Gowing et al. 2003). The micro-catchment rainwater harvesting (water is available at the point of consumption) reduces the need to pump water over long distances and thus saves time and human labor. Poverty alleviation and promotion of socioeconomic well-being of the rural people can be achieved through rainwater harvesting activities (Mbugua and Nissen-Petersen 1995).

Improving nutrient management through mulching and stronger mechanization with animal tracking are the major challenges of rainwater harvesting systems. The effects of RWH on landscape functions are still poorly understood, since there is scanty literature on the impact of RWH practices on biodiversity conservation. Enhanced soil moisture, recharge groundwater, reduced runoff, and increased yield are some of the benefits from adopting RWH techniques that extend beyond rain-fed farming to the whole ecological system.

References

Achouri M (1994) Les techniques d'aménagement des versants. In Séminaire sur la conservation des eaux et du sol en Tunisie 1990–2000. Kasserine

Anschuetz J, Kome A, Nederlof M, de Neef R, van de Ven T (2003) Water harvesting and soil moisture retention. Agromisa Foundation, Wageningen

Critchley W, Siegert K (1991) Water harvesting. A manual for the design and construction of water harvesting schemes for plant production. FAO, Rome

Gowing JW, Young MDB, Hatibu N, Mahoo HF, Rwehumbiza F, Mzirai OB (2003) Developing improved dry land cropping systems for maize in semi-arid Tanzania. Part II. Use of a model to extrapolate and add value to experimental results. Expl Agric 39:293–306

Hatibu N, Mahoo H (1999) Rainwater harvesting technologies for agricultural production: a case for Dodoma, Tanzania. Department of Agricultural Engineering and Land Planning/Sokoine University of agriculture, Tanzania

Itabari JK, Wamuongo JW (2003) Water harvesting technologies in Kenya. Technical note number 16. LWC, 1978. Reduced tillage for soil erosion control. Land and Water Conservation (LWC)

Liniger H, Mekdaschi Studer R, Hauert C, Gurtner M (2011) Sustainable land management in practice. Guidelines and best practices for Sub-Saharan Africa. World Overview of Conservation Approaches and Technologies (WOCAT) and Food and Agriculture Organization of the United Nations (FAO), TerrAfrica

Malesu MM, Oduor AR, Odhiambo OJ (eds) (2007) Green water management handbook. Rainwater harvesting for agricultural production and ecological sustainability. The World Agroforestry Centre, Nairobi

Mazvimavi K, Ndlovu P, Nyathi P (2010) Conservation agriculture practices and adoption by smallholder farmers in Zimbabwe. Poster presented at 3rd AAAE conference, Cape Town, South Africa

Mbugua JK, Nissen-Petersen E (1995) Rain water an underutilized resource. SIDA, Nairobi

Molden D (ed) (2007) Water for food, water for life: comprehensive assessment of water management in agriculture. Earthscan and International Water Management Institute (IWMI), London/Colombo

Ouessar M, Hessel R, Sghaier M, Ritsema C (2012) Water harvesting potential for Africa. An assessment of costs and impacts. Scientific Reports, Report 4, Water Harvesting for Rainfed Africa (WAHARA), Tunisia

Rockström J, Hatibu N, Oweis TY, Wani S, Barron J, Bruggeman A, Farahani J, Karlberg L, Qiang Z (2007) Managing water in rainfed agriculture. In: Molden D (ed) Water for food, water for life: comprehensive assessment of water management in agriculture. Earthscan and International Water Management Institute (IWMI), London/Colombo

Scheierling S (2011) Adapting soil-water management to climate change: insights from a portfolio review (presentation). Learning session: climate-smart agriculture. Agriculture and Rural Development (ARD) Days 2011

Scheierling SM, Critchley WRS, Wunder S, Hansen JW (2013) Improving water management in rainfed agriculture: issues and options in water-constrained production systems. Water Paper, Water Anchor, TWIWA, World Bank

Wani SP, Sreedevi TK, Rockström J, Ramakrishna YS (2009) Rainfed agriculture – past trends and future prospects. In: Wani SP, Rockström J, Oweis T (eds) Rainfed agriculture: unlocking the potential, Comprehensive assessment of water management in agriculture 7. CAB International, Wallingford

Chapter 15
Micro Irrigation

Drip irrigation is the answer to many farmers' woes, especially in water scarce areas

– Ruksan Bose (2004)

Abstract There is a growing interest in the micro irrigation technology that has proven as an efficient way of water use in agriculture. The two systems of micro irrigation include drip irrigation (also called trickle irrigation) and sprinkler irrigation. Even though the micro irrigation system is associated with big investments and large-scale commercial plots, the present trend emphasizes the need for low-cost, simple micro irrigation systems for smallholder farmers. The recurrent water loss is less in micro irrigation and hence suitable for areas with low water supply. The significant water saving and increased crop yields have been obtained using very little water. This method of irrigation can be effectively utilized by farmers in hilly and undulating terrains which can be brought under command areas and drought-prone areas to extend irrigation with meager supplies of water.

Keywords Micro irrigation • Drip irrigation • Sprinkler irrigation • Water saving • Increased yields

15.1 Introduction

Water is a key factor in increasing agricultural production and productivity, and hence irrigation is most important. Irrigation is the application of external source of water to crops rather than rain water to overcome full daily water requirement to complete their growth stages. The basic objective of irrigation is to obtain an optimum yield from the crop grown on the land by supplementing the natural supply of water to land.

Water is important for plants because of the following reasons:

- For transport of minerals and nutrients from the soil to the plants
- To prepare plant food through photosynthesis
- Seed germination

P.P. Reddy, *Sustainable Intensification of Crop Production*,
DOI 10.1007/978-981-10-2702-4_15

Table 15.1 Water losses under various irrigation systems

Method of irrigation	Water losses (%)
Surface	30–45
Gate pipe	15–20
Sprinkler irrigation	5–10
Drip irrigation	3–6

- To provide the appropriate pressure to the plant tissues for maintenance of the plant structure

The losses in water conveyance and application are large (Table 15.1) in traditional surface irrigation methods. These losses can be considerably reduced by adopting micro irrigation systems.

Much of the cultivated area in the world is dependent on rainwater that leaves farmers vulnerable to drought. There is a need for efficient use of water in view of increased competition for scarce water resources and recurrent droughts. Hence, a reevaluation of strategies is required to improve access to water-saving technologies. For many of these problems, water use efficiency is the solution.

The government needs to motivate the farmers to adopt modern irrigation technologies in view of the increasing national or regional water shortage. Modern irrigation technologies must justify the investment by offering sufficient financial return or a reduction in labor demand to the farmer. In order to cover the costs, farmers should grow high-value crops with an assured market. The irrigation technologies selected should be straight forward to operate and maintain and must be comparatively cheap. In the initial years of operation, technical support is a must for farmers.

The micro irrigation systems such as drip irrigation and sprinkler irrigation are the solution to overcome the water shortage and recurrent drought problems.

15.2 Micro Irrigation

The frequent pressurized application of small quantities of water normally as continuous drops through emitters present on the water delivery line, directly on the soil surface, or below the soil surface is called micro irrigation.

15.2.1 Principles

The principles of micro irrigation are as follows:

- Pressurized application of water through a low-pressure network including main, submain, and lateral lines with emitters/drippers.

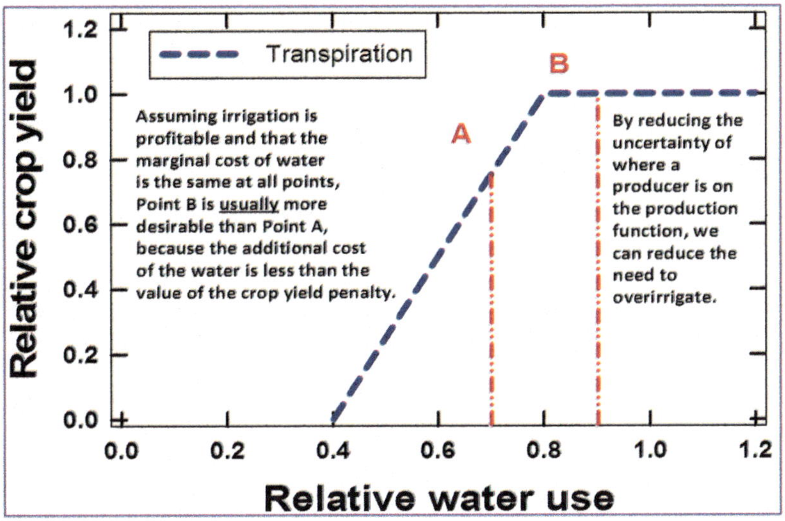

Fig. 15.1 Modern scientific irrigation scheduling

- Frequent application of water in small quantities to fulfill the requirements of the plants.
- Direct application of water in the root zone of the plant.
- Micro irrigation can be used for application of water soluble fertilizers and nutrients (fertigation) and pesticides to control soilborne pathogens (chemigation).
- Suspended impurities which are likely to block the emitters are prevented by passing the water through a filtration system.

15.2.2 Irrigation Scheduling

The determination of when and how much water to apply to achieve high crop water productivity, water conservation, and economic profitability is called irrigation scheduling. It is the process of delaying any unnecessary irrigation so that the cropping season ends before the next irrigation is needed.

A single approach or a combination of weather-, soil-, or plant-based approaches are being used in the modern scientific irrigation scheduling (Fig. 15.1). The daily or hourly evapotranspiration (ET) is calculated from weather data and then use ET as a withdrawal and precipitation and irrigation as deposits in the weather-based approaches. The soil-based approaches normally use changes in soil water levels for estimation of ET and a critical measured soil water level as a trigger for an irrigation event. A measured or remotely sensed characteristic of plant water stress is normally used to trigger an irrigation event in the plant-based approaches.

The economic penalty for over-applying water is usually many times less (8–17 times) than that of under-applying water.

15.2.3 Different Methods

There are two micro irrigation methods which include:

15.2.3.1 Drip Irrigation

Drip irrigation is the application of water slowly to the surface or subsurface soil ideally to match the rate with crop consumption.

15.2.3.2 Sprinkler Irrigation

Similar to the natural rainfall, the irrigation water is applied to the land in the form of spray.

15.3 Drip Irrigation

In arid and semiarid regions with higher water demands, drip irrigation is the latest method of irrigation system. The drip irrigation, also called as trickle irrigation or localized irrigation or micro irrigation, is an irrigation method which saves water (because little water is lost to runoff or evaporation) by allowing water to drip slowly to the roots of plants, either onto the soil surface or directly onto the root zone. It is carried out through a network of valves, pipes, tubing, and emitters. Narrow tubes deliver water directly to the base of the plant (Fig. 15.2). The drip irrigation promotes plant growth, reduces pests and diseases, and controls weeds.

15.3.1 Benefits

15.3.1.1 Increased Water Use Efficiency

- Saving of water to the extent of 30–70 %.
- Direct evaporation losses are reduced.
- Irrigates the full field including edges.
- Superficial drainage is eliminated.
- Covers large area with a specific amount of water.

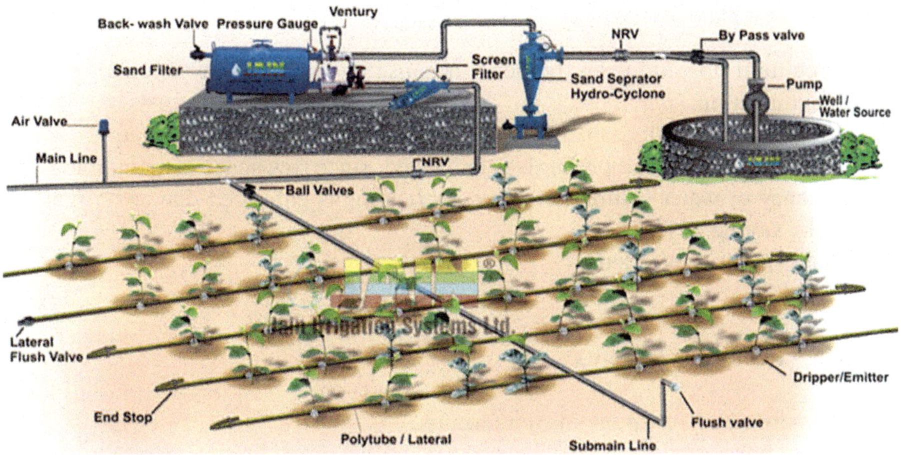

Fig. 15.2 A typical drip irrigation system (Jain irrigation)

- No effect of wind on drip irrigation
- Reduces water consumption due to less grasses and weeds.
- Permits irrigation to the desired depth.

15.3.1.2 Crop Yield and Quality

- Gives uniform yield
- Enhances crop yield per unit of water applied
- Crop quality is improved

15.3.1.3 Root Environment

- Available nutrients are increased.
- Soil erosion is reduced.
- Prevents salt accumulation in the crop root zone.
- Water retention at low tension is conditioned.

15.3.1.4 Pest Management

- Insect and disease incidence is reduced.
- Increases efficiency of pesticide application due to chemigation, thereby saving of pesticides.

15.3.1.5 Agronomic Practices

- Intercultural operations are reduced, because of less weed growth and reduced soil compaction.
- Do not interfere with crop cultivation, plant protection, and harvesting.
- Work efficiency in fruit orchards is enhanced, since row space is maintained dry.
- Efficiency of water soluble fertilizer application is enhanced due to fertigation, thereby saving of fertilizers.
- Soil erosion is controlled.

15.3.1.6 Economic Benefits

- Works efficiently even on uneven terrains.
- Lower labor cost.
- High water application use efficiency (up to 90–95 %).
- Operation and maintenance cost is low.

15.3.2 Limitations

- Damage of components of drip irrigation system by rodents or insects (repellents can be used)
- Physical impurities like soil particles, algae, or chemical impurities like minerals and salts may clog drippers due to poor water quality or poor maintenance by farmers (use of water filters can overcome the problems).
- Higher initial investment in view of pressurized irrigation and use of plasticulture.

15.3.3 Advantages of Drip Irrigation Over Flood Irrigation

The advantages of drip irrigation over flood irrigation are presented in Table 15.2.

15.3.4 Different Types

There are two types of drip irrigation systems which include:

- Surface drip irrigation
- Subsurface drip irrigation

Table 15.2 Advantages of drip irrigation over conventional flood irrigation

	Drip method	Flood method
Water saving	High, between 40 and 100 %	Less. High rates of evaporation, surface runoff, and percolation
Transportation loss	Negligible	High seepages and leakages
Irrigation efficiency	80–90 %	30–50 %
Input cost	Less spent on labor, fertilizers, pesticides, and tilling	Comparatively higher
Weed problem	Almost nil	High
Suitable water	Even saline water can be used	Only normal water can be used
Diseases and pests	Relatively less	High
Efficiency of fertilizer use	Very high since supply is regulated	Heavy losses due to leaching
Water logging	Nil	About 8.5 million ha in India
Water control	Can be regulated easily	Not much control
Cost–benefit ratio	Excluding water savings: 1.3–13.3 Including water savings: 2.8–30.0	Between 1.8 and 3.9
Capital cost/ha	Rs 15,000 to 40, 000 depending on crop spacing	
Yield increase	20–100 % higher than flood method	Less compared to drip

Narayanamoorthy (1996)

15.3.4.1 Surface Drip Irrigation

The surface drip irrigation system uses a thin wall of 8–10 mil drip line with close emitter spacing of 30–45 cm, injecting water at 2.5–15 cm below the surface (Fig. 15.3). This system is often called as "temporary" because the drip line is retrieved and recycled yearly. The submains can be temporary or permanent. In view of the yearly recurring expense on new drip line and the labor cost for installation, this system is typically used on high-value crops.

(i) *Advantages*

- Low cost.
- Weed growth is less.
- Monitoring is easier.

(ii) *Disadvantages*

- Wetting area is limited.
- Emitters are easily clogged.

Fig. 15.3 Wetting area under surface drip irrigation

15.3.4.2 Subsurface Drip Irrigation

The subsurface drip irrigation (SDI) system uses a slightly thicker wall (13–15 mil) drip line with wider emitter spacing of 50–67.5 cm and injected at 20–35 cm below the surface. This system is permanent, making installation and design critical to ensure long life of the system. SDI system is normally used in row crops but is also used in some of the high-value crops.

Application of irrigation water with small emission points (emitters) that are in a series of plastic lines typically spaced between alternate pairs of crop rows applied below the soil surface to the crop root zone to meet water needs is called subsurface drip irrigation. In order to discharge irrigation water directly to the roots, this highly efficient drip irrigation system uses buried drip tubes. The irrigation water is supplied to plants through capillary movement from the bottom. This method can be used for application of small, frequent, just-in-time irrigation water directly to crop root system. In arid, semiarid, hot, and windy areas, it is applicable to almost each and every crop. The subsurface drip irrigation can increase the crop water productivity and stabilize yields (Fig. 15.4) at a greater level than alternative irrigation systems when deficit irrigated.

(i) *System components*: A typical subsurface drip irrigation system layout consists of a settling pond, pumping unit, pressure relief valve, check valve or backflow prevention valves, hydrocyclone separator, chemical/fertilizer injection unit, filtration unit equipped with back flush valves, pressure regulators, air vent valves, and PVC pipes delivering the water to the crop (Fig. 15.5).

The subsurface deficit irrigation enhances productivity of crops and quality of the product. The tomato fruit yield and quality were increased under subsurface deficit irrigation dripper system as compared to the control where water was directly dripped near base of the plants (Xu et al. 2011).

The subsurface drip irrigation has very high water use efficiency since evaporation is reduced almost completely. There is no loss of agricultural land in laying out irrigation system, the implements can work freely, the plant roots do not suffer from waterlogging, and the yields are increased (Table 15.3).

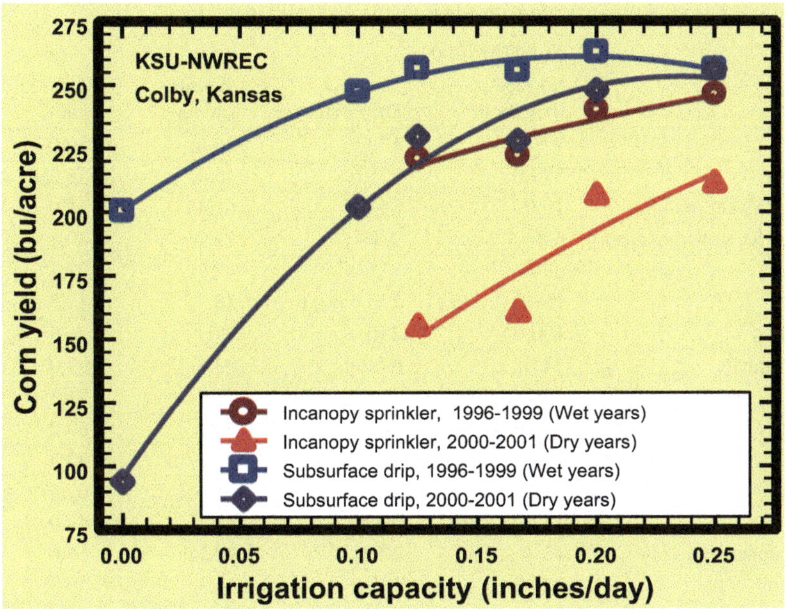

Fig. 15.4 Effect of different irrigation systems on corn yield. The subsurface drip irrigation gives an additional US$ 44/acre compared to the center pivot sprinkler irrigation

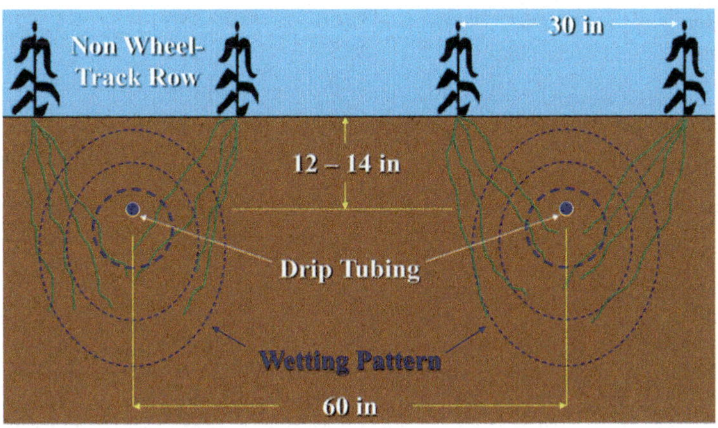

Fig. 15.5 Installation of subsurface drip irrigation system

(ii) *Benefits*

- Elimination of water losses through evaporation and surface runoff (including soil erosion) since the water is applied below the soil surface.
- Uses water more efficiently.
- Highly efficient automated drip irrigation system.

Table 15.3 Effect of drip irrigation on water saving and yield in different crops

Crop	Yield (MT/ha)		% water saving	% increase in yield
	Surface irrigation	Drip irrigation		
Mango	7.5	13.5	34.8	80.0
Banana	575.0	875.0	45.0	52.2
Citrus (Sweet orange)	10.0	15.0	61.4	50.0
Citrus (Kinnow mandarin)	6.8	9.8	21.7	40.1
Grape	264.0	325.0	47.2	23.1
Guava	160.0 (kg/tree)	220.0 (kg/tree)	18.7	37.5
Papaya	130.0	230.0	67.9	76.9
Pomegranate	34.0	67.0	23.8	97.0
Ber	13.7 (kg/tree)	18.0(kg/tree)	18.8	31.3
Potato	172.0	291.0	54.2	69.2
Tomato	61.8	88.7	78.5	43.5
Brinjal	91.0	148.0	61.9	62.6
Chili	42.3	60.9	61.7	44.0
Bell pepper	13.3	22.2	43.1	66.6
Broccoli	140.0	195.0	14.3	39.3
Cabbage	21.4	48.87	32.4	119.2
Cauliflower	171.0	274.0	33.3	60.2
Cucumber	155.0	225.0	55.6	45.2
Gherkin	24.3	48.7	36.2	100.6
Water melon	82.1	504.0	65.3	513.9
Bottle gourd	38.0	55.8	35.7	46.8
Bitter gourd	32.0	43.0	56.6	34.4
Ash gourd	108.0	120	11.9	11.1
Beans	5.6	10.25	36.9	81.8
Okra	100.0	113.1	84.0	13.1
Onion	284.0	342.0	50.0	20.4
Carrot	14.0	26.25	33.8	92.3
Beet root	5.7	8.9	79.1	56.1
Radish	10.5	11.9	76.1	13.3
Sweet potato	42.4	58.9	60.3	38.9
Baby corn	5.7	9.8	43.9	72.4
Sugarcane	75.0	175.0	49.3	133.3
Cotton	2.2	4.2	46.6	88.0
Maize	2.0	3.5	34.9	75.0
Groundnut	4.23	6.1	35.9	43.9
Chickpea	3.1	5.5	42.0	75.4

- Discharges irrigation water directly to the root zone by using buried drip tubes.
- Elimination of surface infiltration, such as crusting, and saturated condition of water ponding.
- Optimum soil moisture content in the root zone is maintained by frequent irrigation.
- Suitable for arid, semiarid, and windy conditions.
- Potential for high uniformity of application.
- More suitable for medium- and large-scale crop production.
- Germination of annual weed seeds is reduced since water is applied directly to the root zone of the crop.
- Plant growth, yield, quality, and crop health are increased.
- Saves water.
- Fertilizers (fertigation) and pesticides (chemigation) can be applied along with drip irrigation.
- Increased productivity.
- Energy needs are reduced.
- Pest and disease problems are reduced.

(iii) *Limitations*

- Initial investment cost is high.
- Planting pattern and water application pattern should match.
- High-quality water is required.
- Regular maintenance cost.
- Needs expert design and trained employees because of its high-tech components.
- Maintenance of the system is time-consuming.
- Lot of repair work due to leakages and plugging. Water should be filtered to minimize the risk of clogging before the water enters the distribution pipes. All the pipes can be flushed periodically to prevent clogging.
- Danger of salt accumulation when saline water is used.
- Costly system which requires expert design and maintenance.

15.3.5 Experiences

Israel has been practicing drip for the past 30 years. During this period, its agricultural output has increased almost fivefold with hardly any increase in the amount of water used. As 60 % of Israel lies in arid and semiarid zones, efficient water use has been the state's guiding principle since it was founded. Two-thirds of the 1.2 billion cubic meters of water used by its farmers comes from fresh water sources.

Jalinder Jamdani from Manerajauri village in southern Maharashtra is one of the many successful drip farmers in India. Jamdani produces about Rs 300,000–400,000 worth of export-quality grapes per acre, whose growth, shape and sweetness he

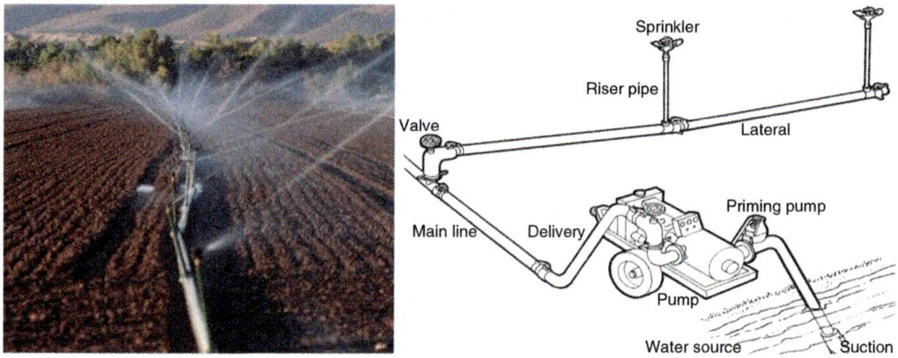

Fig. 15.6 Sprinkler irrigation system

attributes to the measured watering with drip irrigation (Kannan and Gurumurthy 1999).

Under drip irrigation systems in Maharashtra, the sugarcane yield has more than doubled from an average of 27 tons per acre to 40–82 tons per acre with a corresponding decrease in power and water usage. The installation of the drip system in Sangli (Maharashtra) enhanced the sugarcane yield to the extent of 20–25 tons per acre.

Drip irrigation facilities have almost doubled the yield of cotton, saving up to 70% of water.

Grape yield has increased from 15 to 25 tons/ha to 25–40 tons/ha, and water saving has risen by up to 70% under drip irrigation in the Nasik area.

15.4 Sprinkler Irrigation

In sprinkler irrigation system, water is applied through a pressurized system. The pressure causes the water to flow out through the sprinkler nozzle and fly through the air and falls onto the plant and soil surface. The sprinklers are normally located on lateral pipes (Fig. 15.6). This method applies water to plants and soil surface in the form of spray, somewhat similar to rainfall. When the water is sent through small holes with pressure developed from pump, the spray pattern is achieved. Soils that are too steep, rolling, or too shallow having uneven topography (where soil erosion problems are high) can be irrigated efficiently with sprinklers. Since the sandy soils absorb water too fast, this method of irrigation is very much suited. This method of irrigation can also be used in undulating terrains and hilly regions. The crops suitable for sprinkler irrigation include plantation (coffee, tea), and horticultural (fruit orchards, vegetables, flower crops) crops.

15.4.1 Components

In sprinkler irrigation system, water is conveyed under pressure through pipes to the area to be irrigated where it is passed out through sprinklers. The system comprises of the following components (Fig. 15.6):

- Power generator
- Pump unit
- Pipeline—main/submains and laterals
- Sprinkler head
- Couplers
- Other accessories such as water meters, pressure gauge, fertilizer applicator, valves, bends, plugs, and risers

15.4.2 Different Types

The sprinkler systems are of two major types based on the arrangement for spraying irrigation water:

- Rotating head system
- Perforated pipe system

The sprinkler systems are classified into the following types based on portability:

- Portable
- Permanent
    - Center pivot
    - Traveling sprinkler
    - Wheel/reel sprinkler

Application of water at 3.5 cm depth by sprinkler irrigation gave almost equal tomato fruit yield as that of 6 cm application depth by the surface method, thereby saving about 34 % water over 2 years (Table 15.4) in sandy loam soils at Madurai, Tamil Nadu, India (Singh et al. 1999). The sprinkler irrigation system gave 18.6 % higher tomato yield as compared to the ridge and furrow system for the same application depth of 6 cm.

The sprinkler irrigation system is responsible for saving water and in increasing crop yields (Table 15.5).

Table 15.4 Effect of sprinkler irrigation system and depth of irrigation on tomato yield at Madurai, Tamil Nadu, India, 1995–1997

		Yield (t/ha)		
Method of irrigation	Depth of irrigation (cm)	1995	1966	Mean
Sprinkler	6.7	18.39	29.55	23.97
	6.0	16.33	29.50	22.92
	4.9	13.52	27.25	20.39
	3.5	11.90	27.45	19.68
	2.2	10.22	27.65	18.94
Ridge and furrow	6.0	13.84	24.80	19.32
CD (P = 0.05)		1.97	3.56	

Singh et al. (1999)

Table 15.5 The effect of sprinkler irrigation system on water saving and yield increase in different crops

Crops	Water saving (%)	Yield increase (%)	Crops	Water saving (%)	Yield increase (%)
Maize	41	36	Onion	33	23
Sorghum	55	34	Garlic	28	6
Pearl millet	56	19	Okra	28	23
Wheat	35	24	Cowpea	19	3
Barley	56	16	Gram	69	57
Cotton	36	50	Groundnut	20	40
Potato	46	4	Sunflower	33	20
Chilies	33	24	Fenugreek	29	35
Cabbage	40	3	Lucerne	16	27
Cauliflower	35	12			

INCID (1998)

15.4.3 Advantages

- Acts like a rain.
- Larger area is wetted.
- Visual inspection is easy.
- Less clogging due to larger orifice openings.
- Application rate is higher.
- Uniform distribution of water is ensured.
- Suitable for most kinds of soil.
- No hindrance to the use of farm implements.
- Fertilizers and pesticides can be evenly applied through sprinklers.
- Minimum extent of water loss.
- Can irrigate more land.
- No necessity of land leveling operations which are costly.

- Based on the crop requirement (young seedling or mature crops), the amount of water can be controlled.
- Automation is possible.
- Enhancement of yield and quality.

15.4.4 Disadvantages

- Initial investment cost is rather very high.
- High operations and maintenance cost.
- Often trouble from clogged nozzle.
- Failure of sprinklers to revolve.
- More weed growth.
- Can be complex to run.
- High wind interferes with the distribution pattern.
- Labor requirement for moving a pipe and related work.
- Suitable only for high-value crops.
- Evaporation loss when operating under high temperatures.
- High water pressure required in sprinkler (>2.5 kg/cm^2).

15.5 Comparison of Drip and Sprinkler Irrigation Systems

The subsurface drip irrigation system can stabilize corn yields at a greater level than sprinkler irrigation systems when deficit irrigated (Fig. 15.7).

The features of micro irrigation methods are presented in Table 15.6.

15.6 Conclusions

The modern irrigation methods, namely, sprinklers and drip irrigation, can play vital roles in improving water productivity. These irrigation systems are highly efficient in water application and have opened up opportunities to cultivate light-textured soils with very low water-holding capacity and in irrigating undulating farmlands. The technology has also enabled regions facing limited water supplies to shift from low-value crops with high water requirements, such as cereals, to high-value crops with moderate water requirements, such as fruits and vegetables. Implementation of these improved irrigation techniques can be used to save water and energy and increase crop yields. However, currently the use of these improved irrigation methods are limited, primarily due to the high initial cost. Favorable government policies and the availability of credit are essential for popularization of these irrigation methods.

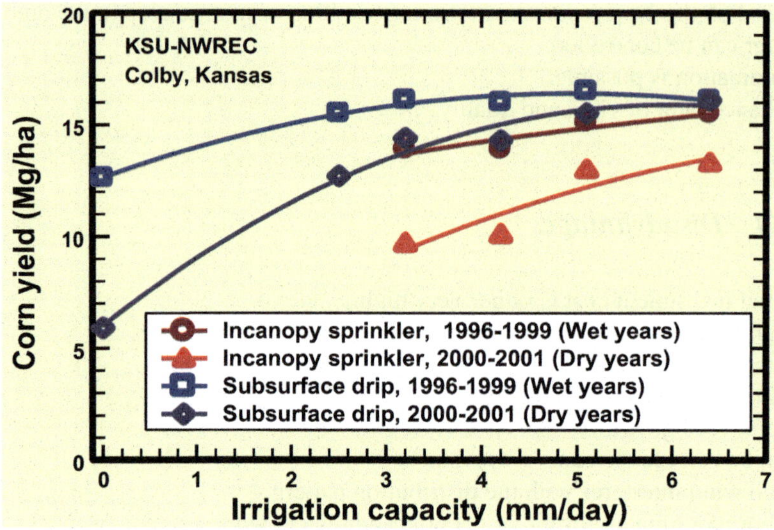

Fig. 15.7 Comparison of drip and sprinkler irrigation systems on stabilization of yield in maize

Table 15.6 Comparison of micro irrigation methods

Features	Drip irrigation	Sprinkler irrigation
Evaporation loss	Minimum	Medium
Wetting of the foliage	Minimum	High
Water consumption by weeds	Minimum	High
Surface drainage	Minimum	Medium
Control of irrigation depth	High	Medium
Crop yield per unit of water applied	High	Medium
Uniformity in crop yield	High	Medium
Soil aeration	High	little
Interference of other tasks by irrigation method	Low	High
Application of fertilizers and pesticides through irrigation water	High	Moderate
Labor cost	Moderate	Moderate
Leveling of land is required	Minimum	Low
Automation of system	High	High
Energy requirements	High	High
Quality of water	High	Moderate
Use of filters	High	Moderate
Control of pests and diseases	High	Moderate

References

INCID (Indian National Committee on Irrigation and Drainage) (1998) Sprinkler irrigation in India. INCID, New Delhi

Kannan S, Gurumurthy S (1999) Drip irrigation and water management. Yojana 2:15–16
Narayanamoorthy A (1996) Evaluation of drip irrigation system in Maharashtra. Mimeograph Series No. 42, Agro-Economic Research Centre, Gokhale Institute of Politics and Economics, Pune
Ruksan B (2004) Yards out of inches. Down to Earth Supplement, July 2004
Singh SR, Chaudhary TN, Bhatta RK, Bhatnagar PR, Saha B, Patil NG (1999) Irrigation water management. In: Singh GB, Sharma BR (eds) Fifty years of natural resource management research in India. Central Soil Salinity Research Institute (CSSRI), Karnal, pp 40–48
Xu HL, Qin FF, Xu QC, Xu RY, Wang TY, Wang R (2011) Applications of xerophytophysiology in plant production: sub-irrigation improves tomato fruit yield and quality. J Food Agric Environ 9:256–263

Chapter 16
Deficit Irrigation

Abstract The emphasis of irrigation management shall shift from production per unit area to maximizing the production per unit of water consumed (water productivity), since insufficiency of water supply for irrigation will be the norm rather than the exception. Maximization of water-use efficiency (WUE) for higher yields per unit of irrigation water applied under scarce water supplies can be achieved through deficit (or regulated deficit) irrigation. There is a potential to enhance WUE by 1.2 times as compared to normal irrigation practices using 25 % deficit irrigation. This chapter reviews yield responses of major field crops such as cotton, maize, potato, sugar cane, soybean, and wheat to deficit irrigation. Integration of deficit irrigation with subsurface drip irrigation (SDI) and conservation tillage in increasing WUE is also discussed.

Keywords Regulated deficit irrigation • Partial root-zone drying • Water-use efficiency • Evapotranspiration • Water productivity

16.1 Introduction

The growing competition for clean water and decreasing water resources for irrigation are the main constraints to meet food requirements in the coming years. About 70–80 % water is withdrawn from the rivers for irrigation purpose. The arid and semiarid regions are already facing serious water shortages as the existing water resources are fully exploited. Therefore the task of increasing food production with less water in the coming decades is a great challenge especially in countries with limited water and land resources. Hence, there is an urgent need to increase the water productivity for food production from rain-fed and irrigated agriculture.

P.P. Reddy, *Sustainable Intensification of Crop Production*,
DOI 10.1007/978-981-10-2702-4_16

16.2 Deficit Irrigation

There is a growing interest in deficit irrigation in order to increase the productivity of water. Enhancing the water-use efficiency for increasing the yield per unit of irrigation water applied is called deficit irrigation. The application of water stress is effected either throughout the crop season or a particular growth period of the crop without appreciable reduction in yield.

The soil water deficit can be to the extent of 60–65 % in the early vegetative growth stage of maize (germinating to 10th leaf) as the root zone at this time is only about one-half to two-thirds of the crop's potential without affecting the plant growth. Withholding irrigation water during this period enhances the opportunity to store rainfall water when it occurs, encourages deeper root growth, and reduces the nutrients leaching. When the crop reaches critical crop growth stage such as the 10–12th leaf stage for maize or early flowering for soybeans (critical periods of high water use), the water deficit level should be maintained at 30–40 %. The soil water deficit level can be increased to higher limits (60–70 %) without appreciable reduction in yield to the crops nearing maturity (e.g., after maize kernels have begun to dent).

16.2.1 Objectives

The objectives of deficit irrigation are as follows:

- To save water by eliminating unwanted irrigations that does not have any or a little effect on yield in order to divert the saved water to irrigate other crops
- To increase the water-use efficiency

16.2.2 Crop Water Requirements

The quantity of water applied at critical crop growth stages to manage soil moisture deficit caused by the evapotranspiration (ET) of the crop is called the crop water requirement that can be measured at different intervals (seasonally, daily, hourly) by using the lysimeter. The evapotranspiration (ET = evaporation + transpiration) can be lowered by reduced soil evaporation and reduced plant transpiration. The soil evaporation is reduced by using efficient irrigation system (drip or trickle), mulching soil with live (cover crops) or dead mulch (organic materials). The plant transpiration can be reduced by shading, saving water during the vegetative stage of growth; crops or crop varieties with a short growing season require less water and are tolerant of drought (Stewart and Musick 1982). In order to ensure successful deficit irrigation, delaying the irrigation until 2 weeks before flowering, adjusting the plant population depending on the water availability, and limiting irrigation during noncritical growth stages is essential.

Table 16.1 Effect of planned evapotranspiration deficit of 25 % on expected relative yield and relative water-use efficiency in different crops

Crop	Stage when ET deficit occurred	Expected relative yield	Relative water-use efficiency	References
Common bean	Vegetative;	0.86	1.14	Calvache and Reichardt (1999)
	Yield formation	0.78	1.04	
Cotton	Whole season;	0.79	1.05	Yavuz (1993)
	Boll formation and flowering	0.88	1.17	Prieto and Angueira (1999)
Groundnut	Flowering	0.82	1.09	Ahmad (1999)
Maize	Whole season	0.82	1.09	Craciun and Craciun (1999)
Potato	Whole season;	0.79	1.06	Kovacs et al. (1999)
	Vegetative	0.90	1.20	Iqbal et al. (1999)
Soybean	Vegetative	0.86	1.14	Kirda et al. (1999)
Sugar beet	Whole season;	0.79	1.05	Bazza andTayaa (1999)
	Mid-season	0.84	1.12	
Sugarcane	Tillering	0.90	1.20	Pene and Edi (1999)
Sunflower	Whole season;	0.77	1.03	Karaata (1991)
	Vegetative yielding	0.79	1.06	
Wheat	Whole season;	0.81	1.08	Madanoglu (1997)
	Flowering and grain filling	0.90	1.20	Waheed et al. (1999)

The crops which are well suited for deficit irrigation applied either at predetermined growth stages or throughout the growing season include wheat, cotton, sunflower, maize, potato, and sugar beet. For example, deficit irrigation imposed during vegetative and yielding stages of sunflower and sugar beet, flowering and boll formation stages in cotton, flowering and grain filling stages of wheat, during vegetative growth of soybean, shall provide feasible and acceptable irrigation options for minimal yield reductions with limited supplies of irrigation water.

The relative evapotranspiration deficit imposed at certain growth stages of field crops such as soybean, groundnut, sugarcane, and common bean showed comparatively less yield reduction (Table 16.1).

16.2.3 Soil Water Measurements

Two common ways of estimating soil water deficit are by the feel/appearance method or use of soil water sensors (tensiometers or electrical resistance blocks) (Fig. 16.1).

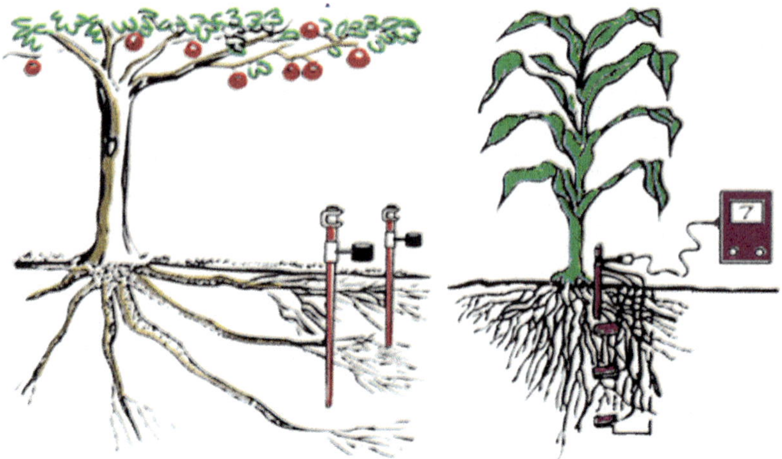

Fig. 16.1 Estimation of soil water deficit using tensiometers (*left*) or electrical resistance meter (*right*)

Table 16.2 Effect of regulated deficit irrigation on fruit quality of peach in different years

	°Brix (%)		
Irrigation regime	2012	2013	2014
T 100	12.2[b]	13.6[b]	14.4[c]
T 75	13.5[a]	15.1[b]	15.6[b]
T 50	13.6[a]	15.2[a]	17.6[a]

[a,b,c]Mean values in a column followed by different letters are significantly different at 5% level

16.2.4 Regulated Deficit Irrigation (RDI)

The planned induction of water stress at specific crop developmental stages that control vegetative growth without negatively affecting yield is called regulated deficit irrigation (RDI). The goals of RDI include reduced water usage and to achieve higher farm income.

With regulated deficit irrigation (RDI), 60 % increase in water-use efficiency was obtained in fruit crops such as pear and peach orchards along with improvement in fruit quality and without appreciable reduction in vegetative vigor and loss in yield in southeastern Australia (Goodwin and Boland 2002). The RDI also significantly increased the fruit quality (Table 16.2).

The early season regulated deficit irrigation in apple controlled vegetative growth, saved water, and suppressed stem water potential, stomatal conductance (Fig. 16.2), and fruit growth rate compared to the trickle and furrow-irrigated controls. The regulated deficit irrigated trees had similar or less vegetative growth and similar or higher yield efficiency than the controls (Ebel et al. 1995). The RDI also improved fruit quality by increasing sugar levels by 9 % (average of 3 years) without any yield reduction (Table 16.3).

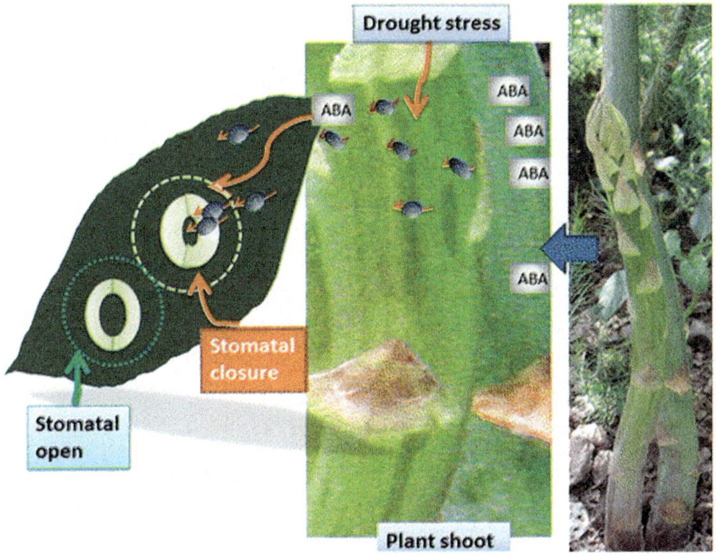

Fig. 16.2 Effect of regulated deficit irrigation on stomatal closure

Table 16.3 Effect of regulated deficit irrigation on fruit yield in different aged apple trees

Age of trees	Water regime	Yield (t/ha)
4	Full irrigation	16.0
	Regulated deficit irrigation	15.8
5	Full irrigation	24.5
	Regulated deficit irrigation	24.3
12	Full irrigation	81.5
	Regulated deficit irrigation	81.2
15	Full irrigation	96.4
	Regulated deficit irrigation	95.5
17	Full irrigation	134.4
	Regulated deficit irrigation	133.6

The regulated deficit irrigation before fruit set to grapevines was responsible for controlling shoot growth and pruning weights and without appreciable reduction in yield at Washington State, USA (Wample and Smithyman 2002).

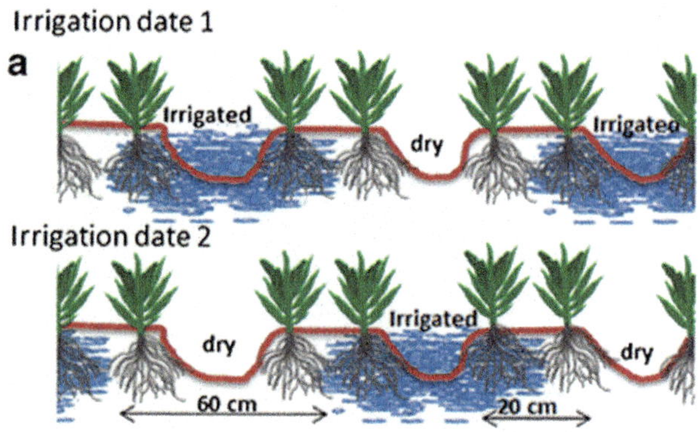

Fig. 16.3 Partial root-zone drying

16.2.5 Partial Root-Zone Drying (PRD)

In addition to RDI, partial root-zone drying (PRD) is also a promising practice for inducing stress tolerance in fruit trees. PRD is a new irrigation technique that subjects one-half of the root system to a dry or drying phase, while the other half is irrigated. The wetted and dried sides of the root system alternate on a 10–14-day cycle (Fig. 16.3) (McCarthy et al. 2002). Both RDI and PRD systems require high management skills. Close monitoring of soil water content is recommended. Both practices improve the WUE of wine grape production. Micro-irrigation facilitates the application of RDI and PRD, for which practical guidelines have been developed.

The benefits of using partial root-zone deficit irrigation are to reduce the amount of irrigation water, to increase WUE, and to increase or maintain crop yields (Table 16.4).

16.2.6 Integration with Other Practices

Deficit irrigation has resulted in increased water productivity and higher crop yields. Similarly, subsurface drip irrigation (SDI) deposits water and fertilizers directly in the crop root zone resulting in high application efficiencies, reduced water evaporation, saving of water, and higher crop yields. Therefore, significant potential exists for water saving and higher crop yields from the combined application of deficit irrigation principles and SDI in crop production.

The RDI practices applied under subsurface drip irrigation system encouraged deeper root development (40–80 cm) in almond than surface treatments (0–40 cm) producing higher root density in the treatments watered the least. Water stress during the pre- and postharvest periods had no visible effect on bud development, bloom, fruit growth, or fruit abscission, without significant reductions in kernel dry weight or kernel percentage (Romero et al. 2004).

Table 16.4 Effects of partial root-zone deficit irrigation on water-use efficiency and crop yields in selected crops

	Partial root-zone deficit irrigation over full irrigation		
Crop	Water-use efficiency	Yield	References
Cotton	Saved 31–33 % of irrigation water, increased WUE by 5–21 % over full irrigation	Increased seed–cotton yield by 5–21 % over full irrigation, due to improved harvest index	Du et al. (2008)
Tomato	Increased WUE on fresh yield by 52 %, mainly due to reduced transpiration rate		Yang et al. (2012)
Potato	Saved water consumption by 32–54 % over full irrigation with early deficit application without yield penalty	Improved the marketable class of potato tubers by as high as 20 %; no yield penalty, with 30 % water reduction	Yactayo et al. (2013), Shahnazari et al. (2007)
Maize	Saved water by 11–32 %; increased canopy WUE by 10–42 %		Li et al. (2010)
Grape		Increased berry yield by 0–43 %, due to promotion of shoot length, pruning weight, berry number per cluster, and cluster weight	De la Hera et al. (2007)

The integration of deficit irrigation along with conservation tillage significantly reduces the water requirement of groundnut crop (USDA 2010).

16.3 Benefits and Limitations

16.3.1 Benefits

The benefits of proper application of deficit irrigation include:

- Reduced risk of nutrient loss by leaching of the root zone resulting in less fertilizer use (Pandey et al. 2000).
- Irrespective of the onset of the rainy season, the crop planning is improved due to control over the date of sowing and length of the growing period (Geerts et al. 2008a).
- In view of stabilization of the harvest in deficit irrigation as compared to rain-fed cultivation, stable income and economic planning can be achieved.
- Improved ground water quality (Unlu et al. 2006).
- The productivity of water and harvest quality of the produce is enhanced.
- Disease incidence (especially of fungal diseases) is reduced due to low humidity.

16.3.2 *Limitations*

The limitations of deficit irrigation are as follows:

- There may be considerable decrease in yield.
- The accessibility to water during periods of high demand (drought-sensitive stages of a crop) is not assured (Zhang 2003).
- Soil salinization risk is more in DI as compared to full irrigation in view of more efficient application of irrigation (Geerts et al. 2008b).
- There is no guarantee for the supply of minimum quantity of water for the crop, below which DI has no significant beneficial effect (Kang et al. 2002).
- The effect of different levels of water stress at various stages of crop growth on crop response is only imperative and not well understood (Kirda 2002).

16.4 Experiences

By applying the deficit irrigation at various growth stages, a water savings of 25–75 % was achieved in winter wheat without any significant loss of yield and profits in the North China Plain (NCP) during 1992 and 2000 (Zhang et al. 2002). The deficit irrigation study showed that three irrigations in dry years, two irrigations in normal years, and a single irrigation in wet years yielded maximum profits. The schedule of irrigation for three irrigations shall be before overwintering, jointing, and heading to milky filling; for two irrigations is at jointing and heading to milky filling; and for the single irrigation is at jointing to booting.

In planned DI, the winter wheat yields have increased by 65 %, and the water-use efficiency has doubled as compared to rain-fed and fully irrigated winter wheat in Turkey (Ilbeyi et al. 2006).

By imposing transient soil moisture deficit stress during the vegetative phase, i.e., 20–45 days after sowing, the water-use efficiency (WUE) and dry matter content were increased in a deficit irrigation study carried out on groundnuts in India (Nautiyal et al. 2002). Withholding of irrigation water during the vegetative growth stage of the crop might have been responsible for enhanced growth of roots resulting in more effective water absorption from deeper soil layers.

The water can be conserved by deficit irrigation on groundnut plants early in the growing season to cause early maturation of the plant with minimal loss in crop yield (English 1990). Similarly, USDA scientists have also reported that by induction of drought in peanut through deficit irrigation early in the season conserved water (USDA 2010). By applying water stress during the early vegetative stage, the groundnut plants physiologically adapt to stressful drought conditions.

A high water productivity and a better nutrient–water balance were obtained in cotton both at India and Turkey, by reducing the irrigation water up to 60 % of the total crop water requirement with minimal yield losses (Raes et al. 2008).

The application of deficit subsurface drip irrigation using the available water resources on the optimal area to cotton plant was found more economical, increased the water-use efficiency, and reduced farming costs in arid East Texas, USA (Henggeler et al. 2002).

By practicing the deficit irrigation during the plant establishment and reproductive stages of quinoa crop, the yields could be stabilized at around 1.6 tons per hectare (Geerts et al. 2008c).

16.5 FAO CROPWAT Model

To optimize crop production under scarce water supply conditions, practical recommendations can be developed using model such as CROPWAT. This model is very useful in predicting the imposed water stress and reduction in crop yield. The negative impact of water stress on crop yield can be predicted based on the relative sensitivity of different growth stages. To predict water stress and crop yield reduction satisfactorily, there is a provision of adjusting standard values in CROPWAT model. To analyze results and to draw conclusions, the researchers can take the help of this model which is a very powerful tool. This model shall help in recording of data in a more uniform manner and enable meaningful comparisons of findings in different studies in several countries. The most useful feature of the CROPWAT model is that it extends the findings and conclusions from studies to conditions not tested in the field. Hence, the practical recommendations can be given to farmers and extension staff on deficit irrigation scheduling under various soil, water supply and crop management conditions.

16.6 Conclusions

The significant savings can be generated in allocation of irrigation water through proper application of deficit irrigation practices. Deficit irrigation improves fruit quality and reduce water supply without negatively affecting the orchard commercial revenues. This is particularly important as growers are entitled to cultivate more land with saved water.

There is an urgent need for making available the guidelines for following deficit irrigation practices for identifying likely critical crop growth stages for imposing reduced evapotranspiration and for evaluating the acceptability and economic feasibility of deficit irrigation by the estimation of expected relative yield decreases.

In arid and semiarid regions, there shall be more widespread adoption of deficit irrigation in view of increasing scarcity and growing competition for water. This method has been tested and found quite useful in several crops such as winter wheat, groundnut, pear, peach, and grapevine. Since the deficit irrigation requires more sophisticated water controls, soil water monitoring, and accurate water management, there is a need to refine and improve the technology continuously as several trees and crops respond differently to water stress. In order to improve the knowledge and management techniques, refining the measuring technology and soil water sensors is required. The farmers should make use of recent major advances in developing and marketing of user-friendly and affordable soil water sensors for deficit irrigation. The improved food security can be achieved by using deficit irrigation scheduling strategies which reduce water demand with less impact on yields and crop quality.

References

Ahmad A (1999) Yield response of groundnut grown under rain-fed and irrigated conditions. In: Kirda C, Moutonnet P, Hera C, Nielsen DR (eds) Crop yield response to deficit irrigation. Kluwer Academic Publishers, Dordrecht

Bazza M, Tayaa M (1999) Contribution to improve sugar beet deficit-irrigation. In: Kirda C, Moutonnet P, Hera C, Nielsen DR (eds) Crop yield response to deficit irrigation. Kluwer Academic Publishers, Dordrecht

Calvache M, Reichardt K (1999) Effects of water stress imposed at different plant growth stages of common bean (*Phaseolus vulgaris*) on yield and N_2 fixation. In: Kirda C, Moutonnet P, Hera C, Nielsen DR (eds) Crop yield response to deficit irrigation. Kluwer Academic Publishers, Dordrecht

Craciun L, Craciun M (1999) Water and nitrogen use efficiency under limited water supply for maize to increase land productivity. In: Kirda C, Moutonnet P, Hera C, Nielsen DR (eds) Crop yield response to deficit irrigation. Kluwer Academic Publishers, Dordrecht

De la Hera ML, Romero P, Gómez-Plaza E, Martinez A (2007) Is partial root-zone drying an effective irrigation technique to improve water use efficiency and fruit quality in field-grown wine grapes under semiarid conditions? Agric Water Manag 87:261–274

Du T, Kang S, Zhang J, Li F (2008) Water use and yield responses of cotton to alternate partial root-zone drip irrigation in the arid area of north-west China. Irrig Sci 26:147–159

Ebel RC, Proebsting EL, Evans RG (1995) Deficit irrigation to control vegetative growth in apple and monitoring fruit growth to schedule irrigation. Hortscience 30(6):1229–1232

English M (1990) Deficit irrigation. I: analytical framework. J Irrig Drain E-ASCE 116:399–412

Geerts S, Raes D, Garcia M, Condori O, Mamani J, Miranda R, Cusicanqui J, Taboada C, Vacher J (2008a) Could deficit irrigation be a sustainable practice for quinoa (*Chenopodium quinoa* Willd.) in the Southern Bolivian Altiplano? Agric Water Manag 95:909–917

Geerts S, Raes D, Garcia M, Mendoza J, Huanca R (2008b) Indicators to quantify the flexible phenology of quinoa (*Chenopodium quinoa* Willd.) in response to drought stress. Field Crop Res 108:150–156

Geerts S, Raes D, Garcia M, Vacher J, Mamani R, Mendoza J, Huanca R, Morales B, Miranda R, Cusicanqui J, Taboada C (2008c) Introducing deficit irrigation to stabilize yields of quinoa (*Chenopodium quinoa* Willd.). Eur J Agron 28:427–436

Goodwin I, Boland AM (2002) Scheduling deficit irrigation of fruit trees for optimizing water use efficiency. In: Deficit irrigation practices, Water reports 22. FAO, Rome, pp 67–78

Henggeler JC, Enciso JM, Multer WL, Unruh BL (2002) Deficit subsurface drip irrigation of cotton. In: Deficit irrigation practices. FAO, Rome, pp 29–38

Ilbeyi A, Ustun H, Oweis T, Pala M, Benli B (2006) Wheat water productivity and yield in a cool highland environment: effect of early sowing with supplemental irrigation. Agric Water Manag 82:399–410

Iqbal MM, Shah SM, Mohammad W, Nawaz H (1999) Field response of potato subjected to water stress at different growth stages. In: Kirda C, Moutonnet P, Hera C, Nielsen DR (eds) Crop yield response to deficit irrigation. Kluwer Academic Publishers, Dordrecht
Karaata H (1991) Kirklareli Kosullarinda Aycicegi Bitkisinin Su-Üretim Fonksiyonlari. PhD thesis, Köy Hizmetleri Arastirma Enst. Report No 24, Kirklareli, Turkey (Turkish)
Kang S, Zhang L, Liang Y, Hu X, Cai H, Gu B (2002) Effects of limited irrigation on yield and water use efficiency of winter wheat in the Loess Plateau of China. Agric Water Manag 55:203–216
Kirda C (2002) Deficit irrigation scheduling based on plant growth stages showing water stress tolerance. In: Deficit irrigation practices. FAO, Rome, pp 3–10
Kirda C, Kanber R, Tulucu K (1999) Yield response of cotton, maize, soybean, sugar beet, sunflower and wheat to deficit irrigation. In: Kirda C, Moutonnet P, Hera C, Nielsen DR (eds) Crop yield response to deficit irrigation. Kluwer Academic Publishers, Dordrecht
Kovacs T, Kovacs G, Szito J (1999) Crop yield response to deficit irrigation imposed at different plant growth stages. In: Kirda C, Moutonnet P, Hera C, Nielsen DR (eds) Crop yield response to deficit irrigation. Kluwer Academic Publishers, Dordrecht
Li F, Wei C, Zhang F, Zhang J, Nong M, Kang S (2010) Water-use efficiency and physiological responses of maize under partial root-zone irrigation. Agric Water Manag 97:1156–1164
Madanoglu K (1997) Water consumption of wheat (Yektay 406) in Central Anatolia. Publication No. 52. Ankara Central TOPRAKSU Research Institute, Ankara, 67 pp
McCarthy MG, Loveys BR, Dry PR, Stoll M (2002) Regulated deficit irrigation and partial root zone drying as irrigation management techniques for grapevines. In: Deficit irrigation practices, Water reports 22. FAO, Rome, pp 79–88
Nautiyal PC, Joshi YC, Dayal D (2002) Response of groundnut to deficit irrigation during vegetative growth. In: Deficit irrigation practices, Water reports 22. FAO, Rome, pp 39–46
Pandey RK, Maranville JW, Admou A (2000) Deficit irrigation and nitrogen effects on maize in a Sahelian environment. I. Grain yield and yield components. Agric Water Manag 46:1–13
Pene CBG, Edi GK (1999) Sugarcane yield response to deficit irrigation at two growth stages. In: Kirda C, Moutonnet P, Hera C, Nielsen DR (eds) Crop yield response to deficit irrigation. Kluwer Academic Publishers, Dordrecht
Prieto D, Angueira C (1999) Water stress effect on different growing stages for cotton and its influence on yield reduction. In: Kirda C, Moutonnet P, Hera C, Nielsen DR (eds) Crop yield response to deficit irrigation. Kluwer Academic Publishers, Dordrecht
Raes D, Geerts S, Vandersypen K (2008) More food, less water. In: Raymaekers B (ed) Lectures for the XXIst century. Leuven University Press, Leuven, pp 81–101
Romero P, Botia P, Garcia F (2004) Effects of regulated deficit irrigation under subsurface drip irrigation conditions on vegetative development and yield of mature almond trees. Plant Soil 260:169–181
Shahnazari A, Liu F, Andersen MN, Jacobsen SE, Jensen CR (2007) Effects of partial root-zone drying on yield, tuber size and water use efficiency in potato under field conditions. Field Crop Res 100:117–124
Stewart BA, Musick JT (1982) Conjunctive use of irrigation and rainfall in semi-arid regions. Adv Agron 1:1–23
Unlu M, Kanber R, Senyigit U, Onaran H, Diker K (2006) Trickle and sprinkler irrigation of potato (*Solanum tuberosum* L.) in the Middle Anatolian region in Turkey. Agric Water Manag 79:43–71
USDA (2010) New farming wrinkle may help peanut growers. USDA Agricultural Research Service
Waheed RA, Naqvi HH, Tahir GR, Naqvi SHM (1999) Some studies on pre-planned controlled soil moisture irrigation scheduling of field crops. In: Kirda C, Moutonnet P, Hera C, Nielsen DR (eds) Crop yield response to deficit irrigation. Kluwer Academic Publishers, Dordrecht

Wample RL, Smithyman R (2002) Regulated deficit irrigation as a water management strategy in *Vitis vinifera* production. In: Deficit irrigation practices, Water reports 22. FAO, Rome, pp 89–100

Yactayo W, Ramírez DA, Gutiérrez R, Mares V, Posadas A, Quiroz R (2013) Effect of partial root-zone drying irrigation timing on potato tuber yield and water use efficiency. Agric Water Manag 123:65–70

Yang L, Qu H, Zhang Y, Li F (2012) Effects of partial root-zone irrigation on physiology, fruit yield and quality and water use efficiency of tomato under different calcium levels. Agric Water Manag 104:89–94

Yavuz MY (1993) Farkli Sulama Yontemlerinin Pamukta Verim ve su Kullanimina Etkileri. PhD thesis, Cukurova University, Faculty of Agriculture, Adana, Turkey

Zhang H (2003) Improving water productivity through deficit irrigation: examples from Syria, the North China Plain and Oregon, USA. In: Kijne JW, Barker R, Molden D (eds) Water productivity in agriculture: limits and opportunities for improvement. International Water Management Institute, Colombo, pp 301–309

Zhang X, Pei D, Li Z, Li J, Wang Y (2002) Management of supplemental irrigation of winter wheat for maximum profit. In: Deficit irrigation practices, Water reports 22. FAO, Rome, pp 57–66

Chapter 17
Supplemental Irrigation

Abstract Water harvesting and supplemental irrigation can be utilized in rain-fed areas to increase the productivity of land and water. Making available of controlled but limited additional water can enable the cultivation of marginal lands with annual rainfall of less than 300 mm. Supplemental irrigation involves the application of small amounts of water during critical crop growth stages that can significantly enhance water productivity (ensuring more sustainable use of groundwater) and enhance yield. In arid and semiarid dry areas, where water is the most important limiting factor, the goal should be to maximize yield per unit of water (water productivity), rather than yield per unit of land. The increased water productivity as a result of the introduction of supplemental irrigation in rain-fed areas is discussed with relevant examples.

Keywords Supplemental irrigation • Water productivity • Critical crop growth stages • Increased yields

17.1 Introduction

The area under rain-fed farming covers about 80 % of the world's agricultural land (95 % in sub-Saharan Africa, 90 % in Latin America, 75 % in West Asia and North Africa, 65 % in East Asia, and 60 % in South Asia) which is responsible for about two-thirds of food production in the world (Rockström et al. 2007). These areas generally receive low annual rainfall of 100–600 mm and renewable water resources are very much limited. The factors which are responsible for limiting the rain-fed agriculture include drought, water scarcity, and land degradation. Even though the productivity is normally low in rain-fed agriculture as compared to irrigated agriculture, the rain-fed agriculture still plays an important role in fulfilling the food and livelihood security to the growing global population. In view of increasing scarcity and competition for irrigation water resources, enhancing the water productivity should be the goal for agricultural productivity increase, environment protection, and poverty alleviation. For sustaining food and livelihood security even at current levels, we should adopt the policy of "more crop and income per drop of water" by increasing the productivity of water in rain-fed agriculture.

P.P. Reddy, *Sustainable Intensification of Crop Production*,
DOI 10.1007/978-981-10-2702-4_17

In view of inappropriate management of soil, water, and nutrients; moisture deficits; and use of inadequate production inputs; the production in rain-fed agriculture is generally low in developing countries. The production and productivity of crops can be enhanced by efficient water management.

To enhance the rain-fed agricultural production, there are three primary ways:

- Improved water management to enhance the productivity of rainwater
- Adequate inputs and agronomic practices (improved cultivars; adequate nutrients; insect, disease, nematode, and weed management)
- Adoption of supplemental irrigation that is a highly efficient and yield-stabilizing technology

The supplemental irrigation (SI) is one of the potential and less utilized technologies for rain-fed agriculture, particularly in Africa and Asia. This method of irrigation has been extensively tested and widely adopted and has significantly improved productivity of different crops in several developed countries. The efficient application of water is very crucial as it can contribute significantly to reduce water losses and enhance water-use efficiency in situations where water is a scarce resource and the amount of water available for supplemental irrigation is normally limited.

Water harvesting (WH) and supplemental irrigation (SI) are the principal technologies to create additional resources of water that can enhance the land and water productivity. Water harvesting technologies can be followed in dry areas where rainfall is around 300–600 mm annually, where crops cannot grow depending only on rainfall. The supplemental irrigation can be utilized in wetter part of the dry areas with annual rainfall of about 300–600 mm. The additional water resources can be used for cultivation of marginal lands with rainfall of less than 300 mm annually.

17.2 Water Harvesting

The surface water and shallow groundwater aquifers are being increasingly used as main sources for supplemental irrigation. The treated sewage water harvesting is one of the nonconventional water resources which is also important and has potential for the future.

The collection of runoff water for its productive use is called water harvesting (Siegert 1994). It is the process of concentrating rainfall as runoff from a large catchment area that can be used in a small target area. The runoff water collected is either stored in some type of (on-farm) storage facility for use as supplemental irrigation of crops or directly applied to an adjacent agricultural field (or plot).

The main features of water harvesting are as follows:

- The "catchment" area is relatively small and is contiguous with the benefiting target area.
- The application is basically uncontrolled to the target area—the objective is simply to capture as much water as possible and store it within the reach of the crop(s), into some type of reservoir or in the soil profile of a cultivated area.

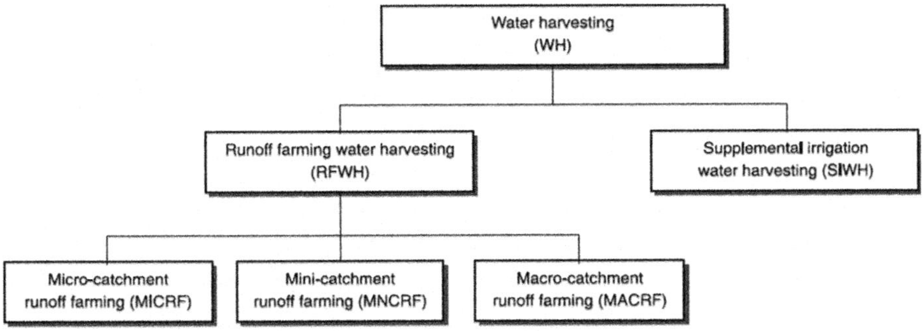

Fig. 17.1 The different types of water harvesting techniques based on the type of storage adoption

- Water harvesting can be used to concentrate rainfall for purposes other than crop production (e.g., recharge of ground water).
- Practiced in areas where surface runoff often has an intermittent character (arid and semiarid regions).
- Requires a runoff producing area and a runoff receiving area for its utilization.
- Water is normally stored directly in the soil profile or in small reservoirs, tanks, and aquifers.

The areas that normally receive an average annual rainfall of at least 250 mm in summer and 100 mm in winter are suitable for water harvesting. The water harvesting serves as an important source of water for irrigation in the drier regions of India, such as Bihar, the western regions of Rajasthan and Gujarat, and the Deccan plateau of central India (Kolavalli and Whitaker 1996).

The four components of water harvesting system include runoff producing catchment, runoff collection scheme, runoff storage facility, and cropped or cultivated area.

The different types of water harvesting techniques based on the type of storage adoption are presented in Fig. 17.1.

17.3 Supplemental Irrigation

In rain-fed agriculture, soil moisture shortage occurs frequently during the most critical stages of plant growth such as sowing, panicle initiation, flowering, and grain filling that can significantly affect the yield. In arid and semiarid dry areas, where water is the most important limiting factor, the goal should be to maximize yield per unit of water, rather than yield per unit of land. In this context, supplemental irrigation which involves the application of small amounts of water during critical crop growth stages can significantly enhance water productivity (ensuring more sustainable use of groundwater) and enhance yield.

The purposes of supplementary irrigation include:

- To optimize yield during water deficit is occasional and of short duration.
- To optimize water-use efficiency during limited availability of water.

The addition of small amounts of water to mainly rain-fed crops to increase and stabilize yields during the periods when rainfall fails to provide sufficient moisture for normal plant growth is called supplemental irrigation. It is effective in alleviating the adverse effects of soil moisture stress on the yield of rain-fed crops during dry spells. The essential characteristic of SI is the supplemental nature of irrigation and rainfall. The water productivity (i.e., the ratio of economic yield of a crop and the total amount of water consumed) of rain and supplemental irrigation exceeds the water productivity of either component if applied alone.

The characteristics of SI in rain-fed areas are as follows:

- Water is applied to rain-fed crops which are generally produced without irrigation.
- It is applied only when rainfall is inadequate.
- Rainfall is the prime source for supplemental irrigation.
- The amount and timing of SI is to provide sufficient water during the critical stages of crop growth to ensure optimal yield in terms of yield per unit of water (Oweis 1997).
- It aims to increase the total farm yield and water-use efficiency by maximizing the area that benefits from the water available (Caliandro and Boari 1992).

SI may potentially have three major effects:

- Yield improvement
- Stabilization of production from year to year (increasing reliability)
- Providing suitable conditions for economic use of higher technology inputs, such as high-yielding varieties, fertilizers, pesticides, and herbicides, irrespective of seasonal rainfall

The supplemental irrigation technology is useful for rain-fed agriculture in dry areas (arid and semiarid regions), particularly where growers are already using some irrigation. It is highly effective and a simple technology which allows farmers to plant and manage crops at the optimal time, without being at the mercy of unpredictable rainfall.

The application of relatively small amounts of water at the time of water shortage during the most critical stages of crop growth (Table 17.1) will assure the survival of an economically valuable crop, which otherwise can lead to death of plants and significant reduction in yields.

In arid and semiarid regions, supplementary irrigation will be desirable because it is mainly used at the critical periods of the crop growth cycle, in order to stabilize and improve yield quantity and quality.

Table 17.1 Critical soil moisture periods for different crops

Crop	Available critical soil moisture limit (%)	Critical soil moisture stage of crops
Cereals	–	Sowing, panicle initiation, flowering, and grain filling
Wheat	–	Crown root initiation and grain-filling stages
Rice	–	Panicle initiation, flag leaf, and milky stages
Sorghum	–	Flowering and grain-filling stages (in rainy season)
Maize	–	Silking and tasseling
Pearl millet	–	Booting and flowering
Finger millet	–	Flowering
Groundnut	–	Peg penetration and pod development
Sunflower	–	Head formation and early grain filling
Sesame	–	Flowering
Soybean	–	Flowering and pod filling
Black gram/Green gram	–	Flowering and early pod development
Cotton	–	Square formation and boll formation and development
Sugarcane	–	Cane formation (up to 120 days after sowing)
Citrus	50	Flowering, fruit setting, fruit growth
Banana	50	Early vegetative period flowering and fruit formation
Mango	50	Start of fruiting to maturity
Pineapple	50	Vegetative growth
Grape	50	Vegetative growth
Guava	50	Period of fruit growth
Jujube (Ber)	50	A drought-resistant plant; supplemental irrigation is required during fruit growth
Olive	–	Vegetative resumption or to improve yield quantity (fruit enlargement stage) and/or quality
Chili	50	Tenth leaf to flowering and fruit development
Potato	65	Stolon formation, tuberization and tuber enlargement
Onion	60	Bulb formation and bulb enlargement
Tomato	60	Flowering and fruit development
Peas	40	Flowering and pod development
Cabbage	60	Head formation and enlargement
Cauliflower	70	Curd formation and enlargement
Eggplant	50	Flowering and fruit development
Cucumber	50	Flowering and fruit development
Okra	45	Flowering and fruit development
Flower crops	–	Bud formation and development
Ornamentals	–	Flowering

Table 17.2 Effect of supplemental irrigation on crop yields at different locations in India

Crops	Supplemental irrigation (cm)	Yield (t/ha)	Yield increase due to irrigation (%)	Location
Short-duration rainy-season crops				
Sorghum	1.6	2.51	560	Hyderabad
Maize	2.0	4.43	40	Jhansi
Finger millet	5.0	2.32	43	Bangalore
Soybean	8.0	2.05	14	Indore
Long-duration rainy-season crops				
Castor	5.0	1.32	31	Hyderabad
Pigeon pea (sole crop)	5.0	0.33	560	Jhansi
Tobacco	4.0	1.30	58	Dantiwada
Post-rainy-season crops				
Wheat	6.0	2.6	123	Dehra Dun
Rape seed	5.0	0.54	116	Ranchi

Vijayalakshmi (1987)

17.4 Benefits

Supplemental irrigation is responsible for giving multiple benefits such as significantly higher water productivity, lower risk of crop failure, and higher and more stable yields (the amount of grain or biomass produced per liter of water).

Application of supplemental irrigation facilitated the farmers to plant their crops early, preventing exposure to terminal heat and drought stress in hot areas and frost in cold areas and increasing yields.

17.4.1 Increased Yields

Among 13 crops from 23 locations tested in India for their response to supplemental irrigation, the increase in grain yield varied from 23 % in the case of sorghum to 345 % in chickpea (Singh et al. 1999). Supplemental irrigation is estimated to increase production by 43 % on an average in developing countries and up by 300 % in southern Africa.

The application of supplemental irrigation gave highest increase in grain yields to the extent of 560 % in rainy-season sorghum and pigeon pea and 337 % in pearl millet, while the least increase in yield by 32 % was observed in groundnut (Table 17.2) (Vijayalakshmi 1987).

In alfisols, considerable benefits have been obtained from supplemental irrigation in both rainy and post-rainy seasons. The water application efficiency (WAE) from supplemental irrigation was higher in sorghum (14.9 kg/mm/ha), while it was lower in pearl millet (8.8–10.2 kg/mm/ha). The response of supplemental irrigation in intercropped pigeon pea was lower. The WAE for both pigeon pea/sorghum and pigeon pea/pearl millet intercropping systems ranged from 5.3 to 6.7 kg/mm/ha,

Table 17.3 Effect of to supplemental irrigation on intercropping systems at International Crops Research Institute for the Semi-Arid Tropics (ICRISAT) Center, Patancheru, India during 1981–1984

Crop/Grain yield (t/ha)[a]	Increase due to irrigation (kg/ha)	WAE[b] (kg/mm/ha)	Crop/grain yield (t/ha)[c]	Increase due to irrigation (kg/ha)	WAE (kg/mm/ha)	Combined WAE (kg/mm/ha)
Intercropping system						
Pearl millet 2.353	403	10.0	Pigeon pea 1.197	423	5.3	6.8
Sorghum 3.155	595	14.9	Pigeon pea 1.220	535	6.7	9.4
Sequential cropping system						
Pearl millet 2.557	407	102	Cowpea 0.735	425	5.3	6.9
Pearl millet 2.215	350	8.8	Tomato 26.250	14,900	186.3	127.1

El-Swaify et al. (1985), Pathak and Laryea (1990)
[a]One irrigation of 40 mm
[b]*WAE* water application efficiency
[c]Two irrigations of 40 mm each

while it was higher (WAE of 186.3 kg/mm/ha) in tomato (Table 17.3) (El-Swaify et al. 1985; Pathak and Laryea 1990).

In cowpea and maize, application of supplemental irrigation at critical crop growth stages enhanced grain yields (Table 17.4) (Fapohunda et al. 1984).

Srivastava et al. (1985) reported that supplemental irrigation was beneficial in vertisols for sequential crops of chickpea and chilies. The WAE was highest for chickpea (5.6 kg/mm/ha), followed by chilies (4.1 kg/mm/ha) and lowest in safflower (2.1 kg/mm/ha) (Table 17.5). The benefit of supplemental irrigation was more in alfisols than in vertisols.

Significant grain yield increases have been obtained under field conditions in wheat and barley with small amounts of supplemental irrigation. The grain yields were increased from 4.6 to 5.8 t/ha in Morocco, from 2.2 to 3.4 t/ha in Iran, and from 1.25 to 3 t/ha in Syria.

The chickpea and sorghum grain yields increased by 345 % and 23 % over control, respectively, when the supplemental irrigation was applied at critical crop growth stages (Singh et al. 1999).

17.4.1.1 Wheat

The most critical crop growth stages (flowering and grain filling) are likely to suffer from shortage of soil moisture in the dry rain-fed areas resulting in reduced crop growth and consequent low yield in wheat. A substantial improvement in wheat grain yield and water productivity (WP) can be achieved by providing supplemental irrigation during sensitive growth stages of crops (Fig. 17.2).

Table 17.4 Effect of supplementary irrigation on yields of cowpea and maize

Crop	Supplemental irrigation (cm)	Grain yield (kg/ha)
Cowpea	Irrigation + rain	1110
	Rain, no irrigation	260
Maize	Irrigation + rain	4825
	Rain, no irrigation	4100

Fapohunda et al. (1984)

Table 17.5 Effect of supplemental irrigation on sequential crops at ICRISAT Center, Patancheru, India during 1981–1985

Sequential cropping system	Mean yield (kg/ha)		Water application efficiency (kg/mm/ha)
	Supplemental irrigation	Increase due to irrigation	
Maize + chickpea	1540	493	5.6
Green gram + chilies	1333	325	4.1
Maize + safflower	1238	165	2.1

Srivastava et al. (1985)

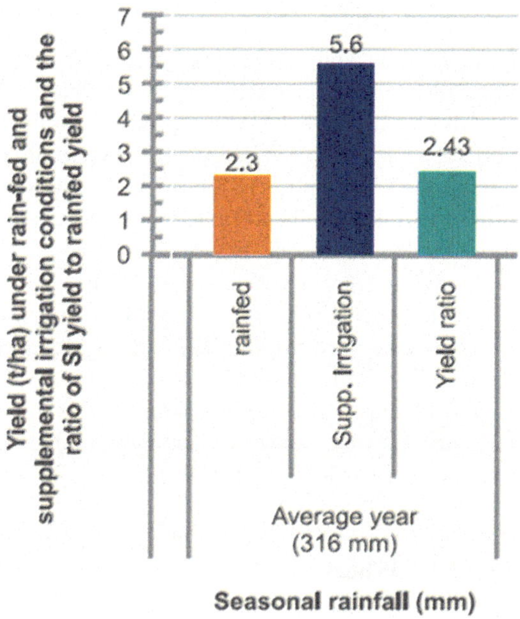

Fig. 17.2 Effect of supplemental irrigation on wheat grain yield under rain-fed conditions

The supplemental irrigation of 120 mm increased rain-fed (316 mm rainfall) wheat yield from 2.3 to 5.6 t/ha at Tel-Hadya, Syria. The mean wheat yield was enhanced from less than 0.8 t/ha under rain-fed conditions to more than 4.8 t/ha with SI in field demonstrations, conducted during 1988–1998 by the International Center for Agriculture Research in the Dry Areas (ICARDA) under farmers' field conditions in Syria.

The rain-fed wheat yield ranged from 0.9 to 2.5 tons/ha, while yield under SI ranged from 3 to 4.5 t/ha in Konya Province of Central Anatolia, Turkey. In Eskisehir Province, yields ranged from 1.1 to 3.2 t/ha (rain-fed) to 2.5 t/ha to 6.25 t/ha with SI (Tenkinel et al. 1992).

The rain-fed wheat yield enhanced from 2.16 to 4.61 t/ha by the application of only 68 mm of irrigation water in the spring during the growing season of 1997/1998 (annual rainfall 236 mm). In northern Iraq, application of 100–150 mm of SI during April and May gave the maximum grain yields (Fig. 17.3) (Adary et al. 2002).

17.4.1.2 Barley

The barley grain yield increased to 1.89 t/ha at 33 % SI, 4.25 t/ha at 66 % SI, and 5.17 t/ha at 100 % SI, as compared to 0.26 t/ha under rain-fed conditions in Mediterranean climate with total rainfall of 186 mm (Oweis and Hachum 2006).

Fig. 17.3 The effect of supplemental irrigation (*left*) on wheat as compared to rain-fed wheat (*right*)

17.4.1.3 Rice

The late transplanted rice in Bangladesh normally suffered from water stress when the rains ended early. Under such condition, application of 60 mm of SI increased the yield by 58 % (Islam and Bhuiyan 1991). In rice, supplemental irrigation during sensitive crop growth stages can increase yields by 30–40 %.

17.4.1.4 Maize

Supplementary irrigation raised maize yield from 0.6 to 2.0 t/ha.

17.4.1.5 Cotton

Providing supplemental irrigation helps the cotton crop to overcome the chances of drought during hot summer days that increase water use. The income of dryland farmers can be further enhanced by introduction of high-value hybrid cotton under protective irrigation.

17.4.1.6 Cowpea

Fapohunda et al. (1984) obtained cowpea yields up to 1.9 t ha^{-1} using 464 mm of supplemental water.

17.4.1.7 Chickpea

The requirement of supplemental irrigation water was reduced by about 27 % with shallow cultivation in moderately cracked furrows before the application of irrigation water, without affecting the chickpea yields (Table 17.6).

Table 17.6 Effect of supplemental irrigation on grain yield of chickpea at ICRISAT Center, Patancheru, India

Treatment	Mean depth of water application (cm)	Grain yield (kg/ha)
No supplemental irrigation	0	690
One supplemental irrigation on uncultivated furrows	6.3	920
One supplemental irrigation on cultivated furrow	4.6	912
SEM		19
CD (P = 0.05)		5.5

Srivastava et al. (1985)

17.4.1.8 Peach

Application of about 100–200 mm of supplemental irrigation to peach orchards during the summer from June to September can overcome severe moisture stress and can improve tree growth, yield, and fruit quality in Mediterranean region.

17.4.1.9 Guava

The highest response to supplemental irrigation was recorded in guava (122.6 %) and the lowest in fig (41.7 %) in a trial conducted in medium deep black soils at Bijapur, North Karnataka, India (Radder et al. 1995).

17.4.2 Improved Water Productivity

The application of a small amount of supplemental water is effective in high water productivity and in alleviating severe moisture stress during the most sensitive stages of crop growth and seed filling. The wheat yield can be increased to 2.5 kg of grain per cubic meter above that of rain-fed production by the application of one cubic meter of water applied at the right time (when the crop is suffering from moisture stress), combined with good management (Fig. 17.4) (Oweis 1997).

On-farm water productivity is higher (2.5 kg/m^3) under supplemental irrigation, lower (0.3 to 1 kg/m^3) under rain-fed conditions, and moderate (0.75 kg/m^3) under full irrigation.

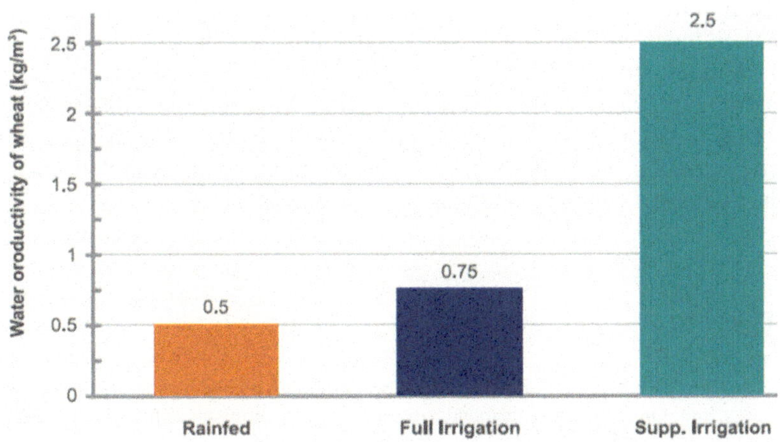

Fig. 17.4 Effect of supplemental irrigation on water productivity in wheat (Oweis 1997c)

Table 17.7 Effect of supplemental irrigation and fertilizer on sorghum grain yield (kg/ha) in Sahel during 1998–2000

Treatment	Mean yield (kg/ha) (1998–2000)*
Control treatment	455[a]
Supplemental irrigation	712[b]
Fertilizer application	975[c]
Supplemental irrigation + fertilizer application	1403[d]

Fox and Rockström (2000)
*Mean values in a column followed by different letters are significantly different at the 5 % level using the Student–Newman–Keuls test

17.5 Integration with Fertilizers

The integration of supplemental irrigation with fertilizer application increased sorghum yield significantly (Fox and Rockström 2000). The supplemental irrigation integrated with fertilizer application gave a grain yield of 1403 kg/ha (which was higher than the farmer's normal practice by a factor of three) as compared to a grain yield of 712 kg/ha with supplemental irrigation alone (Table 17.7).

17.6 Conclusions

In many parts of Africa and Asia, the supplemental irrigation had given significant benefits in the form of both by increasing and stabilizing crop productivity. The beneficial effect increased when the supplemental irrigation (small amounts of 10–15 mm water applied) was given during sensitive crop growth stages. The integration of supplemental irrigation along with the use of other improved inputs, e.g., responsive cultivars, fertilizers, and pesticides, maximized the benefits. The best response on low water-holding-capacity soils (alfisols) can be achieved by providing one or two supplemental irrigations during the post-rainy season; while on high moisture-holding-capacity soils (vertisols), pre-sowing supplemental irrigation was highly beneficial for post-rainy season crops. By cultivation of high-value horticultural crops such as vegetables which are becoming popular (even with poor farmers), the benefits of supplemental irrigation can be enhanced.

References

Adary A, Hachum A, Oweis T, Pala M (2002) Wheat productivity under supplemental irrigation in Northern Iraq, On-farm water husbandry research report series no.2. International Center for Agricultural Research in the Dry Areas (ICARDA), Aleppo

Caliandro A, Boari F (1992) Supplementary irrigation in arid and semi-arid regions. In: International conference on supplementary irrigation and drought water management, vol 1. CIHEAM, Bari

El-Swaify SA, Pathak P, Rego TJ, Singh S (1985) Soil management for optimized productivity under rainfed conditions in the semi-arid tropics. Adv Soil Sci 1:1–64

Fapohunda HO, Aina PO, Hossain MM (1984) Water use – yield relations for cowpea and maize. Agric Water Manag 9:219–224

Fox P, Rockström J (2000) Water harvesting for supplemental irrigation of cereal crops to overcome intra-seasonal dry-spells in the Sahel. Phys Chem Earth, Part B Hydrology, Oceans and Atmosphere 25(3):289–296

Islam J, Bhuiyan LR (1991) Supplemental irrigation—a safeguard technique for successful cultivation of monsoon rice in Bangladesh. Irrig Drain Syst 5:351–352

Kolavalli S, Whitaker ML (1996) Institutional aspects of water harvesting: khadins in western Rajasthan, India. In: VIth annual conference of the international association for the study of common property, Berkeley, California, USA. International Crop Research Institute for the Semi-Arid Tropics, Patancheru, India

Oweis T (1997) Supplemental irrigation: a highly efficient water-use practice. International Center for Agricultural Research in the Dry Areas, Aleppo, 16 pp

Oweis T, Hachum A (2006) Water harvesting and supplemental irrigation for improved water productivity of dry farming systems in West Asia and North Africa. Agric Water Manag 80:57–73

Pathak P, Laryea KB (1990) Prospects of water harvesting and its utilization for agriculture in the semiarid tropics. In: Proceedings of symposium of the SADCC (Southern African Development Coordination Conference), pp 266–278. Land and Water Management Research Program, Gaborone

Radder GD, Itnal CJ, Belgaumi MI (1995) Protective irrigation (life saving) – principles and practices. In: Singh RP (ed) Sustainable development of dryland agriculture in India. Scientific Publishers, Jodhpur

Rockström J, Hatibu N, Oweis TY, Wani SP (2007) Managing water in agriculture. In: Molden D (ed) Water for food, water for life: a comprehensive assessment of water management in agriculture. International Water Management Institute (IWMI), Colombo, pp 315–352

Siegert K (1994) Introduction to water harvesting: some basic principles for planning, design and monitoring. In: Water harvesting for improved agricultural production. Proceedings of the FAO expert consultation, Cairo, Egypt, November 1993, FAO, Rome

Singh SR, Chaudhary TN, Bhatta RK, Bhatnagar PR, Saha B, Patil NG (1999) Irrigation water management. In: Singh GB, Sharma BR (eds) Fifty years of natural resource management research in India. Central Soil Salinity Research Institute (CSSRI), Karnal, pp 40–48

Srivastava KL, Pathak P, Kanwar JS, Singh RP (1985) Watershed-based soil and rain water management with special reference to Vertisols and Alfisols. In: National seminar on soil conservation and watershed management, New Delhi, India

Tenkinel O, Kanber R, Yazar A, Ozekici B (1992) Drought conditions and supplemental irrigation in Turkey. In: International conference on supplementary irrigation and drought water management, vol 1. CIHEAM, Bari

Vijayalakshmi K (1987) Soil management for increasing productivity in semi-arid red soils: physical aspects. In: Pathak P, El-Swaify SA, Singh S (eds) Proceedings of the consultant's workshop on the state of the art and management alternatives for optimizing the productivity of SAT alfisols and related soils, 1–3 December 1983, pp 109–115. International Crops Research Institute for the Semi-Arid Tropics (ICRISAT), Patancheru, India

Part VIII
Agro-ecological Crop Protection

Chapter 18
Agro-ecological Pest Management

Abstract Agro-ecologists opine that interaction between healthy soils and healthy plants is fundamental to agro-ecologically based pest management. They also encourage biodiversification as the primary technique to evoke regulation of pests sustainably. Besides regulation of pest populations through enhanced activity of biological-control agents, biodiversity performs a variety of ecological services including the production of food and recycling of nutrients in the agro-ecosystems. One of the simplest forms of biodiversification such as legume-based crop rotations can simultaneously optimize soil fertility and pest regulation. Crop rotations enhance yields by interrupting insect, disease, and weed life cycles. The microbiological activity of the soil (population of fungal and bacterial antagonists and arbuscular mycorrhizal fungi that better resist pest attack) is also enhanced by crop rotations. Agro-ecologically based pest management makes full use of natural and cultural practices and methods, including host resistance and biological control. In order to stabilize the population of pest species throughout the food web, the new designs should concentrate on managing the farm environment through ecosystem enhancements (i.e., landscape ecology), crop attributes, or other means.

Keywords Agro-ecology • Economic thresholds • Ecosystem enhancements • Pest management

18.1 Introduction

The conventional agricultural intensification is threatened with environmental crisis. The environmental crisis is rooted in genetic homogeneity; soil salinization; ground water depletion; pesticide pollution of soil, water, and food chains; land degradation; and associated vulnerability, which raise serious doubts about the sustainability of modern agriculture. In the foreseeable future, the continuing growth of the human population, which likely reaches 9.5 billion by 2050, requires that we should find ways for sustainable intensification of crop production to increase food production.

Crop pests such as insects, mites, diseases (caused by fungi, bacteria, virus, and nematodes), and weeds are one of the major limiting factors in crop production.

P.P. Reddy, *Sustainable Intensification of Crop Production*,
DOI 10.1007/978-981-10-2702-4_18

They are responsible for more than 30 % crop loss worldwide which is equivalent of food required to feed over one billion people (Birch et al. 2011). One of the possible approaches to increase food production is to reduce crop losses caused by pests and diseases.

The present-day farmers are selecting crops which give high yields and high palatability that are more susceptible to pests by overlooking natural resistance for productivity. The pest management at present is mainly based on chemically based pesticides which give short-term solutions that are harmful to human health and the environment. The negative impact of pest management using pesticides include development of pesticide resistance (requiring more frequent application at higher doses), emergence of secondary pests, and pesticide residues in edible food products responsible for acute poisoning. The pest problems will continue to persist as long as monoculture is practiced which do not possess the necessary ecological defense mechanisms to tolerate pest populations that experience outbreaks. Hence, there is an urgent need to develop alternative strategies for pest management which are sustainable and environment friendly.

The future pest-management strategies should be based on a broad knowledge of the agro-ecosystem that would manage pests rather than eliminate them. The existing agricultural practices that disrupt natural processes should be replaced with agricultural practices that augment natural processes that suppress pests. These practices should be supported with the use of biological-control organisms and products, target-specific chemical pesticides, and pest-resistant plants. This would subsequently lower the risk to users, foster interdisciplinary interaction, and improve the profitability of alternative pest-management methods. The factors that should be kept in mind while developing pest-management strategies include long-term sustainability, protecting human and environmental health, cost-effectiveness, and solving unmet needs of food production.

18.2 Agro-ecological Pest Management

The agro-ecological pest-management strategy involves water and soil conservation, closed energy flows, optimal recycling of nutrients and organic matter turnover, and a balance between pest and natural enemy populations (Gliessman 1998) which provides the use of low-input technologies, biologically mediated soil fertility, sustained yields, a balanced environment, and natural pest regulation through the design of diversified agro-ecosystems. It enhances the immunity of the agro-ecosystem by adopting preventative strategies.

One of the fundamental pillars of ecological pest management is soil health which harbors more active and diverse populations of the soil organisms that compete, antagonize, and ultimately curb soil-borne pests. The crop protection can be achieved by managing pests through a healthy soil, a rich natural enemy biodiversity obtained by a diversified agro-ecosystem, soil nutrient management, soil organic matter management, and reduced tillage and providing field boundaries, borders, and buffers (Thomas et al. 1991).

18.2.1 Goals

The fundamental goals of agro-ecological pest management include safety, profitability, and durability.

18.2.1.1 Safety

The primary criterion of acceptability of novel management systems is to minimize health risks for growers and workers who use them and for consumers of the food.

18.2.1.2 Profitability

Cost-effectiveness (costs and availability of labor, land, equipment, and other production inputs, and crop prices) is the primary criterion for development of effective pest-management strategies which are also easy to implement and readily integrated with other crop production practices.

18.2.1.3 Durability

Long-term pest management, without adverse environmental, economic, or safety consequences, should be the aim of developing pest-management strategies. It is important to address the problem of development of pest resistance in new ecologically based approaches.

18.2.2 Redesign of Agro-ecosystems

There is a need to redesign agro-ecosystems that can contribute to one or more of the following general strategies:

- Producing healthy crops
- Increasing stress on pests
- Enhancing beneficials

18.2.2.1 Producing Healthy Crops

- Building and maintaining soil health
- Reducing tillage for healthy soils
- Maintaining surface residues
- Other practices for building healthy soils – application of FYM and composts and rotation with grasses and legume forages

18.2.2.2 Increasing Stress on Pests

- Discouraging the pests' dispersal or connection with crops
- Disrupting pest populations by destabilizing habitat
- Reducing weeds' access to resources

18.2.2.3 Enhancing Beneficials

- Sowing cover crops between rows of cash crops
- Maintaining beneficial insectary plantings at field edges
- Providing permanent refuge strips or beetle banks for ground beetles, an important group of soil dwelling generalist predators
- Harboring natural predators, parasitoids, and wildlife in perennial grasses, forbs, shrubs, and trees on field edges or in strips
- Conservation tillage, preserving soil structure and complex food webs for ground beetles and other beneficials
- Encouraging root disease-suppressing microbes with life-sustaining organic matter by means of cover crops, animal manures, and composts

18.3 Pillars

In order to build farm's natural defenses, the agro-ecological pest management relies primarily on the preventive practices such as aboveground and belowground principles and practices (crop and soil management). The pest management in the agro-ecosystem is based on the use of multiple tactics rather than a "silver bullet" (Fig. 18.1).

18.3.1 Reduced Tillage

Growing of crops without disturbing the soil through tillage from year to year is called no-tillage farming (zero tillage or direct drilling) or reduced-tillage farming. This practice is helpful to the farmer in many ways including reduction or

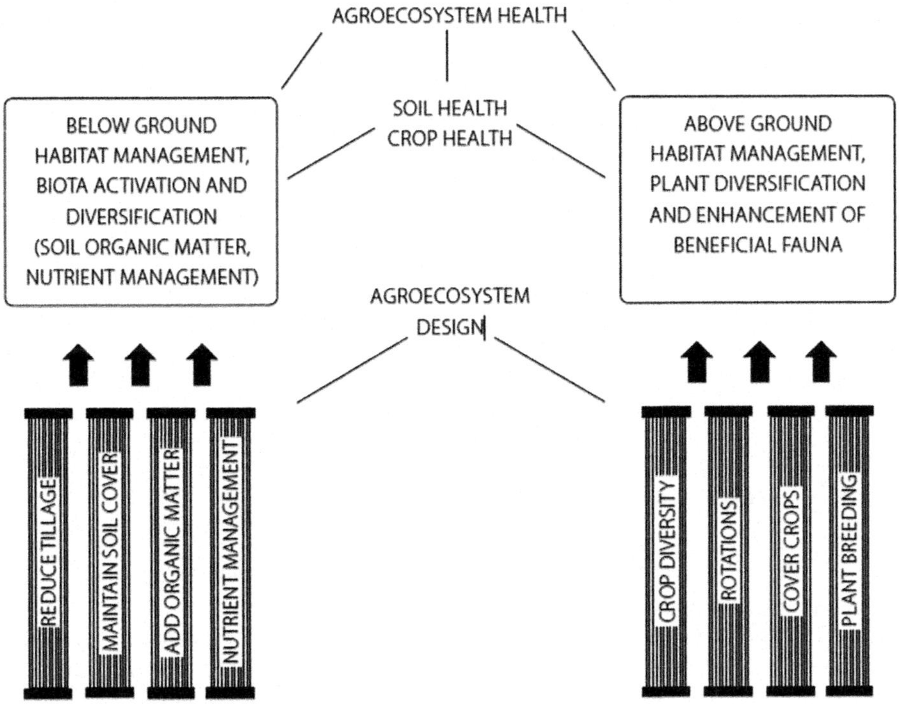

Fig. 18.1 The pillars of agro-ecological pest management

elimination of soil erosion, recycling of nutrients in the soil, increased amount of water infiltration into the soil, increased organic matter retention in soil, and increased amount and variety of life in the soil including disease-suppressing organisms.

The well-managed conservation agriculture systems through reduced tillage contain high biodiversity below and above the ground which provide a stable habitat for natural enemies of insect pests and pathogens, including antagonistic microorganisms. In tillage-based systems, most of these natural enemies would not exist at a level to give a beneficial effect. By adopting good conservation agriculture systems especially with diversified crop rotations, pest and disease pressures are significantly reduced over time resulting in saving on pesticide costs.

The reduced and minimum tillage in groundnut decreased the incidence of tomato spotted wilt tospovirus by 42 % over conventional tillage (Table 18.1). This practice has now become recommended strategy to manage the disease.

Some virus diseases vectored by aphids are reduced under no-till farming because winged aphids are more often attracted to barren ground (conventional tillage system), compared to ground covered by residue (conservational tillage system).

The occurrence of charcoal rot in maize is decreased under no-till farming in many rotations due to conservation of moisture in the soil.

Table 18.1 Effect of tillage treatment on the incidence of tomato spotted wilt tospovirus on groundnut

Type of tillage	Tomato spotted wilt tospovirus incidence (%)
Conventional tillage	24.6
Minimum tillage	14.2
Reduced tillage	14.2
CD ($P = 0.05$)	3.9

Leser (1994) reported that the aphid population was always lower in conservation tillage or crop rotation systems compared to continuous-cotton conventional tillage systems.

18.3.2 Maintaining Soil Cover

The maintenance of a protective cover of vegetative residues such as straw, maize stalks, palm fronds, and stubbles on the soil surface at all times is called mulch farming system. At the time of year when soil erosion risk is greatest, mulching practice is an invaluable tool. For mulching to be effective for prevention of soil erosion, about 70 % of the soil surface must be covered by the mulch. This requires about 4 to 6 t/ha^{-1} of straw mulch in the tropics. A mulch rate of 12 t/ha^{-1} has been observed to improve soil physical properties (Lal 1975). The benefits of mulching surface soil include:

- Reduced soil erosion
- Improved water infiltration in soil
- Decreased inflow water velocity
- Reduced weed germination and emergence
- Controls pests and diseases
- Soil surface protection against raindrop impact
- Enhanced soil burrowing activity of earthworms (e.g., *Hyperiodrilus* spp. and *Eudrilus* spp.) (Lal 1975)
- Shade and physically hindering emerging weeds
- Enhance crop growth and competitiveness
- Soil moisture conservation
- Modify soil temperature
- Reduced soil surface crusting
- Improved transport of water in soil profile (Aina 1984)
- Improved soil moisture storage in root zone

The annual saving of 32 % of rainfall in water runoff has been reported by Lal (1976) from mulching in humid Western Nigeria. Surface soil mulching was responsible for drastic reductions in runoff and erosion from a mulched pineapple field on a 20 % slope (Table 18.2) (Roose 1988).

Table 18.2 Effect of residue mulch on runoff and soil erosion

	Runoff (% rainfall)		Soil erosion (t ha^{-1} $year^{-1}$)		
Country	Bare	Mulched	Bare	Mulched	Reference
Ghana	49.8	1.40	313.0	0.42	Mensah-Bonsu and Obeng (1979)
Côte d'Ivoire	36.4	0.33	18.3	1.90	Roose (1988)
Nigeria	29.0	0.10	410.0	1.00	Lal (1976)
Nigeria	42.1	2.40	232.6	0.20	Lawes (1966)

The elimination of weeds or retardation of their growth is the primary reason for using surface mulch. The annual weed suppression by surface mulch is due to limiting the light, moisture, and gas exchange required for weed establishment.

Teasdale et al. (2004) reported that the high levels of cover crop mulch inhibited weed germination and emergence. The annual weeds with small seeds are especially susceptible to suppression by mulches.

The soil mulched with hairy vetch reduced annual grass weeds in maize to the extent of 83 % as compared to maize grown in unmulched soil (Fig. 18.2).

A thick cover crop (hairy vetch and rye) mulch controlled the Colorado potato beetle (CPB) and early blight disease (prevented splashing of pathogen onto the leaves) in tomato. The straw mulch creates a microenvironment which increases the population of predators such as lady beetles, ground beetles, and lacewings, reduces the CPB's ability to locate potato plants, and thereby suppresses CPB activity by reducing defoliation and increasing the yields by one-third compared to plots with no mulch.

For nocturnal predators like spiders and ground beetles, the straw mulch provides humid, sheltered hiding places. The straw mulch also decreases the visual contrast between foliage and bare soil and makes it harder for flying pests like aphids and leafhoppers to "see" the crops they attack.

In apple orchard ecosystems, the use of compost as a mulch should be encouraged for sustainable management of arthropod pests (spotted tentiform leaf miner, *Phyllonorycter blancardella,* and migrating wooly apple aphid, *Eriosoma lanigerum* nymphs) by encouraging predator population (Brown and Tworkoski 2004).

18.3.3 Addition of Organic Matter

The addition of organic amendments such as FYM, green manures (single and mixed species), seed meals, oil cakes, saw dust, coir pith, crab shells, dried plant material, good quality compost, organic waste, and peats to the soil is responsible for the management of crop diseases, nematode pests (Akhtar and Alam 1993), and weeds. The use of special composts containing particular antagonistic microorganisms is the recent trend. The soil-borne diseases of potato (including common scab

Fig. 18.2 Effect of hairy vetch mulch in reducing annual grass weeds in maize

and Verticillium wilt) and plant nematodes were controlled by the application of organic amendments to the soil.

The release of toxic compounds during the decomposition of organic materials, increase in nematode antagonists, and improved crop nutrition and plant growth following soil amendment with organic matter may lead crops to tolerance from plant-parasitic nematodes.

The addition of organic amendments to the soil was found to be very effective in the management of diseases caused by pathogens such as *Fusarium* spp. (Klein et al. 2011), *Pythium* spp. (McKellar and Nelson 2003; Veeken et al. 2005), *Rhizoctonia solani* (Diab et al. 2003), and *Sclerotinia* spp. (Lumsden et al. 1983). It also helps to improve soil structure and water and nutrient holding capacity of the soil and enhanced microbial activity which contributes to biological control (Cook and Baker 1983).

The population of actinomycetes in tomato rhizosphere was higher in organically managed soil than in conventionally managed soil resulting in suppression of corky root severity (Workneh and Van Bruggen 1994). The organically managed soils contain higher population of antagonistic fungi (*Trichoderma* spp.) and bacteria (fluorescent pseudomonads) (Van Bruggen 1995).

18.3.4 Nutrient Management

In order to make plants increase the growth rate, stay more vigorous, make better use of soil water, and to produce greater numbers of roots, allowing more surface area for root absorption and nutrient uptake, and shortening susceptible stages of plant growth, maintaining balanced soil nutrition is essential. The plant nutrition indirectly affects pathogens and diseases, since mineral nutrients are essential metabolic regulators for plant growth. The population of antagonistic microorganisms in the rhizosphere can be enhanced by the quality of root exudates secreted by healthy plants. The antagonistic microorganisms produce toxins that directly affect pathogen development and survival and compete with pathogens for nutrients (Palti 1981) or produce antibiotics (Tjamos et al. 1992).

The incidence of diseases such as *Phytophthora* and *Fusarium* and outbreaks of insects like aphids and leafhoppers are increased with enhanced nitrogen fertilizer levels (Campbell 1989). In organically grown crops, the diversity of insect pests is lower due to lower nitrogen content in low-input systems (Magdoff 1992).

The infection by disease-causing organisms such as *Pythium* and *Rhizoctonia* was reduced in organic-rich soils which generally exhibited good soil fertility as well as complex food webs and beneficial organisms.

In organic fertilization regimes, the incidence of key insect pests of broccoli, cabbage aphid (*Brevicoryne brassicae*), and flea beetle (*Phyllotreta cruciferae*) was reduced due to low leaf nitrogen content, and yield increased as compared to conventional inorganic fertilization (Altieri 1994).

Kajimura (1995) reported that in organic rice fields, the density of immigrant plant hopper, *Sogatella furcifera*, and the settling rate of female adults and survival rate of immature stages of ensuing generations were significantly lower. The density of nymphs and adults in the ensuing generations decreased in organically farmed fields due to lower number of eggs laid by a female of the invading and following generations, and the percentage of brachypterous females in the next generation was also lower.

The infestation of the aphid, *Metopolophium dirhodum,* was higher in conventional winter wheat fields (where inorganic fertilizers were used) as compared to organic farms. Similarly, the European corn borer females preferred to significantly lay more eggs in chemically fertilized plants as compared to organically fertilized soils (Phelan et al. 1995).

18.3.5 Crop Diversity

The plant diversity is responsible for minimizing the economic damage from insect pests by facilitating the populations of existing or naturally occurring beneficial organisms by supplying them with appropriate habitat and alternative food sources (Altieri and Letourneau 1984). The pollen and nectar present in the flowering plants are required as alternative food for natural enemies (for normal fecundity and longevity) which are the key elements of diversified agro-ecosystems. In annual crops and orchards with rich floral undergrowth, increased parasitism occurs due to the presence of parasitoids (Leius 1967). The population of natural enemies (predators, parasitoids, and pathogens) is diverse and abundant in diverse agro-ecosystems, where fewer pesticides are used, than in monocultures or in fields routinely treated with pesticides.

18.3.5.1 Field Margins with Biodiversity

The field margins can be converted into reservoirs of natural enemies by providing habitats which supply pollen, nectar, and other resources. These natural enemies move from field margins into crops and provide biological control (more in border than in center of fields) of several crop pests:

- The parasitism of the rape pollen beetle is about 50 % higher at the edges of fields than in the center, in Germany.
- The parasitism of European corn borers in Michigan by the ichneumonid wasp, *Eriborus terebrans*, is higher at the outskirts of fields.
- The numbers and efficiency of the sugarcane weevil parasite, *Lixophaga sphenophori,* are enhanced in Hawaii at field margins due to nectar-bearing plants in field margins.
- In California, the leafhopper population in grape vineyards adjacent to French prunes is reduced by the egg parasite *Anagrus epos* (a parasitic wasp), since the prunes harbor an economically insignificant leafhopper whose eggs provide *Anagrus* with its only winter food and shelter.

18.3.5.2 Creation of Natural Enemies Corridors

The refuge strips of anise flowering plants (50–100 m) (increase the diversity and abundance of natural enemies) can be provided amid soybean crop for multiplication and spread of natural enemies into field centers (Fig. 18.3).

The bean aphid can be controlled by syrphids in sugar beet fields by providing corridors of tansy leaf (*Phacelia tanacetifolia*) every 20–30 rows. Likewise, the cabbage aphid can be managed by planting strips of buckwheat and tansy leaf in Swiss cabbage fields which increased populations of a small parasitic wasp that attacks the aphid.

Fig. 18.3 Effect of providing refuge strips of anise flowering plants amid soybean crop to attract beneficial insects that prey on insect pests (Photo by Doug Landis)

Plants from the aster, carrot, and buckwheat families are especially useful. Examples of flowering plant mixes might include species from the daisy or sunflower family (Compositae) and from the carrot family (Umbelliferae). The flowering plant families that attract natural enemies useful for biological control of crop pests are presented in Table 18.3.

Table 18.3 The flowering plant families that attract natural enemies useful for biological control of crop pests

Family	Flowering plants	Scientific name
Umbelliferae (carrot family)	Caraway	*Carum carvi*
	Coriander	*Coriandrum sativum*
	Dill	*Anethum graveolens*
	Fennel	*Foeniculum vulgare*
	Flowering ammi (Bishop's flower)	*Ammi majus*
	Queen Anne's lace (wild carrot)	*Daucus carota*
	Toothpick ammi	*Ammi visnaga*
	Wild parsnip	*Pastinaca sativa*
Compositae (Aster family)	Blanket flower	*Gaillardia* spp.
	Coneflower	*Echinacea* spp.
	Coreopsis	*Coreopsis* spp.
	Cosmos	*Cosmos* spp.
	Goldenrod	*Solidago* spp.
	Sunflower	*Helianthus* spp.
	Tansy	*Tanacetum vulgare*
	Yarrow	*Achillea* spp.
Legumes	Alfalfa	*Medicago sativa*
	Big flower vetch	*Vicia grandiflora*
	Fava bean	*Vicia faba*
	Hairy vetch	*Vicia villosa*
	Sweet clover	*Melilotus officinalis*
Brassicaceae	Basket of gold alyssum	*Aurinium saxatilis*
	Hoary alyssum	*Berteroa incana*
	Mustards	*Brassica* spp.
	Sweet alyssum	*Lobularia maritima*
	Yellow rocket	*Barbarea vulgaris*
	Wild mustard	*Brassica kaber*
Other species	Buckwheat	*Fagopyrum esculentum*
	Cinquefoil	*Potentilla* spp.

18.3.5.3 Weeds Attracting Natural Enemies

The weeds contribute to the biodiversity of agro-ecosystems, camouflage crops from colonizing pests, making the crops less apparent to their prospective attackers, and attract natural enemies by producing flowers which supply pollen and nectar to the beneficials.

The farmers should be emphasized about the weed management over weed control, since a growing appreciation for the complex relationships among crops, weeds, pests, and natural enemies exists.

18.3.5.4 Boosting Beneficials from Beetle Banks

Several grasses can provide habitats for high densities of overwintering predators (predaceous ground beetles, *Demetrias atricapillus* and *Tachyporus hypnorum*) with temperature moderation. The "beetle banks" can be established at the centers of cereal fields by sowing orchard grass on earth ridges. The establishment of beetle banks is cost-effective, since the natural enemies they harbored were quite effective in preventing cereal aphid outbreaks that pesticide savings outweighed the labor and seed costs required to establish them.

The following aspects need to be considered while improving the habitats for beneficials:

- Sowing cover crops between rows of cash crops
- Maintaining "beneficial insectary plantings" at field edges
- Providing permanent refuge strips or beetle banks for generalist predators (ground beetles)
- Harboring natural predators, parasitoids, and wildlife in perennial grasses, forbs, shrubs, and trees on field edges or in refugee strips
- Maintaining soil structure and complex food webs for ground beetles and other beneficials through conservation tillage
- Supplying root disease-suppressing microbes with life-sustaining organic matter by means of cover crops, animal manures, and composts

18.3.6 Cover Cropping

The cover cropping (growing of pure or mixed stands of legumes and cereals) is responsible for improving soil structure and water infiltration, enhancing soil fertility, preventing soil erosion, modifying the microclimate, and suppressing pests including insects, pathogens, nematodes, and weeds. The cover crops can also break disease cycles and reduce populations of bacterial and fungal diseases (Everts 2002), and parasitic nematodes (Potter et al. 1998; Vargas-Ayala et al. 2000), and increase the biological control of insect pest populations by harboring predators and parasitoids. Inclusion of mustards or other species of the Brassicaceae family as cover crops suppresses fungal disease populations through biofumigation (release of naturally occurring toxic chemicals during the degradation of glucosinolate compounds in their plant cell tissues) (Lazzeri and Manici 2001).

The leafhopper pests in grape vineyards can be managed by planting clover and other legume ground covers that attract beneficial wasps and spiders and provide them with habitat and food source (Hanna et al. 2003).

The allelopathic compounds produced by some cover crops can suppress weeds and plant-parasitic nematodes. The allelopathic effects of cover crops such as crimson clover and hairy vetch are enhanced by their incorporation in soil rather than left on the surface in no-till management (Teasdale and Daughtry 1993). The weed emergence is also delayed by cover crops, giving a competitive edge to the primary crop.

The allelopathic effects of cover crops on certain weeds are presented in Table 18.4.

The integration of cover cropping along with minimum tillage is effective for the management of cotton pests by increasing the beneficial insect populations (carabid beetles and spiders) and increased the net returns to the extent of $60 per acre.

Combining minimum tillage with cover crops and cover crop mulch creates enough biological diversity to control cotton pests. Compared to tilled fields, fields planted with winter cover crop like vetch, winter grains, or clover using conservation tillage enhanced beneficial insect populations. In fact, overall seasonal densities of certain types of carabid beetles and spiders in the "conservation" fields were a full 14 times higher than in the conventional fields. Input costs were nearly identical, but average yields in the conservation fields were about 100 pounds higher per acre than conventional yields. Moreover, net returns were $60 per acre higher in the conservation plots.

Cover crops and surface crop residues can be used to control or inhibit weeds in subsequent cash crops in three basic ways:

Table 18.4 The allelopathic effects of cover crops on certain weeds

Cover crop	Weeds suppressed	Reference
Hairy vetch	Lambs quarters, yellow foxtail, yellow nut sedge, pitted morning glory	Teasdale and Daughtry (1993), and White et al. (1989)
Crimson clover	Pitted morning glory, wild mustard, Italian ryegrass	Teasdale and Daughtry (1993), and White et al. (1989)
Cereal rye	Lambs quarters, redroot pigweed, common ragweed	Barnes and Putnam (1986), and Masiunas et al. (1995)
Wheat	Morning glory, prickly sida	Liebl and Worsham (1983)
Velvet bean	Yellow nutsedge, chickweed	Hepperly et al. (1992)
		Fujii et al. (1992)
Sorghum Sudan grass	Annual ryegrass	Forney and Foy (1985)

- By smothering and shading them so they do not receive adequate air and light
- By outcompeting them for nutrients
- By producing an effect known as allelopathy, the toxic effect on weed seed germination and seedling growth that occurs as residues of some cover crops decomposes

The cover crops listed in Table 18.5 have shown allelopathic effects on certain weeds. It should be noted that the allelopathic effects of crimson clover and hairy vetch are more apparent if the cover crop is incorporated rather than left on the surface in no-till management (Teasdale and Daughtry 1993).

The gray-seeded Mucuna and sunn hemp green manure cover crops gave the best weed control with reductions in infestation of 95 and 93 %, respectively, as compared to the check plot without green manure cover crops that had 2676 weed plants/m^2.

The use of cover crops to attract natural predators of pests by providing their habitat is a form of biological control known as habitat augmentation (Table 18.6) (Bugg and Waddington 1994).

Table 18.5 Effect of different summer cover/green manure crops (CGMC) on weed population at Choré Experimental Station (60 days after sowing)

	Weeds		Species (%)	
Treatments	Plants/m^2	Relative %	Narrow-leaved[a]	Broad-leaved[b]
Without CGMC	2676	100	19	81
Cowpea var. Tupí	782	29	1	99
Lablab (white seed)	675	25	20	80
Pigeon pea	518	19	15	85
Calopo	425	16	4	96
Jack bean	382	14	3	97
Black-seeded Mucuna	300	11	7	93
Sunn hemp	191	7	9	91
Gray-seeded Mucuna	130	5	9	91

Florentín (1997)
[a]Predominant narrow-leaved weeds: sandbur (*Cenchrus echinatus*), sour grass (*Digitaria insularis*)
[b]Predominant broad-leaved weeds: dayflower (*Commelina* sp.), painted spurge (*Euphorbia heterophylla*), hispid starburr (*Acanthospermum hispidum*)

Table 18.6 The effect of common cover crop species that attract beneficial insects

Cover crop	Beneficial insects
Buckwheat	Parasitic wasps, ladybugs, tachinid and hover flies, lacewings
Clovers	Parasitic wasps, big-eyed bugs, minute pirate bugs, ladybugs, tachinid flies, and aphid midges
Hairy vetch	Minute pirate bugs, ladybugs, predatory and parasitic wasps
Cereals	Ladybugs

The populations of some plant-parasitic nematodes are effectively reduced by warm-season legume cover crops:

- The integration of cover crops such as hairy indigo and joint vetch along with mulching with clippings of cowpea was highly effective in maintaining low populations of root-knot (*Meloidogyne incognita*) and sting (*Belonolaimus longicaudatus*) nematodes (Rhoades and Forbes 1986).
- In greenhouse and field tests, the cover crop velvet bean was effective in lowering population densities of several root-knot species present simultaneously (Rodriguez-Kabana et al. 1992).
- The use of some cover/green manure crops (pigeon pea, crotalaria, hairy indigo, velvet bean, and joint vetch) in rotation reduced root-knot nematode damage in a subsequent snap bean crop.

18.3.7 Crop Rotation

Crop rotation is an important tool in organic agriculture for the management of insect and mite pests, diseases, nematodes, and weeds. The inclusion of crops that do not support growth and reproduction of the pathogen in rotation reduces the amount of soil-borne inoculum. The crops in the same family generally should not follow one another in the field. For example, cucumbers should not follow cantaloupes, since both belong to the same family. Similarly, a cucumber–melon–squash rotation obviously invites disease problems. A minimum of at least 2 years should lapse between the crops from same families. The typical crop rotation may include Brassicaceae (cole crops), followed by Asteraceae (lettuce, cut flowers), followed by Solanaceae (tomatoes, potatoes, peppers, eggplants), and followed by Cucurbitaceae (squashes, cucumbers, melons).

In order to optimize pest regulation and soil fertility, the simplest forms of biodiversification include the legume–cereal crop rotations. The yield increases in crop rotations are mainly due to the action of interrupting insect, disease, and weed life cycles and enhancing the growth and activity of soil microorganisms, including vesicular–arbuscular mycorrhizae, which facilitate crops to utilize soil nutrients and water more efficiently. The soil-borne diseases which are difficult to control can be managed by following a 2–8-year succession of crops not susceptible to the same pathogen (Table 18.7).

Some common characteristics of the crop pests that can be controlled by rotations (Flint and Roberts 1988) include:

- Fairly narrow host range of the pest.
- Source of pest should be within the field.
- Pest's incapability to survive long periods without a living host.
- Pest should be as immobile as possible.

Table 18.7 Effect of duration of crop rotation for the management of soil-borne diseases

Vegetable crop	Disease	Years without a susceptible crop
Asparagus	Fusarium rot	8
Cabbage	Club root	7
	Blackleg	3–4
	Black rot	2–3
Muskmelon	Fusarium wilt	5
Parsnip	Root canker	2
Peas	Root rots	3–4
	Fusarium wilt	2
Pumpkin	Black rot	2
Radish	Club root	7

The spring lettuce crops will have less weed competition by following crop rotation with a fast growing, vigorous summer cover crops, such as Japanese millet, cowpea, or soybean and then planting fall broccoli.

18.3.8 Plant Breeding

Plant breeding is responsible for strengthening numerous plant defenses against pests such as insect and mite pests, diseases, nematodes, and weeds. The plant defenses may be either structural (tough leathery leaves, leaf hairs, and spines) or toxic plant chemicals that deter feeding (allelopathic chemicals that deter weed growth and leaf waxes that form barriers for pests and diseases).

The selection for a single resistant gene (monogenic or oligogenic in vertical resistance) could inadvertently eliminate other genes affecting other pests or genes that play a role in attracting the natural enemies of pests, since the focus of plant breeding for pest resistance is frequently limited to a specific pest/plant interaction. The development of new races of the pest can easily break down the resistance, since single gene is responsible for resistance. The initial amount of inoculum is reduced in vertical resistance which can delay the epidemic.

The horizontal resistance (also called polygenic resistance, minor gene resistance, or field resistance) is governed by many genes, and it is effective against many races of the pathogen. It slows down the rate at which disease increases in the field.

Our knowledge is steadily expanding on the role of plants in their own defense that can be used to breed and engineer plants whose defenses work harmoniously with natural systems. There is a need for more research as well as plant breeding programs which focus on enhancing natural defenses. More emphasis should be placed on open-pollinated crops rather than hybrids for their adaptability to local environments and greater genetic diversity.

18.4 Conclusions

The major constraints to agricultural production throughout the world include insect and mite pests, pathogens (fungal, bacterial, viral, and nematodes), and weeds. Ecologically based pest management (EBPM) promotes the interactions between cropping systems, pests, and naturally occurring biological-control agents to reduce damage associated with pests. Safety, profitability, and durability are the three fundamental goals essential for developing and implementing EBPM strategies that are long-lasting and have a positive impact on crop protection. There is a need for coordinated interdisciplinary and multidisciplinary research support to develop and implement EBPM. The following aspects need emphasis in future:

- Crop protection strategies involving ecological principles
- Interdisciplinary and multidisciplinary institutional research
- Managed ecosystems through ecological research
- Resources necessary for EBPM to be conserved

The EBPM recommends agricultural practices that enhance natural processes, supported by biological-control organisms and products, resistant plants, and narrow-spectrum pesticides.

References

Aina PO (1984) Contribution of earthworms to porosity and water infiltration in tropical soil under forest and long-term cultivation. Pedobiologia 26:131–136

Akhtar M, Alam MM (1993) Utilization of waste materials in nematode control: a review. Bioresour Technol 45:1–7

Altieri MA (1994) Biodiversity and pest management in agroecosystems. Haworth Press, New York, 185 pp

Altieri MA, Letourneau DK (1984) Vegetation diversity and insect pest outbreaks. CRC Crit Rev Plant Sci 2:131–169

Barnes JP, Putnam AR (1986) Evidence for allelopathy by residues and aqueous extracts of rye (*Secale cereale*). Weed Sci 86:384–390

Birch ANE, Begg GS, Squire GR (2011) How agro-ecological research helps to address food security issues under new IPM and pesticide reduction policies for global crop production systems. J Exp Biol 62:3251–3261

Brown MW, Tworkoski T (2004) Pest management benefits of compost mulch in apple orchards. Agric Ecosyst Environ 103:465–472

Bugg RL, Waddington C (1994) Using cover crops to manage arthropod pests of orchards – a review. Agric Ecosyst Environ 50:11–28

Campbell R (1989) Biological control of microbial plant pathogens. Cambridge University Press, Cambridge, 199 pp

Cook RJ, Baker KF (1983) The nature and practice of biological control of plant pathogens. American Phytopathological Society, St. Paul

Diab H, Hu S, Benson DM (2003) Suppression of *Rhizoctonia solani* on impatiens by enhanced microbial activity in composted swine waste amended potting mixes. Phytopathology 93:1115–1123

Everts KL (2002) Reduced fungicide applications and host resistance for managing three diseases in pumpkin grown on a no-till cover crop. Plant Dis 86:1134–1141

Flint ML, Roberts PA (1988) Using crop diversity to manage pest problems: some California examples. Am J Altern Agric 3:163–167

Florentín M (1997) Efecto residual de los abonos verdes sobre la infestación de malezas y la producción de cultivos comerciales. In: Viedma L (ed) Coord. Curso sobre Siembra Directa, MAG/DIA, PROCISUR/BID. Centro Gráfico, Encarnación, Paraguay, pp 69–84

Forney DR, Foy CL (1985) Phytotoxicity of products from rhizospheres of a sorghum-sudangrass hybrid (*Sorghum bicolor* x *Sorghum sudanese*). Weed Sci 33:597–604

Fujii Y, Shibuya T, Yasuda T (1992) Allelopathy of velvet bean: Its organic production – cover crops for organic farms 18 discrimination and identification of LDOPA as a candidate of allelopathic substances. Japan Agric Res Q 25:238–247

Gliessman SR (1998) Agroecology: ecological processes in sustainable agriculture. Ann Arbor Press, Chelsca

Hanna R, Zalom FG, Roltsch WJ (2003) Relative impact of spider predation and cover crop on population dynamics of *Erythroneura variabilis* in a raisin grape vineyard. Entomol Exp Appl 107:177–191

Hepperly P, Aguilar EH, Perez R, Diaz M, Reyes C (1992) Pigeon pea and velvet bean allelopathy. In: Rizvi SJH, Rizvi V (eds) Allelopathy: basic and applied aspects. Chapman and Hall, London, pp 357–370

Kajimura T (1995) Effect of organic rice farming on plant hoppers 4. Reproduction of the white backed plant hopper, *Sogatella furcifera* (Homoptera: Delphacidae). Res Popul Ecol 37:219–224

Klein E, Katan J, Gamliel A (2011) Soil suppressiveness to Fusarium disease following organic amendments and solarization. Plant Dis 95:1116–1123

Lal R (1975) Role of mulching techniques in tropical soil and water management. IITA Technical Bulletin No. 1. Ibadan

Lal R (1976) Soil erosion problems on an alfisol in Western Nigeria and their control. IITA Monograph No. 1. Ibadan

Lawes DA (1966) Rainfall conservation and the yields of sorghum and groundnuts in northern Nigeria. Exp Agric 2:139–146

Lazzeri L, Manici LM (2001) Allelopathic effect of glucosinolate-containing plant green manure on *Pythium* sp and total fungal population in soil. Hortscience 36:1283–1289

Leius K (1967) Influence of wild flowers on parasitism of tent caterpillar and codling moth. Can Entomol 99:444–446

Leser JF (1994) Management of cotton aphids: texas style. In: Proceedings Beltwide Cotton conference National Cotton Council, vol I. Memphis, pp 137–141

Liebl RA, Worsham AD (1983) Inhibition of pitted morning glory (*Ipomoea lacunose* L.) and certain other weed species by phytotoxic components of wheat (*Triticum aestivum* L.) straw. J Chem Ecol 9:1027–1043

Lumsden RD, Lewis JA, Millner PD (1983) Effect of composted sewage sludge on several soil-borne pathogens and diseases. Phytopathology 73:1543–1548

Magdoff FR (1992) Building soils for better crops: organic matter management. University of Nebraska Press, Lincoln, 176 pp

Masiunas JB, Weston LA, Weller SC (1995) The impact of rye cover crops on weed populations in a tomato cropping system. Weed Sci 43:318–323

McKellar ME, Nelson EB (2003) Compost induced suppression of *Pythium* damping-off is mediated by fatty-acid metabolizing seed-colonizing microbial communities. Appl Environ Microbiol 69:452–460

Mensah-Bonsu, Obeng HG (1979) Effects of cultural practices on soil erosion and maize production in the semi-deciduous rainforest-savanna transitional zones of Ghana. In: Greenland DJ, Lal R (eds) Soil physical properties and crop production in the humid tropics. Wiley, Chichester, pp 509–519

Palti J (1981) Cultural practices and infectious crop diseases. Springer, New York, 243 pp

Phelan PL, Mason JF, Stinner BR (1995) Soil fertility management and host preference by European corn borer, *Ostrinia nubilalis*, on *Zea mays*: a comparison of organic and conventional chemical farming. Agric, Ecosyst Environ 56:1–8

Potter MJ, Davies K, Rathjen AJ (1998) Suppressive impact of glucosinolates in Brassica vegetative tissues on root lesion nematode *Pratylenchus neglectus*. J Chem Ecol 24:67–80

Rhoades HL, Forbes RB (1986) Effects of fallow, cover crops, organic mulches, and fenamiphos on nematode populations, soil nutrients, and cash crop growth. Nematropica 16:141–151

Rodriguez-Kabana R, Pinochet J, Robertson DG, Weaver CF, King PS (1992) Horse bean (*Canavalia ensiformis*) and crotalaria (*Crotalaria spectabilis*) for the management of *Meloidogyne* spp. Nematropica 22:29–35

Roose EJ (1988) Soil and water conservation lessons from steep-slope farming in French-speaking countries of Africa. In: Moldenhauer WC, Hudson NW (eds) Conservation farming on steeplands. SWCS, Ankeny, pp 129–139

Teasdale JR, Daughtry CST (1993) Weed suppression by live and desiccated hairy vetch. Weed Sci 41:207–212

Teasdale JR, Abdul-Baki AA, Mills DJ, Thorpe KW (2004) Enhanced pest management with cover crop mulches. In: Bertschinger L, Anderson JD (eds) Proceedings XXVI IHC – sustainability of horticultural systems Acta Horticulturae, 638, ISHS

Thomas MB, Wratten SD, Sotherton NW (1991) Creation of "islands" habitats in farmland to manipulate populations of beneficial arthropods: predator densities and emigration. J Appl Ecol 28:906–917

Tjamos EC, Papavizas GC, Cook RJ (1992) Biological control of plant diseases: progress and challenges for the future, NATO ASI Series. Plenum Press, New York, 462 pp

van Bruggen AHC (1995) Plant-disease severity in high-input compared to reduced-input and organic farming systems. Plant Dis 79:976–984

Vargas-Ayala R, Rodriguez-Kabana R, Morgan-Jones G, McInroy JA, Kloepper JW (2000) Shifts in soil micro-flora induced by velvet bean (*Mucuna deeringiana*) in cropping systems to control root-knot nematodes. Biol Control 17:11–22

Veeken AHM, Blok WJ, Curci F, Coenen GCM, Temorshuizen A, Hamelers HVM (2005) Improving quality of composted biowaste to enhance disease suppressiveness of compost-amended, peat based potting mixes. Soil Biol Biochem 37:2131–2140

White RH, Worsham AD, Blum U (1989) Allelopathic potential of legume debris. Weed Sci 37:674–679

Workneh F, Van Bruggen AHC (1994) Microbial density, composition, and diversity in organically and conventionally managed rhizosphere soil in relation to suppression of corky root of tomatoes. Appl Soil Ecol 1:219–230

Chapter 19
Cultural Approaches

Abstract The cultural control practices frequently involve variations of standard horticultural and agricultural practices that often prove to be effective methods for pest management. The primary aim of cultural control is prevention and reduction of pest outbreaks. They create an environment which is favorable for the crop and unfavorable for the pests. Several cultural practices, such as sanitation, crop rotation, cover cropping, trap cropping, and crop diversification, water and mineral nutrition management, and alteration of planting and harvesting times, are mainly used for enhancing soil quality and plant health. Cultural practices are also less commonly known as ecological control methods, as they modify the relationships between a pest population and its natural environment. The primary advantages of cultural control tactics include simplicity and low cost. Even though cultural practices alone may not be able to give complete pest controls, they can form potential components in minimizing pest injury and should be considered in any integrated control program.

Keywords Crop rotation • Fertilizer • Irrigation • Sanitation • Soil quality • Insect pests • Plant parasitic nematodes • Pathogenic fungi • Cultural practices

19.1 Introduction

The pests (insect and mite pests, diseases, nematodes and weeds) are one of the important limiting factors in crop production and productivity. They are responsible for significant crop losses to the extent of 26– 40 % of the attainable yield every year in major food and cash crops (Oerke 2006). The present modern agricultural systems are reliant on agrochemical inputs for pest management. There is a consumer demand for pesticide-free food products and a social concern for reduced dependence on toxic chemicals to solve pest problems.

Overreliance on pesticides disrupts parasitoid and predator populations, causes outbreaks of secondary pests, exposes farmers to serious health risks, has negative consequences for the environment, causes development of pesticide resistance, leaves pesticide residues in food products, decreases effectiveness of many pesticides, adds to increased costs, and pollutes the air, soil, and water (Pimentel and

P.P. Reddy, *Sustainable Intensification of Crop Production*,
DOI 10.1007/978-981-10-2702-4_19

Levitan 1986). Hence, there is an immediate need for alternative pest management strategies to overcome the above limitations and to provide sustainable and eco-friendly production system. In this respect, several beneficial cultural practices can be utilized for pest management and to minimize the use of toxic chemicals.

19.2 Cultural Approaches

In order to avoid pest injury to crops and to reduce pest populations, the deliberate alteration of either the cropping system itself or specific crop production practices can be adopted in the agro-ecosystem. They reduce pest establishment, growth and reproduction, dispersal, and survival by making the environment less attractive. Cultural control practices are the modifications of standard horticultural and agricultural practices that changes pest's habitat or environment. These practices are also called as ecological pest management strategies because they are dependent on detailed knowledge of the bioecology of the crop–pest–natural control–environment relationships. The main advantages of cultural control practices include low cost and simple to adopt by the farmers. Prevention is better than cure is the ideology of the cultural control practices. The main objective of cultural control practices is to bring the pest populations below economic threshold levels.

The cultural practices although reduce pest damage to a considerable extent; they alone may not be in a position to provide effective pest control. Hence, they are good candidates that can be used as components in the integrated pest management.

19.2.1 Strategies

The strategies on which cultural control is based are as follows:

- Enhance its natural enemies or alters the crop's susceptibility to the pest thereby reducing pest survival on the crop
- Interfere with oviposition preferences, host plant discrimination, or location by both adults and immatures by making the habitat or crop unacceptable to pests
- Manipulate the environment in order to increase the population of natural enemies
- Multiplication of the pest on crop is prevented.
- Reduce pest populations by creating adverse biotic conditions and modifying the crop.
- Utilize knowledge of the pest's life cycle, particularly its dispersal and overwintering habits so that the crop is not made available to the pest in space and time

19.2.2 *Different Types*

There are two types of cultural control practices such as primary and secondary cultural controls.

19.2.2.1 Primary Cultural Control

These practices are utilized specifically to control insect pests. The examples of primary cultural control include:

- Avoiding certain pests (e.g., chafer beetle, stem borer, rice root weevil) by adjusting the land preparation or time of planting.
- Feeding by ants or birds is compensated by increasing the seeding rate.
- Rice caseworm control by draining the field.
- Rice whorl maggot damage prevention by transplanting older seedlings.

19.2.2.2 Secondary Cultural Control

These practices are used for general crop health but which also prevent the pest buildup. The examples of secondary cultural control are as follows:

- Avoiding buildup of certain pests (e.g., brown plant hoppers, gall midge) by split application of nitrogen fertilizer
- Controlling stem borers by turning over the stubble and plowing to prepare the soil for planting
- Preventing mole crickets, ants, and other soil pests by maintaining water in the field
- Weed management

19.2.3 *Benefits and Constraints*

19.2.3.1 Benefits

- Compatibility with other pest management practices
- Cost-effective
- Delayed development of resistance as compared to chemical control
- Does not need extra labor
- Less impact on the environment
- Low cost
- Makes the environment less favorable for pests
- Modifications of normal farming practices

- No detrimental side effects
- Preventive strategy rather than curative
- Suitable for low-value crops

19.2.3.2 Constraints

- Certain cultural controls may cause soil erosion problems.
- Community-wide adoption is required.
- Do not always provide complete economic control of pests.
- Effective for one pest but may be ineffective against a closely related species.
- Long-term planning is required.
- May increase some pests, while suppressing others.
- Need careful timing.
- Slow acting and not useful for controlling pest outbreaks.

19.3 Cultural Practices

The key cultural practices used for pest management include:

- Altering planting and harvesting dates (coordination of planting and harvest dates to avoid certain pests, e.g., sweet corn and corn earworm)
- Crop rotation – good for insects with long life cycles (e.g., corn rootworms)
- Fertilization level (a well-fertilized plant will grow and mature more quickly, thereby decreasing likelihood of pest damage)
- Fertilizer management (fertilizer applications and crop nutrition)
- Habitat manipulation/diversity
- Intercropping
- Plant spacing and density
- Planting trap crops (a trap crop is more attractive to the pest than the main crop; attracts pests; then the trap crop is destroyed or treated with insecticide)
- Sanitation (removal of dead or diseased wood, old rachises, dead leaves, mummified fruits, and crop residues)
- Selecting pest-resistant varieties (host plant resistance, including genetically modified cultivars)
- Use of mulches
- Use of quality and pest-free seeds
- Use of sterile growing media when potting
- Water management (irrigation, drainage)

19.3.1 Crop Rotation

The pest population builds up to damaging levels by growing a single crop year after year in the same field since they get sufficient time to become established. The pest's life cycle can be broken by rotating the field to a different type of nonhost crops by starving pests. The wireworms (Elateridae) and rootworms (*Diabrotica* spp.) population in maize fields can be reduced by rotation with wheat, oats, or legumes. Similarly, the population of clover root curculio (*Sitona hispidula*) that feeds exclusively on legumes can be eliminated by rotation with maize or small grains. The Verticillium wilt in tomatoes can be managed by rotating with mustard or broccoli. Crop rotation of soybean with maize is effective against soybean cyst nematode, *Heterodera glycines*.

The diversity of a pest's environment and creating discontinuity in its food supply are the main principles on which crop rotation operates. Rotations are effective against pests under the following situations:

- A life cycle of 1 year or longer.
- Pest attack on annual or biennial crops.
- Pests that are fairly immobile (cannot move easily from one field to another).
- Pests which do not survive for more than one or two seasons without their hosts.
- Presence of pest before the crop is planted.
- Soil-inhabiting pest species with relatively narrow host range and limited dispersiveness.

The green manure legume cover crops such as soybeans or cowpeas can be utilized in crop rotation as they help to reduce soil erosion and increase soil fertility. The use of pest-resistant crops is also an alternative. For the rotation to be effective, crops from different families which are nonhosts to the pest targeted should be used (Table 19.1). The most common rotations include grasses, legumes, and root crops.

Crop rotations can also act as insurance against the introduction of unknown pathogens (Bruehl 1987; Katan 2003).

Table 19.1 Different plant families and their representative crops

Family	Representative crops
Alliaceae	Onion, garlic, shallot, leek
Apiaceae	Carrot, parsley, celery, parsnip
Brassicaceae	Cabbage, cauliflower, mustard, radish, broccoli, Brussels sprout, turnip, kale, rutabaga
Cucurbitaceae	Cucumber, watermelon, pumpkin, cantaloupe, gourd, squash
Fabaceae	Bean, pea, cowpea, lablab
Solanaceae	Potato, tomato, chili, eggplant

19.3.1.1 Benefits

The benefits of crop rotations are as follows:

- Soil quality improvement.
  - Soil organic matter increase by cover crops and green manures.
  - Loosening of compacted soils with crops having different rooting depths and optimization of use of water and nutrient resources.
- Soil fertility enhancement.
  - Growing leguminous crops.
- Manage plant diseases.
  - Pest inoculum is reduced.

19.3.1.2 Duration

The duration of crop rotations depends upon the pests to be controlled. Most of the foliar pathogens can be managed by 2–3-year rotations out of the susceptible crop, while many soilborne pathogens require rotations of 4 years and above. Rotations of longer duration are required for some soilborne pathogens such as Fusarium wilt of muskmelon (5 years), clubroot of cabbage (7 years), and Fusarium rot of asparagus (8 years).

19.3.2 Intercropping

The simultaneous growing of two or more different crops (e.g., grain + legume – nitrogen-fixing plant) at the same time in the same field or in alternate rows is called as intercropping (also known as mixed cropping). This is another way of increasing environmental diversity to reduce pest populations. When host and nonhost plants are mixed together in a single planting, it lowers the pest's overall attractiveness to the environment.

Intercropping reduces pest incidence by encouraging the population of natural enemies due to:

- Affects the pest's ability to find host plants as the nonhost plant masking odors of the host plant by conferring associational resistance
- Greater distribution (temporal and spatial) of nectar and pollen resources
- Increased ground cover, particularly important for diurnal enemies
- Increased prey, offering alternative food sources when the pest species are scarce or at an appropriate time in the predator's life cycle

19.3.3 Trap Cropping

The utilization of more attractive food sources (trap crops) to divert insect pest attack away from the main crop at risk is called trap cropping. The trap crop is planted in infested land so that the pest is stimulated to attack, then the trap crop is normally destroyed before the completion of the pest's life cycle, or it will not provide all the requirements necessary for the completion of the pest's life cycle. Alternatively the trap crop may be preferentially attacked in the presence of the main crop to be protected.

The trap crops can also be grown simultaneously with more valuable main crops. The efficiency of trap crops in reducing the spread of pathogens depends on the plant and its growth habit differences including height differentials and position relative to the "protected" crop.

The trap crops are effectively utilized for the management of several insect pests. The Mexican bean beetle and the bean leaf beetle are controlled by planting (in small areas) early maturing bean varieties 10 days to 2 weeks prior to planting the main bean crop. The early maturing bean varieties attract adult beetles which can be killed by spraying with insecticides.

The cotton is intercropped with strips of alfalfa which attracts *Lygus* bugs more strongly than cotton, which are killed before they move into adjacent cotton fields by spraying with an insecticide. The same strategy has also been adopted for the management of *Lygus* bugs (Miridae) on strawberries using alfalfa as a trap crop.

The hatching of nematode eggs or the germination of resting structures of pathogens are stimulated by root exudates of trap crops, and the pathogens eventually die because the pathogens are not in a position to establish a compatible relationship with the trap crop host resulting in reduction of pathogen inoculum level in soil. The incidence of clubroot of crucifers (*Plasmodiophora brassicae*) is reduced in infested soil by the ryegrass (*Lolium* spp.) trap crop. Similarly, the incidence of powdery scab of potatoes (*Spongospora subterranea*) is reduced by the poisonous weed thorn apple (*Datura stramonium*) trap crop.

The tomato fruit borer, *Helicoverpa armigera*, is managed by using of African marigold (*Tagetes erecta*) as a trap crop. It involves integration of planting of one row of 45-day-old marigold seedlings after every 16 rows of 25-day-old tomato seedlings along with spraying of *Ha* NPV at 250 LE/ha or 4 % neem seed kernel extract (NSKE) or 4 % neem seed pulverized extract (NSPE), at 28 and 45 days after planting coinciding with peak flowering (Srinivasan et al. 1994) (Fig. 19.1). The IPM practice gave 26 % more yield and 28 % more returns than farmers' practice (Khaderkhan et al. 1998a). IPM fields recorded higher tomato fruit yields of 74.038, 14.250, and 22.293 MT/ha as compared to 45.056, 13.000, and 18.772 MT/ha in non-IPM fields, respectively (Amerika Singh et al. 2004).

The integrated pest management of cruciferous pests was developed using Indian mustard as a trap crop. It involves planting of paired rows of mustard after every 25 rows of cabbage/cauliflower (Fig. 19.2) and spraying of 4 % NSKE at primordial formation. Two more sprays of 4 % NSKE may be given at 10–15 day interval after

Fig. 19.1 Integrated management of tomato fruit borer using African marigold as a trap crop

Fig. 19.2 IPM of cabbage and cauliflower pests using Indian mustard as a trap crop

the first spray. IPM controls diamondback moth (*Plutella xylostella*), leaf webber (*Crocidolomia binotalis*), stem borer (*Hellula undalis*), aphids (*Brevicoryne brassicae, Hyadaphis erysimi*), and bug (*Bagrada cruciferarum*) (Srinivasan and Krishna Moorthy 1991). The IPM gave 60 % more yield and 152 % more returns than pure cabbage crop (Khaderkhan et al. 1998b).

19.3.4 Fertilizer Management

The plants develop strong immune systems and are more tolerant when they are grown with proper nutrition which responds aggressively under pests attack. The well-balanced plant nutrition is a good preventative measure for addressing the pest problems before they become severe.

The damage due to some pests, especially sucking pests such as certain aphids and mites can be reduced by spraying with mineral nutrients and plant hormones (e.g., from seaweed extracts). Proper fertilization enhances vigorous plant growth, shortens the susceptible stages, give better tolerance to, and opportunity to compensate for, pest damage.

The major nutrient elements such as nitrogen, phosphorous, and potassium can have direct and indirect effects on pest population. In general, the increased pest incidence, especially of sucking pests such as mites and aphids, was observed with high concentrations of nitrogen. On the other hand, the incidence of certain pests is reduced by applications of phosphorous and potassium (e.g., wireworm populations often tend to increase in low-phosphorous soils).

Some plants are predisposed to the foliage diseases when the nitrogen is outbalanced with other nutrients. The rice plants grown in high-nitrogen soils are found more susceptible to blast (*Pyricularia oryzae*) and scald (*Rhyncosporium oryzae*).

The delay in onset and lesser severity of take-all disease in barley (*Gaeumannomyces graminis*) and reduced incidence of potato scab (*Streptomyces scabies*) was observed in soils fertilized with phosphates that enhances the crop maturity, thereby young susceptible tissues are not exposed to the inoculum of obligate parasites for long periods of time.

A wide range of fungal parasites (including *Fusarium oxysporum* f. sp. *vasinfectum*), bacteria (*Corynebacterium insidiosum* and *Xanthomonas* spp.), several viruses, and various nematodes are managed by application of adequate amounts of potassium to the soil. Some of the disadvantages of excessive nitrogen applications are also compensated by the application of potassium fertilizers. The penetration of the host by the pathogen, its multiplication, survival, aggressiveness, and rate of establishment in the host are reduced by potassium. It enhances healing of wounds, reduces infection by wound parasites such as *Botrytis*, increases resistance to frost injury, delays maturity in some crops, postpones senescence, and reduces the risk of infection by facultative saprophytes.

The cell wall's resistance to penetration by facultative pathogens such as species of *Rhizoctonia, Sclerotium, Botrytis*, and *Fusarium* is enhanced by adequate application of calcium.

The host susceptibility is deceased by the applications of some mineral nutrients. The maize downy mildew is reduced by the application of zinc, the occurrence of *Cercospora* leaf spot in groundnuts is inhibited by fertilization with sulfur, and the incidence of take-all disease in wheat is reduced by the application of copper.

19.3.5 Water Management

The plants can be kept healthy, vigorous, and resistant to pest injury by proper irrigation and good fertilization. The infestations of two spotted spider mites (*Tetranychus urticae*) in tree fruits can be considerably reduced by overhead irrigation. Rice pests such as stem borers, gall midge, *Hispa*, black bugs, and plant hoppers can be suppressed by alternate draining and flooding. Other rice pests like whorl maggots, root-feeding midges, water weevils, and caseworms can be controlled by draining the field for 1–2 days.

In order to retard the disease development and to reduce the level of inoculums, the irrigation can be adopted. The incidence of apple scab is reduced by overhead sprinkler irrigation of dormant fruit trees by reducing or inactivating airborne inoculum by washing it out of the atmosphere. The short-lived ascospores of the scab pathogen are released in response to temperature changes, while the tree is dormant and cannot survive until leaves are present. The germination of powdery mildew spores is encouraged by short daily watering because the plants do not stay wet for long enough for the fungus to penetrate. The activity of microorganisms that destroy sclerotia can be encouraged by alternate wetting and drying.

The rate of water application through drip or trickle irrigation which supplies water directly to the root zones of individual plants is not sufficient to disperse plant pathogens. In addition, the drip irrigation provides islands of soil wetting areas rather than uniform wetting conditions, which probably prevents the spread of soilborne pathogens.

On the other hand, irrigation avoids the water stress which predisposes the plants to some diseases. The severity of *Phytophthora* root rot in safflower (Dunniway 1977) and on disease caused by *Macrophomina phaseolina* (Ghaffar and Erwin 1969) is increased under water stress conditions, and the disease incidence is reduced by an adequate irrigation regime.

The soilborne fungal pathogens causing southern blight and some wilt diseases are prevented by planting on raised beds, which improves the soil drainage. This practice of planting on raised beds is also helpful for the management of soilborne pathogens in vegetable crops and leguminous crops such as peanuts, soybeans, and cluster bean when grown in compact and poorly drained soils. The soil drainage can also be improved by incorporation of organic matter (cover crop, compost, etc.) into soil.

19.3.6 Altering Planting or Harvesting Dates

The simple cultural practice of altering the time of planting or harvesting can be utilized to manage pests by creating discontinuity in the pest's food supply. For example, the Hessian fly (*Mayetiola destructor*) in the Midwest can be managed by delayed planting of winter wheat until after the "fly-free date." The planting dates are estimated based on the fact that wheat is not mature enough to attract egg-laying flies, yet still has enough warm weather to grow sufficiently before winter. By planting in early spring and harvesting before larvae mature, the sweet corn can escape most injury from corn earworms (*Helicoverpa zea*).

For the management of bark beetle (Scolytidae) infestations in pine plantations, the time of harvesting is the most effective method. The good bark beetle management strategy involve that timber stands be harvested for lumber before the trees reach full maturity, since trees become more susceptible to beetle outbreaks as they grow older. Similarly, removal of the alfalfa edible foliage (by adjusting harvest time) before the larvae of alfalfa weevil (*Hypera postica*) and alfalfa caterpillar (*Colias eurytheme*) have completed development can cause high mortality of pests.

The strip harvesting system can be employed to manage *Lygus* bugs in alfalfa. As the older strips are harvested, the pest moves into the younger hay strips, instead of flying to other crops as they do if the entire field is cut at the same time. There shall be no increase in the *Lygus* bug population, since the natural enemies of the pest also move from older strip to newer strip. The *Lygus* bug adults deposit eggs when they move to the new half-grown hay strip, which is cut in about 15 days that destroys many of the eggs and newly hatched nymphs at harvest.

By altering the time for planting and early maturing crop varieties, the seedlings emerge from soil early and the crop matures faster than the time period when disease pressure is greatest due to the environment. Similarly, the cotton crop escapes from root rot by early planting in spring.

The wheat plants escape from streak mosaic virus by delayed planting. Likewise, the cotton plants escape from damping-off caused by *Pythium* sp. by late planting.

Time of planting is used largely to:

- Avoid oviposition period of particular pests, invasion by migrants, and/or disease transmission by insect vectors.
- Destroy the crop before the pest enters diapause.
- Synchronize the pest attack and presence of its natural enemies, with weather conditions that are adverse for the pest or with the abundance of an alternative host.

19.3.7 Mulching

Covering of the soil surface with any material is called mulching. Crop residues, saw dust, leaves, grass, Spanish moss, farm yard manure, cereal straw and stalks, compost, weeds, reeds, and various aquatic plants are some of the natural materials used for mulching the soil.

Besides pest control, reduction of soil moisture and soil erosion, smothering weeds, increasing water absorption and infiltration, lowering soil temperature by as much as 5 °C, enriching soil by supplying organic matter and nutrients, and protecting seedlings against the impact of rain, hail, and wind are some of the functions of mulching.

Mulching is responsible for reducing the spread of soilborne bacterial and fungal propagules by preventing water splashing (which contains pathogen propagules) on leaves, flowers, or fruits. Soil mulching with crop residues encourages competition between soil microorganisms for carbon, nitrogen, or both resulting in fewer problems with soilborne pathogens.

A number of fruit crop insect pests can be managed by mulching soil with crop residues which provide habitats for natural enemies.

Mulching of soil surface with rye residues prevent germination of small-seeded weeds (weed suppression is effective for about 30–60 days) due to the release of allelopathic chemicals. The effects of soil mulching with several cereal cover crop residues on weed population are presented in Table 19.2.

Table 19.2 The effect of soil mulching with several cereal cover crop residues on weed population

Tillage	Cover crop mulch	Weeds/sq. ft.	Weed weight lbs./sq. ft.
Conventional	None	12.0	0.22
None	None	5.0	0.14
None	Rye	0.9	0.10
None	Barley	0.8	0.09
None	Wheat	0.3	0.07

Putnam et al. (1983)

19.3.8 Sanitation

Sanitation includes an array of cultural practices aimed at removing food and shelter from pests during critical life stages, or the physical destruction of pests through cultural practices. Prevention of inoculum introduction (by whatever means, into the field, greenhouse, farm, or community) and reduction or elimination of the inoculum (which is already present at these sites) are the two main goals of sanitation (Palti 1981).

The next season's population of apple maggots (*Rhagoletis pomonella*), codling moths (*Cydia pomonella*), and plum curculio (*Conotrachelus nenuphar*) is reduced by collecting dropped fruits from beneath an apple tree. The overwintering populations of cotton pink bollworms (*Pectinophora gossypiella*), European corn borers (*Ostrinia nubilalis*), and sugarcane borers (*Diatraea saccharalis*) can be eliminated by removal of crop debris from cotton fields after harvest. Shot hole borers (*Scolytus rugulosus*) and lesser peach tree borers (*Synanthedon pictipes*) are managed by burning or shredding or pruning the wood from a peach orchard which would otherwise emerge and reinfest the orchard.

The spread of destructive diseases such as bacterial wilt of cucurbits and virus diseases of stone fruits can be managed by removal (roguing) of diseased plants as they appear. Similarly, the removal and destruction of infested, diseased, or dead wood can greatly reduce overwintering stages of pest populations (e.g., mite eggs, aphids, scale, and fire blight infested terminals) and their spread during the next year. The pruning of water sprouts, sucker growth, or foliage (during the dormant phase of the orchards) that is preferred by aphids helps to control these pests in apples.

Some of the other sanitation practices followed for the pest management include:

- Destruction of cull piles in production fields
- Keeping leaves as dry as possible
- Planting on raised beds
- Preventing water splashing
- Provision of good air circulation
- Removal or destruction of plant pathogens from farm equipment and tools
- Removing infected plant parts

Fig. 19.3 Staking of tomato plants to prevent leaves and fruits touching the soil

- Use of clean stakes for trellising (wash soil and debris from stakes and disinfect with a 10 % bleach solution)
- Preventing leaves and fruits from touching soil by staking (Fig. 19.3)

19.3.9 Planting Density and Spacing

The behavior of the insect pest in searching for food or for an oviposition site, and the relative rate of growth of the plant and its pest population per unit of time, are affected by the planting density and spacing. The maximization of yield per unit area without reducing crop quality, so that the yield advantages override the reduction of pest incidence, is the primary objective of this cultural method. The increased plant densities may be used to compensate for expected crop losses from pathogens. The pest population as well as damage can also be reduced by this method.

It is based on the following observations:

- High-density crops also increase some pest populations.
- Low-density planting attracts some insect pests because they are silhouetted against bare ground (e.g., more aphids are attracted at low plant density).
- Promotion of vigorous and strong plants (which in itself can be a good cultural control measure) can be achieved by using proper plant spacing (e.g., a good protection for corn against corn stalk borer).
- Rapid crop maturation which could also provide a means of encouraging early fruiting and harvesting of crops of indeterminate flowering plants is encouraged by plant spacing resulting in management of cotton boll weevils and pink bollworms.

- The attractiveness of the crop to host-seeking corn earworm moths is reduced by using narrow row spacing, which results in quicker closing the plant canopy over the soil.
- The effectiveness of natural enemies for the pest management is enhanced with close spacing.

The crop density and disease development are directly correlated. In general, as the crop density increases, the disease incidence also increases. For example, the serious epidemics of damping-off caused by fungal species such as *Pythium, Fusarium, Rhizoctonia,* and *Cylindrocladium* occurred by the practice of crowding of seedlings in the nursery due to the following reasons:

- Splash dispersal of inoculum is easier due to increased contact between the leaves or roots of neighboring seedlings.
- Temperatures are more uniform, the relative humidity increases, and leaves stay wet longer after rain or dew (the microenvironment within the crop is altered).
- The distance to travel by pathogens or their vectors from one host to the next is reduced.
- When the seedlings are close together, the transfer of inoculum from one plant to another is easy.
- With denser plantings, more plants may be wounded during cultivation creating more opportunities for weak pathogens to infect.

All these conditions favor the development of disease.

The corn earworm infestations can be reduced by using narrow row spacing which helps in closing the canopy over the soil quicker and reducing the attractiveness of the crop to host-seeking corn earworm moths.

On the other hand, the reduced disease incidence and increased yields result from dense stands. For example, the groundnut rosette virus is reduced by dense stands because the aphid vector is inhibited from landing (aphid vector of the groundnut rosette virus is attracted to broken ground cover rather than continuous ground cover). Hence, the disease incidence increases with weeding in this case.

The plant densities can be manipulated by varying sowing or planting rates, by pruning, thinning, trellising, fertilization, water management, staking, and harvesting plants or plant parts, depending on the types of crop. The increased plant densities may be used to compensate for expected crop losses from pathogens.

Farmers can manipulate plant densities by varying sowing or planting rates and may use increased rates to compensate for expected crop losses from pathogens. Crop density can also be manipulated by pruning, thinning, trellising, fertilization, water management, staking, and harvesting plants or plant parts, depending on the types of crops. Prevention of foliage or root contact is the principal reason for all these practices.

19.3.10 Pest-Resistant Varieties

This is a simple, cost-effective practice that is especially useful in areas where disease is already prevalent or the soils are infected. The disease resistant labeling is not a guarantee that the plant is resistant to all the invading pathogens. For example, certain tomato cultivars are resistant to fungal diseases such as *Fusarium* and *Verticillium*, while they are susceptible to foliar diseases such as gray leaf spot and leaf blight. The disease resistance may involve a single gene (vertical resistance) or many genes (horizontal resistance).

19.3.10.1 Benefits

The benefits of resistant cultivars are as follows:

- Affordable by farmers
- Practical method of pest control
- Preventative measure
- Strong line of defense

19.3.10.2 Drawbacks

The drawbacks of resistant cultivars include:

- Desirable traits may be found in susceptible varieties.
- Genetic resistance is not always available.
- Resistance may be overcome by pathogen.

19.3.11 Disease-Free Seed and Planting Stock

Many pests are seed-borne; found on the exterior or interior of seed. The seed-borne pests may be managed by using pest-free are certified seeds if possible. Exterior seed-borne pathogens can be removed with disinfectants (bleach, hydrogen peroxide, ethanol).

The selection of disease-free planting material forms a very important control measure, since many plants are vegetatively propagated. Some diseases such as black scurf of potato and red rot of sugarcane can be controlled by planting healthy planting stock in disease-free fields.

19.3.12 Disease-Escaping Varieties

Due to their inherent characteristics, some early maturing crop varieties escape the onslaught of the pathogen and resist their attack (e.g., early maturing varieties of pea and wheat escape damage due to powdery mildew, *Erysiphe polygoni*, and black stem rust, *Puccinia graminis tritici*, respectively).

19.4 Conclusions

There is an increasing interest in cultural practices for pest management in order to reduce dependence on the use of pesticides. These practices improve the soil health and reduces pest incidence in a sustainable manner. Each and every cultural practice should be reevaluated and modified if necessary as a potential tool for pest management. These practices which are simple and less costly should also be used as components in developing bio-intensive integrated pest management strategies.

References

Amerika Singh, Trivedi TP, Sardana HR, Sabir N, Krishna Moorthy PN, Pandey KK, Sengupta A, Ladu LN, Singh DK (2004) Integrated pest management in horticultural crops – a wide area approach. In: Chadha KL, Ahluwalia BS, Prasad KV, Singh SK (eds) Crop improvement and production technology of horticultural crops. Horticultural Society of India, New Delhi, pp 621–636

Bruehl GW (1987) Soil-borne plant pathogens. MacMillan, New York

Dunniway JM (1977) Predisposing effect of water stress on the severity of Phytophthora root rot in safflower. Phytopathology 67:884–889

Ghaffar A, Erwin DC (1969) Effect of soil water stress on root rot of cotton caused by *Macrophomina phaseolina*. Phytopathology 59:795–797

Katan J (2003) Role of cultural practices for the management of soilborne pathogens in intensive horticultural systems. Acta Hortic 635:11–18

Khaderkhan H, Nataraju MS, Nagaraja GN (1998a) Economics of IPM in tomato. In: Parvatha Reddy P, Krishna Kumar NK, Verghese A (eds) Advances in IPM for horticultural crops. AAPMHE, Bangalore, pp 151–152

Khaderkhan H, Nataraju MS, Nagaraja GN (1998b) Integrated pest management in cabbage: a cost benefit-constraint analysis. In: Parvatha Reddy P, Krishna Kumar NK, Verghese A (eds) Advances in IPM for horticultural crops. AAPMHE, Bangalore, pp 153–154

Oerke EC (2006) Crop losses to pests. J Agric Sci 144:31–43

Palti J (1981) Cultural practices and infectious crop diseases. Springer-Verlag, Berlin

Pimentel D, Levitan L (1986) Pesticides: amounts applied and amounts reaching pests. Bioscience 36:86–91

Putnam AR, DeFrank J, Barnes JP (1983) Exploitation of allelopathy for weed control in annual and perennial cropping systems. J Chem Ecol 9:1001–1010

Srinivasan K, Krishna Moorthy PN (1991) Indian mustard as a trap crop for management of major Lepidopterous pests on cabbage. Trop Pest Manag 37:26–32

Srinivasan K, Krishna Moorthy PN, Raviprasad TN (1994) African marigold as a trap crop for the management of the fruit borer *Helicoverpa armigera* on tomato. Int J Pest Manag 40:56–63

Chapter 20
Habitat Management

Abstract Due to high levels of disturbance, many agro-ecosystems are unfavorable environments for natural enemies of pests. The habitat management is a form of conservation biological control (an ecologically based approach) that provides resources for natural enemies, such as food and shelter, in enhancing biological control in agro-ecosystems. Maintaining different cropped and non-cropped corridors within the agricultural landscape in the form of hedgerows, corridors for natural enemies, flowering weeds to attract beneficials, pollen and nectar resources, and beetle banks provide alternative prey resources for pest predators and hosts for parasitoids, as well as shelter from adverse climatic conditions. The conservation and performance of natural enemies can be achieved through high diversity of habitats and vegetation and reduced disturbance.

Keywords Conservation of natural enemies • Habitat management • Encouraging beneficials • Pollen and nectar sources • Supplemental foods

20.1 Introduction

The natural enemies such as predators, parasitoids, and pathogens play a major role in managing pests of crop plants. They contribute an estimated sum of US$ 400 billion per year globally to the economic value to the ecosystem services (Van Lenteren 2006). In spite of this, the field of biological control has been neglected since the past five decades, giving more importance and resources to chemical pesticides, which are responsible for decreased biodiversity, acute human toxicity, pollution of groundwater, and reduced resilience. Hence, there is a public outcry for the alternative pest management strategies for the effective and eco-friendly pest management. The research on pests should include associations with other crops, host plants, and adjacent plant communities (habitats) that are very important in the life cycle of pests and their management.

In view of the recent progress in our knowledge on conservation biology, there is a need to changeover from the traditional biological control to the conservation biological control by conserving the habitats for increasing the population buildup of natural enemies (Letourneau 1998; Pickett and Bugg 1998). The use of strategies

P.P. Reddy, *Sustainable Intensification of Crop Production*,
DOI 10.1007/978-981-10-2702-4_20

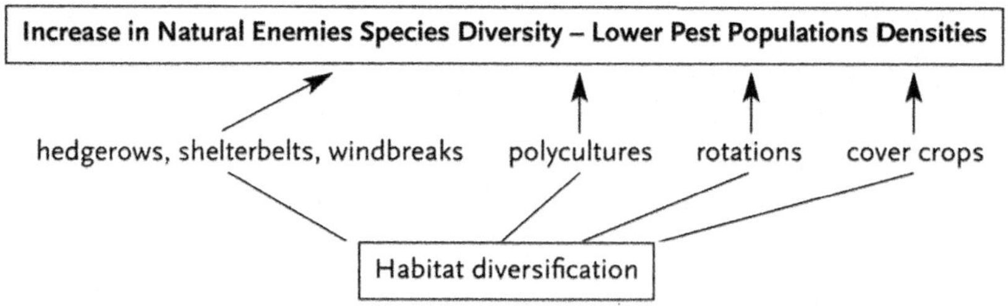

Fig. 20.1 Effect of habitat diversification through agricultural practices on natural enemy population

that increases the effectiveness of natural enemies by manipulation of their habitat to enhance their survival and/or physiological and behavioral performance is called the conservation biological control. This strategy can be applied to both indigenous (native) and exotic (i.e., introduced) natural enemies (Barbosa 1998).

20.2 Habitat Management

The effects of habitat manipulation for enhancing the population of natural enemies are beginning to receive attention in crop protection. It is generally believed that the large, interconnected habitats are favorable for biological diversity and population stability (Kruess and Tscharntke 2000). These resources provided by habitat manipulation operate to reduce the pest populations through enhancement of natural enemies. The pest suppression resulting in this manner from natural enemies is referred to as "top-down" control ("enemies hypothesis"), while suppression of herbivores within-crop habitat manipulation tactics (cover crops and green mulches) provide "bottom-up" control ("resource concentration hypothesis") (Root 1973).

The availability of overwintering sites and the ability of an insect to locate suitable habitats and food resources during its lifetime are determined by the composition of agricultural landscape (Perrin 1980).

The agricultural practices that enhance the abundance and diversity of above- and below-ground organisms should be encouraged in order to provide key ecological services to agro-ecosystems (Fig. 20.1).

Some of the methods that are employed to conserve and enhance the population of natural enemies include:

- *Enhancing natural enemy population*
  - Minimizing insecticide applications
  - Maintaining ground covers, standing crops (cover crops), and crop residues
- *Providing supplementary resources*
  - Need more alternative prey and preferred hosts

  - Need refuge sites (shelter)
  - Erecting artificial nesting structures
  - Nutritional requirements
    - Flowering plants
    - Pollen and nectar resources
    - Mixtures of protein hydrolysate, sugar, and water
- *Managing vegetation in field margins*
  - Overwintering sites for the predator
  - Perimeter trap cropping
  - Growing flowering plants
  - Flowering field margins
  - Push–pull systems
  - Beetle banks
  - Conservation headlands
  - Provision of refugee (for mating or overwintering)
- *Creating corridors for natural enemies*
  - Strip highways of habitat [with multiple flowering shrubs – sunflower families (Compositae) and from the carrot family (Umbelliferae)]
- *Weeds to attract beneficials*
  - Provide alternative resources to support beneficials
  - Understory flowering weeds

20.3 Enhancing Natural Enemy Population

The important component of IPM programs is to increase the populations of natural enemies of pests (beneficials) by maintaining ground covers, standing crops (cover crops), and crop residues and minimizing pesticide applications. The succession of plant biodiversity shall encourage beneficials by maintaining stable, varied, and natural habitats.

20.3.1 Cover Crops

Cover crops can be regulated to provide habitat for natural enemies. Providing a diverse array of vegetation is the simplest tactics to manage insect pests. They attract beneficial insects for pest control and improve pollination. By providing food and habitat, the cover crops encourage the beneficials. The flowering cover crops (clovers, legumes, buckwheat) are particularly helpful for attracting the beneficial

Fig. 20.2 *Left,* the cover crop crimson clover (Photo: KJ Graham). *Right,* the predator *Geocoris punctipes* (Photo: H Pilcher)

Fig. 20.3 The effect of cover crop crimson clover on predator population

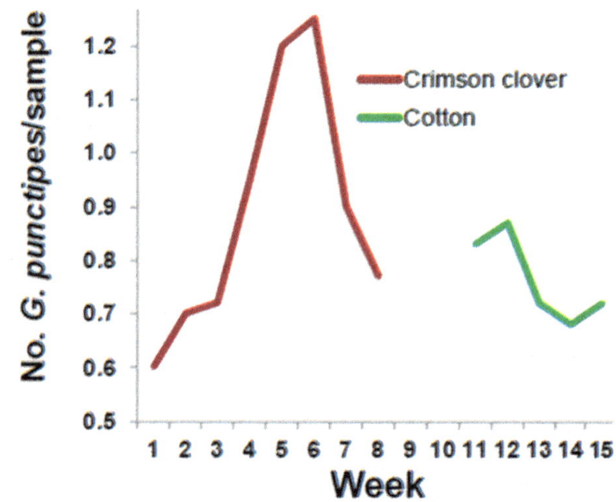

animals and insects. They are also useful to attract pollinators, which shall enhance pollination rates of crops in the field.

Cover cropping of non-Bt cotton with crimson clover (Fig. 20.2) encouraged biological control of *Heliothis* pests and reduced the need for pesticides without affecting yield. The populations of the predator (*Geocoris punctipes*) (Fig. 20.3) built up on crimson clover in the early spring and moved onto cotton and subsequently preyed on bollworm pests in early season cotton (Fig. 20.3). The crimson clover also provided nectar to predators, honey bees, and other insect pollinators.

Conservation tillage in rye cover crop promoted biological control of *Heliothis* pests in non-Bt cotton and in addition minimized the need for insecticides without sacrificing yield.

Conservation tillage in the rye cover crop conserved the habitat of fire ants and thus enhanced predation of *Heliothis* pests in cotton (Fig. 20.4). Fire ants are predators of eggs and larvae of *Heliothis* pest insects in cotton.

Fig. 20.4 The fire ant predator on *Heliothis* pests

Fig. 20.5 *Left*-Thrips damage on cotton. *Right*-The wheat cover crop in cotton for control of thrips

The wheat cover crop in cotton reduced the thrips population (Figs. 20.5 and 20.6).

The enhanced biological control of leafhoppers (*Erythroneura elegantula*) (Fig. 20.7) and thrips (*Frankliniella occidentalis*) in grape vineyards can be achieved by habitat diversification using summer understory cover crops such as buckwheat (*Fagopyrum esculentum*) and sunflower (*Helianthus annuus*) which flower during most of the growing season that support large numbers of parasitoids (*Anagrus epos*) and predators, including spiders (Nicholls et al. 2001). The vegetation of summer cover crops harbored pollen, nectar, and neutral insects which served as alternate food and hosts for important predators along with the parasitic wasp *Anagrus*.

The cover crop can be used in rotations for controlling soilborne pathogens. Incorporation of green manure *Brassica* cover crop in soil acts as biofumigant in managing the soilborne pathogens such as *Aphanomyces*, *Pythium*, *Rhizoctonia*, and *Fusarium* root rot in peas; cavity spot and *Fusarium* in carrot; *Pythium*,

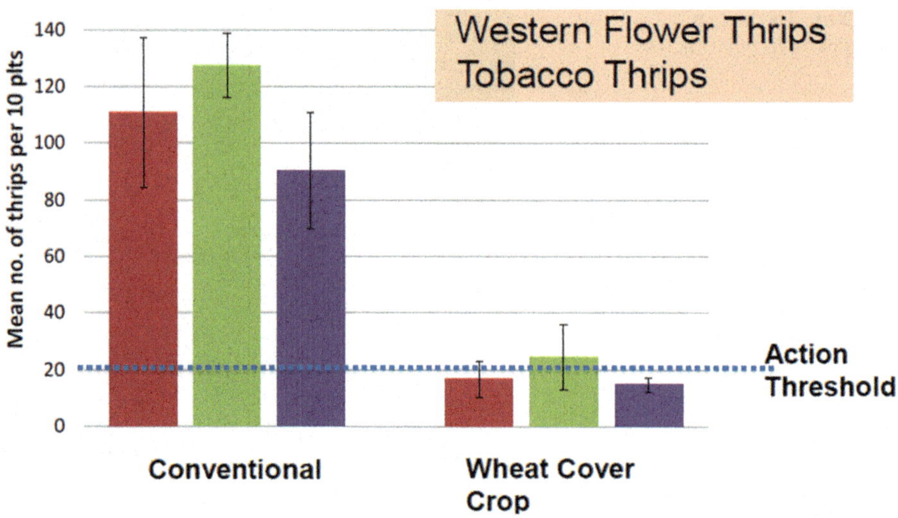

Fig. 20.6 Effect of wheat cover crop on thrips population (per 10 plants) in three management zones

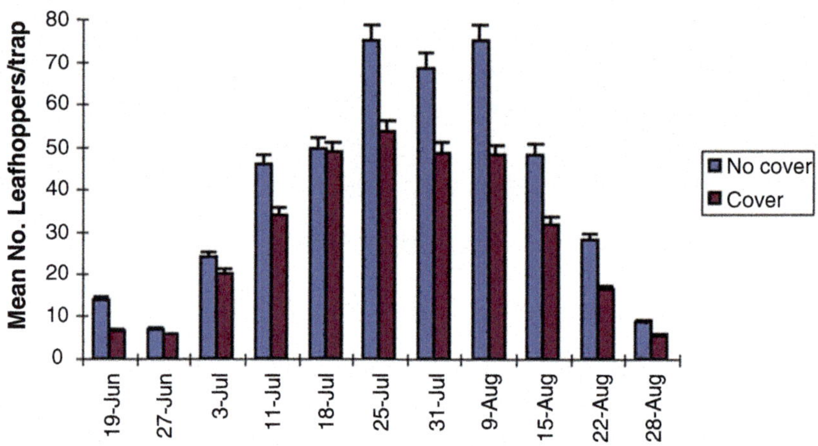

Fig. 20.7 Effect of summer cover crops on population densities of adult leafhoppers, *Erythroneura elegantula* in grape vineyards in Hopland, California

Fusarium, and *Rhizoctonia* root rots in beans; *Verticillium* wilt in potato; pink root in onion; and *Pythium* in lettuce (Sanders 2005).

The weeds in crop plants can also be managed by cover crops that release allelopathic compounds that inhibit the germination or growth of weeds and by competing for light and nutrients. Some of the cover crops which can suppress weeds include cereal grains and grasses (suppress fall and winter weeds), Buckwheat, spring cereals, Japanese millet, sorghum Sudan grass hybrids, and annual ryegrass and fall rye (smother crops). Some cover crops such as legume, cereal, and *Brassica* can reduce weed populations by releasing allelopathic compounds. The spread of cover crop residues on soil surface suppress weed seed germination by physically modifying the amount of natural light, soil temperature, and soil moisture. Mulching of soil surface with rye residues inhibit seedling growth of many annual small-seeded broadleaf weeds, such as lambs quarters and pigweed by releasing allelochemicals.

The cover crops which contain compounds toxic to plant-parasitic nematodes can be grown in a field and incorporated into the soil for preventing their population buildup.

20.3.2 Intercrops

The abundance and diversity of beneficial insects in the agro-ecosystems can be increased by intercropping with flowering species such as clover which provide a favorable habitat (Fig. 20.8) (Theunissen 1994). The natural agro-ecosystems with larger diversity tend to be more stable and less subject to variations in pest and disease populations (Altieri and Nicholls 2004). Considerable reduction in pest population and increase in the ratio of natural enemies to pests was noticed when the pear orchard was intercropped with aromatic plants as compared to orchards with only natural grass or clean tillage (Beizhou et al. 2011).

20.3.3 Minimizing Pesticide Applications

Along with pests, pesticides (neonicotinoids, carbamates, organophosphates, and pyrethroids) are responsible for killing beneficial natural enemies such as parasitoids and predators. Hence, the pesticides should be used carefully only when the

Fig. 20.8 Effect of intercropping on increase in natural enemy abundance and a potential decrease in insect pest abundance

pest population exceeds established economic thresholds. The use of selective pesticides rather than broad-spectrum and persistent insecticides and minimizing their use help in preserving the natural enemy populations in the agro-ecosystems (El-Wakeil et al. 2013). The nonpersistent pesticides do less harm to natural enemies that migrate in after the application and can provide better long-term control of the pest.

Instead of chemical pesticides, eco-friendly biopesticides have been recognized as an important alternative for pest management (Ravensberg 2015). The microbial insecticides such as strains of the bacterium *Bacillus thuringiensis* (*Bt*) are toxic only to black fly larvae, caterpillars, and certain beetles that consume treated foliage.

The use of botanical pesticides (neem products, ryania, pyrethrums, or insecticidal soaps) that function as stomach poisons is relatively less harmful to natural enemies and provides effective protection against targeted arthropod pests (Birch et al. 2011). They act as stomach poisons and do not harm parasitoids or predators directly because these compounds are toxic only when consumed along with treated foliage. The natural pyrethrums or insecticidal soaps kill fewer beneficial as they breakdown quickly on treated surface.

20.4 Providing Supplementary Resources

Providing alternative resources such as food for adult natural enemies, alternative prey or hosts, and shelter from adverse conditions increase the abundance and diversity of natural enemies appears to be a simple approach for enhancing biological control. The resources that can be utilized to attract natural enemies include growing plants that provide nectar or pollen resources, using "banker plants" to provide supplementary prey, and growing plants that produce volatiles which attract natural enemies.

20.4.1 Alternative Prey and Preferred Hosts

The survival and reproduction of natural enemies can be increased by maintaining considerable populations of alternative prey to fluctuate below damaging threshold levels. This can be achieved by planting the hosts for alternative prey around fields or even as strips within fields. In order to augment the populations of natural enemies on ornamental and food crops, the banker plant systems (or open rearing systems for non-pest herbivore to serve as alternative prey for beneficials) can be used (Van Driesche et al. 2008). For natural enemies to survive and reproduce for long periods even when no pests are present in the crop, banker plant systems are used to provide alternative food or hosts (Frank 2010). The effectiveness of the general predators that consume diamondback moth larvae depends on the relative abundance of aphids on cabbage. Likewise, when the preferred prey, western flower thrips, is scarce in field, anthocorid bugs benefit from alternative prey.

Another tactic that can be employed to augment the population of natural enemies is to provide their preferred host. The two parasitic wasps *Trichogramma evanescens* and *Apanteles rubecula* prefer cabbage butterflies as their preferred host. The population of cabbage butterflies can be increased to tenfold in spring by continual releases of fertile females, which helps to build up the population of two parasitic wasps all along the season at effective levels.

20.4.2 Refugee Sites (Shelters)

The provision of appropriate shelters from environmental hazards, such as periods of heat, cold, rain, or pesticide application, is required for natural enemies to promote their survival (foraging, resting, overwintering, or nesting). The shelters that can provide protection to beneficials from windy areas include hedgerows, windbreaks, or shelterbelts. Many natural enemies use flowers' pollen and nectar from windbreaks and hedgerows as their food (Beane and Bugg 1998). The herbaceous and woody plants in crop fields often serve as overwintering and resting shelters for various natural enemies (Beane and Bugg 1998).

20.4.3 Artificial Nesting Structures

The predation of cotton leaf worms and tobacco hornworms from the red wasp (*Polistes annularis*) increased by erecting artificial nesting structures. The biological control in crop fields can be enhanced by building houses to lacewings which provide shelter during harsh weather conditions (McEwen and Sengonca 2001). Some predatory mites which feed on pest mites use domatia (naturally occurring shelters from certain plants) as shelters. The population of predatory mites is higher in plant leaves with greater domatia structures (Loughner et al. 2008). The lower population of leaf-feeding mites and foliar fungal pathogens (powdery mildew) in certain grape cultivars is due to the presence of leaf domatia (English-Loeb et al. 2005).

20.4.4 Nutritional Requirements

20.4.4.1 Flowering Plants

The flowering plants such as yarrow, dill, angelica, anthemia, aster, caraway, cosmos, fennel, hyssop, lovage, mint, showy goldenrod, sweet goldenrod and tansy provide habitat for beneficial insects and spiders. The plants belonging to the family

Table 20.1 The flowering plants that attract beneficial insects

Plant	Beneficial insects attracted
Berseem clover	Big-eyed bugs
Black locust	Lady beetles
California lilac	Hoverflies
Caraway	Lacewings, hoverflies, insidious flower bugs, spiders, parasitic wasps
Common knotweed	Big-eyed bugs, hoverflies, parasitic wasps, soft-winged flower beetles
Cowpea	Parasitic wasps
Crimson clover	Minute pirate bugs, big-eyed bugs, lady beetles
Flowering buckwheat	Hoverflies, minute pirate bugs, predatory wasps, tachinid flies, lacewings, lady beetles
Hairy vetch	Lady beetles, minute pirate bugs, predatory wasps
Queen Anne's lace	Lacewings, predatory wasps, minute pirate bugs, tachinid flies
Soap bark tree	Hoverflies, green lacewings, brown lacewings
Spearmint	Predatory wasps
Sweet alyssum	Tachinid flies, hoverflies, chalcids
Subterranean clover	Big-eyed bugs
Sweet fennel	Parasitic wasps, predatory wasps
Toothpick ammi	Hoverflies, minute pirate bugs, soft-winged flower beetles, tachinid flies
Tansy	Parasitic wasps, lady beetles, insidious flower bugs, lacewings
White sweet clover	Tachinid flies, bees, predatory wasps
Wild buckwheat	Hoverflies, minute pirate bugs, tachinid flies
Yarrow	Lady beetles, parasitic wasps, bees

Compositae are very effective in attracting natural enemies. Among the composites, by far the highest numbers of insects (including beneficials) are attracted to tansy, a vigorous spreading perennial. The access to flowering resources is required to the natural enemies that belong to a broad range of orders such as Hymenoptera, Diptera, Coleoptera, Heteroptera, Neuroptera, Araneae, and Acari (Wäckers et al. 2005). The efficacy of biological control of insect pests is dependent on the availability of flowering resources essential for natural enemies (van Rijn and Sabelis 2005). Some of the flowering plants like berseem clover, cowpea, crimson clover, buckwheat, hairy vetch, and white clover can be used as cover crops.

Other flowering plants which attract beneficials are presented in Table 20.1.

20.4.4.2 Pollen and Nectar Resources

The non-prey food, such as pollen and nectar, is required by most of the natural enemies as part of their diet. Significant increase in the activity, longevity, and fecundity of the predators and parasitoids is dependent on the availability of pollen and nectar resources (Wäckers et al. 2008; Hogg et al. 2011). The best nectar sources suitable for small parasitoids and also larger predators include plants having very small flowers (Fig. 20.9). Adult hoverflies, which are aphid predators, use pollen necessary for eggs to mature and produce their young. In order to attract and nourish natural enemies, readymade seed mixes of flowering plants are available.

There is a need to select plants which provide non-prey food to the beneficials without providing resources to pest insects. For example, lacy phacelia (*Phacelia tanacetifolia*) and *Nasturtium* (*Tropaeolum majus*) provided resources for natural enemies without enhancing moth pests (Baggen et al. 1999). Similarly, three types of flowering resource plants sown under the grapevines gave higher egg parasitism of light brown apple moth (*Epiphyas postvittana*) (Begum et al. 2006). The integration of non-prey food along with attraction of natural enemies (Fig. 20.10) in an attract-and-reward approach has potential for enhancing biological control synergy (Simpson et al. 2011a, b).

Fig. 20.9 Nectar-bearing plants with small attractive flowers – wild carrot, dill, and goldenrod

Fig. 20.10 Integration of non-prey food and attraction of natural enemies into the crop to enhance biological control

20.4.4.3 Artificial Food Supplements

In order to attract or aggregate lady beetles, syrphid flies, and adult lacewings, artificial food supplements containing yeast, whey proteins (casein hydrolysate), and sugars may be required for egg production in the absence of abundant prey. Some of the artificial foods available from suppliers of natural enemies include Wheast, Bug Pro, and Bug Chow.

20.5 Managing Vegetation in Field Margins

The field margins can be manipulated into habitats for natural enemy reservoirs by providing overwintering sites, with pollen, nectar, and additional food resources. The predators and parasitoids move from field corridors into crops and provide effective biological control of crop pests which is usually more effective in crop rows near wild vegetation than in field centers due to:

- European corn borers at the margins of fields are more prone to parasitism by the ichneumonid wasp, *Eriborus terebrans*.
- Parasitism of rape pollen beetle was about 50 % at the edge of the fields, while the parasitism decreased to 20 % at the center (Thies and Tscharntke 1999).
- Egg parasitic wasp *Anagrus epos* reduces leafhopper populations in grape vineyards adjacent to French prunes which harbor an economically insignificant leafhopper whose eggs provide *Anagrus* with its only winter food and shelter.
- Nectar-bearing plants in field edges improve the numbers and efficiency of the sugar cane weevil parasite, *Lixophaga sphenophori*.

Some of the practices that can be used to increase the natural enemies by managing field margins are as follows:

- Beetle banks
- Hedgerows
- Perimeter trap cropping

20.5.1 Beetle Banks

In order to provide overwintering sites for natural enemies, the raised beetle banks are normally established in the margins or middle of bigger fields so that beneficials can reach center of the field more quickly (Thomas et al. 1992) (Fig. 20.11). The installation of beetle banks can be ideally located in less productive land to negate the loss of productive land.

The grasses such as *Lolium, Dactylis, Agrostis,* and *Holcus* can be sown on raised beetle banks. The pollen- and nectar-bearing flowering plants can also be grown on

Fig. 20.11 The beetle bank

these banks to attract beneficials belonging to Hymenoptera and Syrphidae. The polyphagous herbivore predators can reach populations to the extent of 1500/m^2 (Thomas et al. 1992).

The beetle banks provide not only the predatory beneficial fauna to the farmer but also the potential additional economic advantage of less pesticide required in the main cropped area and increased yield due to pest management. For example, the aphid population kept below damage threshold by enhanced natural enemy populations through installation of beetle banks can save a sum of £300 per year in labor and insecticide costs and prevent an aphid-induced yield loss of 5 % to save £660 for a 20 ha field (Thomas et al. 1991).

20.5.2 Hedgerows

The hedgerows provide habitats for overwintering natural enemies which can give more effective biological control of crop pests. In maize crop embedded in hedgerows, the parasitism of the armyworm, *Pseudaletia unipuncta*, was significantly higher as compared to maize fields without hedgerows. Similarly, the parasitoid *Eriborus terebrans* gave higher parasitism of *Ostrinia nubilalis* larvae in edges of maize fields with hedgerows than without them (Marino and Landis 1996).

The "mini hedgerows" are raised narrow strips at field margins planted with suitable vegetation (grasses or flowering plants as adult food sources for beneficials). The suitable grasses for planting on mini hedgerows include *Holcus lanatus* and *Dactylis glomerata* to increase predators and parasitoids in whole field (Thomas 1990). These grasses were found to provide overwintering habitats for cereal aphid predators such as caradids, *Bembidion lampros*; rove beetles, *Tachyporus* spp.; and several species of linyphiid spiders; and increased their population densities (Chiverton 1989).

20.5.3 Perimeter Trap Cropping

Planting of a trap crops specifically attractive to target pests, completely encircling the main crop, is called perimeter trap cropping. This type of planting system repels pests away from the main crop, significantly reduce pesticide applications, and preserve the beneficials in the main crop. It functions by aggregating the pests in the border area which helps in the multiplication of natural enemies, killing the pests.

Planting of trap crops such as cowpeas and snap beans repels Mexican bean beetles and stink bugs away from soybeans. The unsprayed trap crop attracts 92 % of pepper maggot population resulting in effective protection (98–100 %) of sweet bell peppers inside. The pepper maggot on the trap crop can be managed by spraying pesticide. The attack of diamondback moth larvae was prevented on cabbage by using perimeter trap cropping with collard that reduced insecticide applications by 56 %. Similarly, the damage of flea beetles was controlled by planting mustard (cv. "southern giant") around fields of cruciferous crops (cabbage, cauliflower, and broccoli).

The diamondback moth (DBM) on cabbage can be managed by planting two rows of collards trap crop around the cabbage that attracts and concentrates the pest (Fig. 20.12). The DBM on collards is controlled by the naturally occurring parasitic wasp, *Diadegma insulare*. This system reduced the pesticide use by 56 % and increased net savings to $117–$156/ha.

20.5.4 Provision of Refugee

Conserving non-cropped areas like hedgerows in field margins, leaving part of crop areas relatively unmanaged (perennial field margins), and using cover crops (Altieri and Letourneau 1982; Thomas et al. 1991) provide refugee for beneficials. The refugee provides habitats for natural enemies for their survival, persistence, and overwintering.

The grape vineyards located within 5.6 km of an established blackberry refuge benefitted immediately from the immigrant parasitic wasps (*Anagrus epos,* an egg

Fig. 20.12 Perimeter trap cropping of cabbage with collards

parasitoid) for the management of grape leafhopper, *Erythroneura elegantula*. The wasp overwinters in the eggs of the blackberry leafhopper, *Dikrella californica*, and population of the parasitoid increases enormously on the eggs by late March and early April. There is a widespread dispersal of newly produced *Anagrus* wasps that move to grape vineyards to manage leafhoppers (Doutt et al. 1966).

20.6 Corridors for Natural Enemies

20.6.1 Strip Highways of Habitat

The highways of habitat to predators and parasitoids can be provided by planting diverse flowering plants into strips that cut across crop fields every 50–100 m. The natural enemies use this habitat for multiplication and dispersal into field centers. The bean aphids (*Brevicoryne brassicae*) on wheat, sugar beets, and cabbage can be managed by planting tansy leaf (*Phacelia tanacetifolia*) every twenty to thirty rows to enhance the population of syrphid predators. Likewise, the cabbage aphids (*Brevicoryne brassicae*) can be controlled by planting strips of tansy leaf and buckwheat in Swiss cabbage fields which enhanced small parasitic wasp populations (White et al. 1995). Similarly, the aphids on lettuce and cruciferous crop fields can be managed by planting strips of *Alyssum* every 50–100 m in order to attract syrphid flies (Fig. 20.13).

Fig. 20.13 Effect of *Alyssum* corridor for the management of aphids on lettuce by attracting syrphid flies (Photo by Miguel Altieri, Univ. of Calif.)

20.7 Weeds to Attract Beneficials

20.7.1 Providing Alternative Resources

The flowering weed plants which contribute to the biodiversity of agro-ecosystems are the best suited to attract natural enemies, since they supply essential shelter and food resources. The likelihood of certain pest outbreaks appearing in weed-diversified crop fields are less likely than in weed-free cropping systems because weeds camouflage crops from colonizing pests, making the crops less apparent to their potential pests. Hence, the emphasis should be laid on weed management than weed control.

Plowing the land at different times of the year modifies the weed species composition of surrounding crop fields (Altieri and Whitcomb 1979). The diversity and numbers of predaceous insects increased with the increase in diversity and numbers of certain weeds. The plots plowed in December harbored most abundant coccinellid predators because of increased abundance of Mexican tea *(Chenopodium ambrosioides)* and goldenrod (*Solidago* sp.), which in turn provided suitable food (aphids and other insect pests) and habitat for predators (Altieri and Whitcomb 1979).

20.7.2 Understory Cover Cropping

The presence of flowering weed undergrowth that bloom most of the growing season increases the biological control of insect pests under perennial orchard systems. Maintaining floral diversity throughout the growing season in California vineyards, in the form of summer cover crops such as buckwheat (*Fagopyrum esculentum*) and sunflower (*Helianthus annuus*), supports large numbers of predators and parasitoids, thereby favoring enhanced biological control of leafhoppers and thrips in grape vineyards (Nicholls et al. 2001).

20.8 Conclusions

Maintaining different cropped and non-cropped habitats at the farm level has a great potential to conserve and sustain many natural enemies that also contribute to a rapid colonization of new habitats for biological control of crop pests. In order to qualitatively (number of species) and quantitatively (abundance of individuals) maintain and conserve natural enemies at the farm level, different habitat types should be provided to supply adequate resources and conditions for sustainable pest management.

References

Altieri MA, Letourneau DK (1982) Vegetation management and biological control in agroecosystems. Crop Prot 1:405–430

Altieri MA, Nicholls CI (2004) Biodiversity and pest management in agroecosystems, 2nd edn. Food Products Press, New York

Altieri MA, Whitcomb WH (1979) Manipulation of insect patterns through seasonal disturbance of weed communities. Prot Ecol 1:185–202

Baggen LR, Gurr GM, Meats A (1999) Flowers in tri-trophic systems: mechanisms allowing selective exploitation by insect natural enemies for conservation biological control. Entomol Exp Appl 91:155–161

Barbosa P (ed) (1998) Conservation biological control. Academic Press, San Diego/London

Beane KA, Bugg RL (1998) Natural and artificial shelter to enhance arthropod biological control agents. In: Pickett CH, Bugg RL (eds) Enhancing biological control: habitat management to promote natural enemies of agricultural pests. University of California, Berkeley, pp 239–254

Begum M, Gurr GM, Wratten SD, Hedburg PR, Nichol HI (2006) Using selective food plants to maximize biological control of vineyard pests. J Appl Ecol 43:547–554

Beizhou S, Jie Z, Jinghui H, Hongying W, Yun K, Yuncong Y (2011) Temporal dynamics of the arthropod community in pear orchards intercropped with aromatic plants. Pest Manag Sci 67(9):1107–1114

Birch ANE, Begg GS, Squire GR (2011) How agro-ecological research helps to address food security issues under new IPM and pesticide reduction policies for global crop production systems. J Exp Biol 62:3251–3261

Chiverton PA (1989) The creation of within-field overwintering sites for natural enemies of cereal aphids. Proceedings of the 1989 brighton crop protection conference weeds, vol 3, pp 1093–1096

Doutt RL, Nakata J, Skinner FE (1966) Dispersal of grape leafhopper parasites from blackberry refuge. Calif Agric 20(10):14–15

El-Wakeil N, Shalaby S, Abdou G, Sallam A (2013) Pesticide-residue relationship and its adverse effects on occupational workers. In: Stanislav T (ed) Insecticides – development of safer and more effective technologies, pp 57–81. DOI: 10.5772/54338. InTech

English-Loeb G, Norton AP, Gadoury D, Seem R, Wilcox W (2005) Tri-trophic interactions among grapevines, a fungal pathogen, and a mycophagous mite. Ecol Appl 15:1679–1688

Frank SD (2010) Biological control of arthropod pests using banker plant systems: past progress and future directions. Biol Control 52:8–16

Hogg BN, Bugg RL, Daane KM (2011) Attractiveness of common insectary and harvestable floral resources to beneficial insects. Biol Control 56:76–84

Kruess A, Tscharntke T (2000) Effects of habitat fragmentation on plant-insect communities. In: Ekbom B, Irwin ME, Robert Y (eds) Interchanges of insects between agricultural and surrounding landscapes. Kluwer Academic Publishers, Dordrecht, pp 53–70

Letourneau DK (1998) Conservation biology: lessons for conserving natural enemies. In: Barbosa P (ed) Conservation biological control. Academic Press, San Diego/London, pp 9–38

Loughner R, Goldman K, Loeb G, Nyrop J (2008) Influence of leaf trichomes on predatory mite (*Typhlodromus pyri*) abundance in grape varieties. Exp Appl Acarol 45:111–122

Marino PC, Landis DL (1996) Effect of landscape structure on parasitoid diversity and parasitism in agroecosystems. Ecol Appl 6:276–284

McEwen PK, Sengonca C (2001) Artificial overwintering chambers for *Chrysoperla carnea* and their application in pest control. In: McEwen PK, New TR (eds) Lacewings in the crop environment. Cambridge University Press, Cambridge, pp 487–496

Nicholls CI, Parrella M, Altieri MA (2001) The effects of a vegetational corridor on the abundance and dispersal of insect biodiversity within a northern California organic vineyard. Landsc Ecol 16(2):133–146

Pickett CH, Bugg RL (eds) (1998) Enhancing biological control: habitat management to promote natural enemies of agricultural pests. University of California Press, Berkeley
Perrin RM (1980) The role of environmental diversity in crop protection. Prot Ecol 2:77–114
Ravensberg WJ (2015) Commercialization of microbes: present situation and future prospects. In: Lugtenberg B (ed) Principles of plant-microbe interactions. Springer Publishing, Cham, pp 309–317
Root RB (1973) Organization of a plant-arthropod association in simple and diverse habitats: the fauna of collards (*Brassica oleracea*). Ecol Monogr 43:95–124
Sanders D (2005) Growers guidelines. American Vegetable Grower, October 2005. University of California SAREP Online Cover Crop Database: Mustards. http://www.sarep.ucdavis.edu
Simpson M, Gurr GM, Simmons AT, Wratten SD, James DG, Leeson G, Nicol HI, Orre GUS (2011a) Field evaluation of the 'attract and reward' biological control approach in vineyards. Ann Appl Biol 159:69–78
Simpson M, Gurr GM, Simmons AT, Wratten SD, James DG, Leeson G, Nicol HI, Orre GUS (2011b) Attract and reward: combining chemical ecology and habitat manipulation to enhance biological control in field crops. J Appl Ecol 48:580–590
Theunissen J (1994) Intercropping in field vegetable crops: pest management by agrosystem diversification – an overview. Pestic Sci 42:65–68
Thies C, Tscharntke T (1999) Landscape structure and biological control in agroecosystems. Science 285:893–895
Thomas MB (1990) The role of man-made grassy habitats in enhancing carabid populations in arable land. In: Stork NE (ed) The role of ground beetles in ecological and environmental studies. Intercept Ltd, Andover, pp 77–85
Thomas MB, Wratten SD, Sotherton NW (1991) Creation of 'island' habitats in farmland to manipulate populations of beneficial arthropods: predator densities and emigration. J Appl Ecol 28:906–917
Thomas MB, Wratten SD, Sotherton NW (1992) Creation of 'island' habitats in farmland to manipulate populations of beneficial arthropods: predator densities and species composition. J Appl Ecol 29:524–531
Van Driesche RG, Lyon S, Sanderson JP, Bennett KC, Stanek EJ III, Zhang R (2008) Greenhouse trials of *Aphidius colemani* (Hymenoptera: Braconidae) banker plants for control of aphids (Hemiptera: Aphididae) in greenhouse spring floral crops. Fla Entomol 91:583–591
Van Lenteren J (2006) Ecosystem services to biological control of pests: why are they ignored? Proc Neth Entomol Soc Meet 17:103–111
van Rijn PCJ, Sabelis MW (2005) The impact of plant-provided food on herbivore–carnivore dynamics. In: Wäckers FL, van Rijn PCJ, Bruin J (eds) Plant-provided food for carnivorous insects: a protective mutualism and its applications. Cambridge University Press, Cambridge, pp 223–266
Wäckers FL, Rijn PCJ, Bruin J (2005) Plant-provided food for carnivorous insects: a protective mutualism and its applications. Cambridge Univsersity Press, Cambridge
Wäckers FL, van Rijn PCJ, Heimpel GE (2008) Honeydew as a food source for natural enemies: making the best of a bad meal? Biol Control 45:176–184
White AJ, Wratten SD, Berry NA, Weighmann U (1995) Habitat manipulation to enhance biological control of Brassica pests by hover flies (Diptera: Syrphidae). J Econ Entomol 88:1171–1176

Chapter 21
Push–Pull Strategy

Abstract The "push–pull strategy" is based on a novel cropping system approach involving behavioral manipulation of insect pests and their natural enemies for integrated pest and weed management in farming systems. This technology works through integration of stimuli that act to make the main crop unattractive or unsuitable to the pests, the intercrop repels the pest (push) while luring them toward the border crop (pull) from where the pests are subsequently concentrated and managed by pesticides or natural enemies. The components of push and pull strategy are generally nontoxic, reduce the use of insecticides, and provide high-value animal fodder, facilitating milk production and diversifying farmers' income resources. This strategy is very useful for smallholder farmers who practice mixed cropping systems. It effectively manages insect pests and weeds, enhances main crop yields, and is economical as it uses locally available plants and non-expensive external inputs. The push–pull strategies for the management of stem borers in maize and sorghum, boll worms in cotton, pea leaf weevil in beans, Colorado potato beetle in potatoes, pollen beetle in oilseed rape maggot on onions, and thrips on chrysanthemums have been discussed.

Keywords Attractant • Repellent • Semiochemicals • Behavioral manipulation • Stimulo-deterrent diversionary strategy • Integrated pest and weed management • Push–pull

21.1 Introduction

The novel "push–pull" strategy (also called as "stimulo-deterrent diversion") uses repellent intercrop and attractive trap plant, deployed in tandem, to manipulate the distribution and abundance of insect pests and/or natural enemies. By using a repellent intercrop, the insect pests are repelled away from the main crop (push) and are attracted (pull) simultaneously by a trap crop to adjacent/border areas where they are collected and controlled. The feature of this strategy is that both the trap and the repellent plants are of economic importance.

Pyke et al. (1987) from Australia first conceived this "push–pull" strategy to reduce reliance on insecticides for the management of boll worm, *Helicoverpa* spp.,

P.P. Reddy, *Sustainable Intensification of Crop Production*,
DOI 10.1007/978-981-10-2702-4_21

Fig. 21.1 Stem borer damage on maize

in cotton to which the moths were becoming resistant. Miller and Cowles (1990) termed this strategy as "stimulo-deterrent diversion" and utilized it for the control of the onion fly, *Delia antiqua*, as an alternative to chemical control.

The "push–pull" strategy is being widely adopted by resource-poor farmers in Africa for the successful management of stem borers on cereal crops resulting in significant impact on increased food production and productivity in the region. As part of IPM strategies for the future, this strategy is likely to address the global demand for reduction of toxic chemicals in the environment. This technology is economical (it is based on locally available plants, not expensive external inputs), fits well with traditional mixed cropping systems, and is suitable for smallholder farmers for effectively solving the major pest problems of crop plants.

21.2 Management of Insect Pests

The "push–pull" strategy can be adopted for the management of several insect pests in subsistence farming (stem borers in maize and sorghum), in intensive arable agriculture (boll worms in cotton, pea leaf weevil in beans, Colorado potato beetle in potatoes, and pollen beetle in oilseed rape), and in horticulture (maggot on onions, and thrips on chrysanthemums).

21.2.1 Stem Borers in Maize and Sorghum

Stem borers such as *Chilo partellus* (Fig. 21.1), *Eldana saccharina*, *Busseola fusca*, and *Sesamia calamistis* are important limiting factors in the production of maize and sorghum and are responsible for yield losses to the extent of 30–40 % (Khan

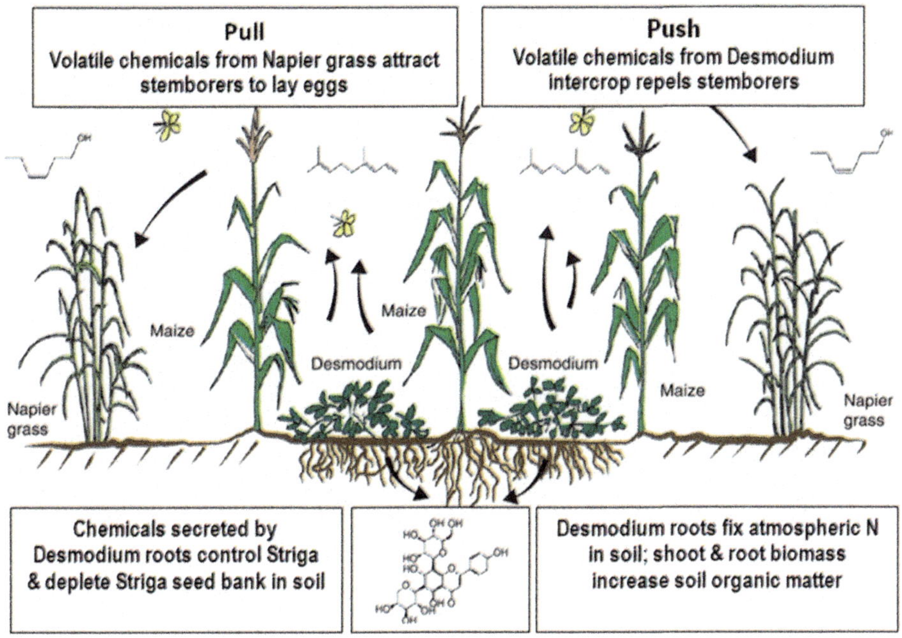

Fig. 21.2 A diagrammatic presentation on how push–pull strategy works for the management of cereal stem borers

and Pickett 2004). These pests are difficult to control as the immature stages of the pests are present inside the stem of host crops. The chemical control of stem borers is not economical, since maize and sorghum are low-value crops cultivated by resource-poor, smallholder farmers. Hence, there is an urgent need to develop improved management strategies involving low inputs.

The scientists at the International Center of Insect Physiology and Ecology (ICIPE), Kenya, and Rothamsted Experiment Station, the United Kingdom, have developed a push–pull strategy for the management of stem borers and *Striga* weed on maize and sorghum. They used repellent intercrops such as molasses grass (*Melinis minutiflora*), leguminous silver leaf Desmodium (*Desmodium uncinatum*), and green leaf Desmodium (*Desmodium intortum*) for driving the stem borers away ("push") and attractive trap crops like Napier grass (*Pennisetum purpureum*) and Sudan grass (*Sorghum vulgare sudanense*) in the crop border to attract female moths ("pull") to lay eggs (Figs. 21.2 and 21.3). When the eggs hatch, 80 % die as the Napier grass also produces a sticky substance that traps the larvae (restricts larval development, causing few to survive) (Fig. 21.4) (Khan et al. 2006b). Besides controlling stem borers, Molasses grass also increased stem borer parasitism by a natural enemy, *Cotesia sesamiae*, when intercropped with maize (Fig. 21.5) (Khan et al. 1997).

The volatile chemicals, such as (E)-ß-ocimene and (E)-4,8-dimethyl-1,3,7-nonatriene (Fig. 21.6) (Khan et al. 2000; Kimani et al. 2000), are produced by the *Desmodium* spp. which repel the stem borer moths from the maize ("push"), fix up

Fig. 21.3 Molasses grass (*Melinis minutiflora*) intercropped with maize at 1:1 ratio and surrounded by trap crop Sudan grass (*Sorghum vulgare sudanense*) in push–pull strategy at Suba district (Kenya)

to 100 kg N ha^{-1} yr^{-1}, and release root allelochemicals that induce abortive germination of the parasitic weed, *Striga hermonthica*. Similarly, the Napier grass releases volatile chemicals, such as octanal, nonanal, naphthalene, 4-allylanisole, eugenol, and linalool (Fig. 21.6) (Khan et al. 2000), which attract female moths ("pull") to lay eggs. The chemicals, such as 4″,5″-dihydro-5,2′,4′-trihydroxy-5″-isopropenylfurano-(2″,3″;7,6)-isoflavanone, produced by Desmodium roots stimulate *Striga* seed germination; and other chemicals, such as 4″,5″-dihydro-2′-methoxy-5,4′-dihydroxy-5″-isopropenylfurano-(2″,3″;7,6)-isoflavanone (suicidal germination), inhibit their attachment to maize roots, thereby reducing Striga seed bank (Khan et al. 2006a).

Leguminous desmodium maintains soil stability and improves soil fertility through enhanced soil organic matter content and nitrogen fixation. All the three crops (main, inter, and border crops) are of economic importance to farmers as livestock fodder. This technology increased maize yield from 1 to 3.5 t ha^{-1} and sorghum yield from 1 to 3 t ha^{-1} (Khan et al. 2011). This strategy in maize planted along with the push–pull plants gave a return of $US2.30 for every dollar invested, compared to only $US1.40 obtained by planting maize as a monocrop (Khan et al. 1997).

21.2.2 Cotton Boll Worms

The cotton bollworm *Helicoverpa armigera* (also called as tomato fruit borer, gram pod borer, and American bollworm) (Fig. 21.7) is one of the most important pests of cotton responsible for considerable losses. The pest has developed resistance to

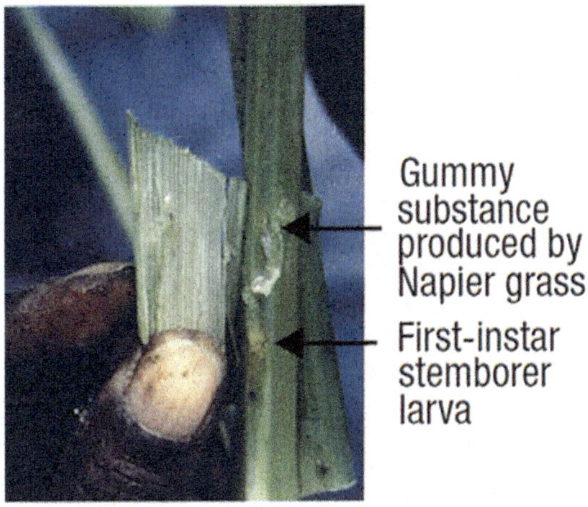

Fig. 21.4 Production of sticky substance by Napier grass tissues in response to penetration by first and second instar stem borer larvae

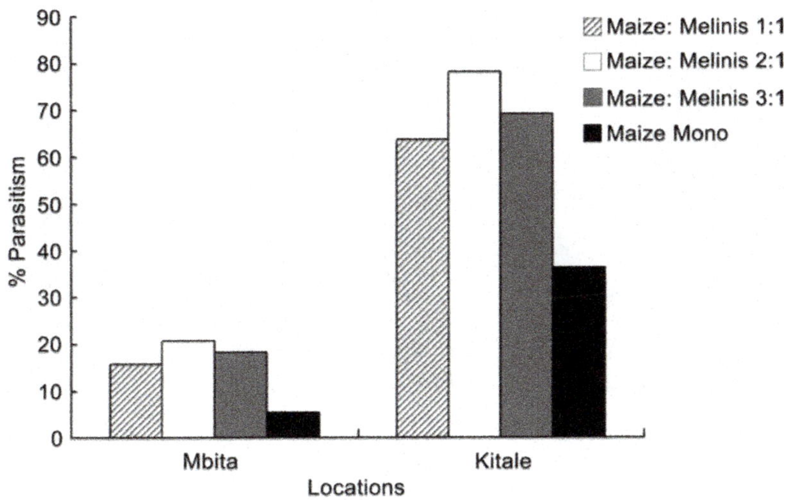

Fig. 21.5 Parasitism of stem borer larvae by *Cotesia sesamiae* in maize – molasses grass (*Melinis minutiflora*) intercrops planted in various ratios

most of the chemical insecticides normally used for its management due to indiscriminate use of insecticides over several years (Kranthi et al. 2004). Hence, there is a need to develop a novel strategy for the effective management of cotton bollworm using eco-friendly components such as cultural methods, botanicals, and biocontrol agents.

Duraimurugan and Regupathy (2005) developed the "push–pull strategy" using neem products to "push" colonizing bollworms away from cotton and to concentrate the pests on trap crops, such as okra and pigeon pea, so that a selective biological control agent (HaNPV) could be used effectively and economically to bring down the pest population. The cotton bollworm preferred to feed on okra and pigeon

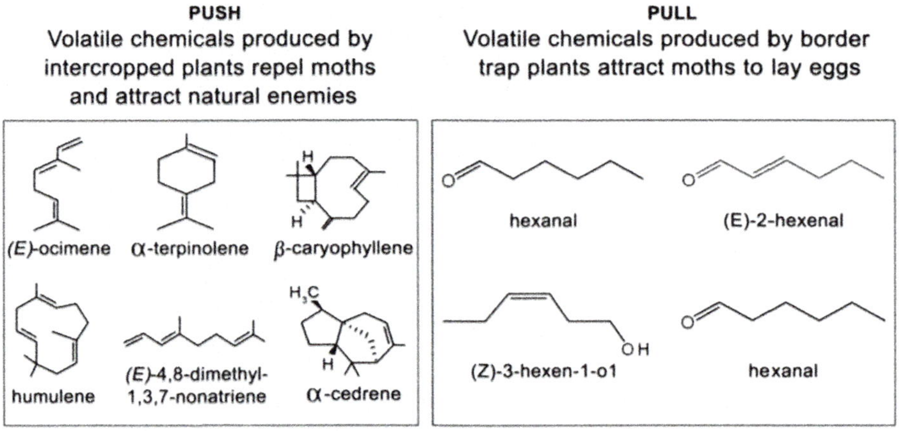

Fig. 21.6 Volatile chemicals produced by molasses grass (push) and Napier grass (pull)

Fig. 21.7 The bollworm damage on cotton leaves and boll

pea than on cotton. Spraying of cotton plants with 5 % neem seed kernel extract (NSKE) repelled *H. armigera* toward untreated okra and pigeon pea. Application of HaNPV on okra and pigeon pea gave effective control of *H. armigera* and reduced damage to boll, locule, and interlocules as compared to cotton sole crop (untreated control). The combined use of trap crops, application of NSKE on cotton and NPV on trap crops, was superior in reducing the incidence of *H. armigera* and damage by 61–72 % and increasing the kapas yield by 85–119 % over control (Table 21.1). The recovery of HaNPV-infected *H. armigera* larvae varied from 37.5 to 47.5, 32.8–39.2, and 14.2–20.2 % on okra, pigeon pea, and cotton, respectively.

Table 21.1 Effect of push–pull strategy with combined use of trap crops, neem, and HaNPV on bollworm incidence in cotton at two locations

Treatments	Vaigaidam (summer 2003)		Bhavanisagar (summer 2003)	
	% boll damage (open boll basis)*	Kapas yield (t/ha)	% boll damage (open boll basis)*	Kapas yield (t/ha)
Cotton (NSKE treated) + okra (NPV treated)	11. 2 (19.5)[a]	1.83[a]	14.8 (22.5)[a]	1.60[a]
Cotton (NSKE treated) + okra (NPV untreated)	14. 1 (22.0)[b]	1.76[ab]	21.5 (27.5)[c]	1.45[abc]
Cotton (NSKE untreated) + okra (NPV treated)	29. 1 (32.6)[ef]	1.43[def]	28.0 (31.9)[e]	1.23[cde]
Cotton (NSKE untreated) + okra (NPV untreated)	29. 7 (33.0)[ef]	1.38[ef]	33.1 (35.1)[f]	1.10[e]
Cotton (NSKE treated) + pigeon pea (NPV treated)	23. 1 (28.7)[c]	1.72[abc]	17.2 (24.4)[b]	1.55[ab]
Cotton (NSKE treated) + pigeon pea (NPV untreated)	24. 4 (29.5)[de]	1.58[b–e]	22.1 (28.0)[cd]	1.50[ab]
Cotton (NSKE untreated) + pigeon pea (NPV treated)	28. 4 (32.2)[de]	1.35[f]	28.3 (32.1)[e]	1.25[cde]
Cotton (NSKE untreated) + pigeon pea (NPV untreated)	30.9 (33.7)[f]	1.28[fg]	29.1 (32.6)[e]	1.20[de]
Cotton (NSKE and NPV treated)	23.9 (9.2)[c]	1.61[a–d]	22.2 (28.1)[cd]	1.36[a–d]
Cotton (NSKE treated)	26.9 (31.2)[d]	1.56[cde]	23.7 (29.1)[d]	1.35[bc]
Cotton (NPV treated)	36.9 (37.4)[g]	1.12[gh]	35.2 (36.3)[g]	0.90[f]
Cotton sole crop (untreated check)	40.0 (39.19)[f]	0.99[h]	38.0 (38.0)[g]	0.73[g]

[a-h]Means in a column followed by the same letter(s) are not significantly different (P=0.05) by DMRT

*Figures in parentheses are arcsine transformed values

21.2.3 *Pea Leaf Weevil in Beans*

The pea leaf weevil, *Sitona lineatus*, is a pest of field peas and faba beans. The adults feed on the terminal (clam) leaf of seedlings (reduces leaf area) which can reduce stand density, and the larvae feed on nitrogen-fixing root nodules that contain *Rhizobium leguminosarum* bacteria (Fig. 21.8).

The "push–pull strategy" was developed for the management of pea leaf weevils using neem antifeedant (push) to repel the pest and border planting of winter peas as trap crops (pull) to attract the pest (Smart et al. 1994). The application of neem antifeedant was also effective in reducing the abundance of weevils, but repeated applications were necessary to maintain the efficacy. The border trap crop (winter peas) can be treated with the aggregation pheromone to enhance the concentration of pest insects (semiochemically assisted trap crop). Instead of winter pea, clover

Fig. 21.8 Feeding damage of pea leaf weevils on leaves and nodules

Fig. 21.9 Colorado potato beetle damage on leaves

can also be used as a trap crop which would act as a refugee for predators of *S. lineatus* (Cook et al. 2007).

21.2.4 Colorado Potato Beetle

The Colorado potato beetle (*Leptinotarsa decemlineata*) (Fig. 21.9) is a major pest of Solanaceous crops in general and potato (*Solanum tuberosum*) in particular. The "push–pull strategy" was developed for the management of this pest using neem-based antifeedant (push) to repel the pest and early perimeter planting of potato as trap crops (pull) to attract the pest (Martel et al. 2005b). The early-planted potato trap crop was sprayed with a slow-release attractant [formulation comprising of (*Z*)-3-hexenyl acetate, (*R*,*S*)-linalool, and methyl salicylate] at weekly intervals that significantly attracted more adult beetles, eggs, and larvae of the Colorado potato beetle (Martel et al. 2005a) where they can be treated with insecticides. Alternatively, the Colorado potato beetle aggregation pheromone (*S*)-3,7-dimethyl-2-oxo-6--octane-1,3-diol can also be used (Dickens et al. 2002) that can enhance attraction and subsequent biocontrol by predators such as the spined soldier bug (*Podisus maculiventris*) (Dickens 1999).

Fig. 21.10 Pollen beetle (*Meligethes aeneus*) damage on oilseed rape

21.2.5 Oilseed Rape Pollen Beetle

In order to manage the pollen beetle (*Meligethes aeneus*) (Fig. 21.10) on oilseed rape (*Brassica napus*), a "push–pull" strategy has been developed based on turnip rape (*Brassica rapa*) as an attractive trap crop (Potting et al. 2005). Turnip rape is a much preferred host to the pollen beetle than oilseed rape when it is used as a perimeter trap crop that considerably reduced the population of pollen beetle in oilseed rape under field conditions. The factors responsible for effectiveness of turnip rape to attract pollen beetle include growth stage-related olfactory and visual stimuli (high proportions of alkenyl glucosinolates) (Cook et al. 2006). The insecticides, parasitoids, or pathogens (*Metarhizium anisopliae*) can be used to reduce pest populations in the trap crop.

21.2.6 Onion Maggots

The maggot *Delia antiqua* (Fig. 21.11) is a major pest of onion (*Allium cepa*) in temperate regions. A push–pull strategy has been developed by manipulating the ovipositional behavior of this pest by treating onion seedlings with chemical deterrents (50 % cinnamaldehyde formulated in activated charcoal) to repel the pest (push) and simultaneously providing deeply planted onion culls (small or sprouting unmarketable bulbs) as trap crop on which onion flies prefer to lay eggs (pull) (Miller and Cowles 1990). Both the components were effective in reducing the oviposition considerably.

This strategy of behavioral manipulation has the following advantages:

- Overriding of the pest by deterrents and concomitantly avoiding severe deprival of pest to its host
- Integrating the effects of "push" and "pull" multiplicatively
- Enhanced biological control opportunities in sites where the pest becomes concentrated

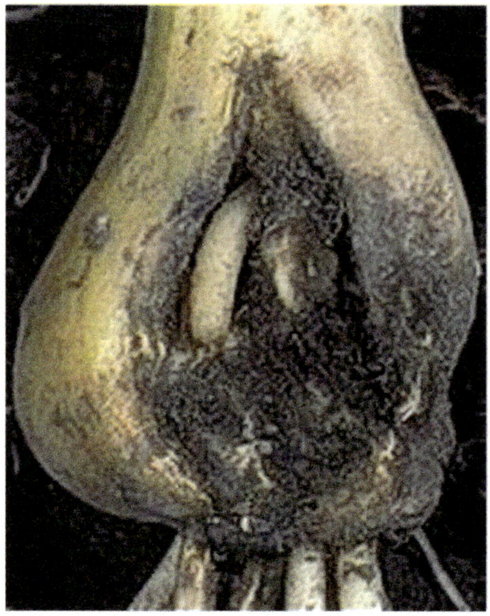

Fig. 21.11 Onion maggot damage on bulb

21.2.7 Chrysanthemum Thrips

The production of greenhouse-grown chrysanthemums is severely affected by Western flower thrips (*Frankliniella occidentalis*) that cause direct damage by feeding (Fig. 21.12), indirect damage by transmitting viruses, and unacceptability of flowers in market due to their presence.

A push–pull strategy has been developed to push thrips from target plants by spraying the antifeedant polygodial (extracted from *Tasmannia stipitata*) on the main crop and concentrating them onto trap plants (chrysanthemum cultivar 'Springtime' that is most attractive) (Bennison et al. 2001a) where the predators are released or maintained. The trap crop also provided pollen for the maintenance of predators in the absence of thrips. The attractiveness of trap crop was increased by baiting with the attractive host plant volatile *E*βf, reducing infestations on the antifeedant-treated main crop (Bennison et al. 2001a, b).

21.3 Benefits and Limitations

21.3.1 Benefits

Some of the common benefits of push–pull strategies are presented in Fig. 21.13.

- Synergic effects due to integration of two or more push and pull components (Martel et al. 2005b).

Fig. 21.12 Thrips damage on leaves and flower of chrysanthemum

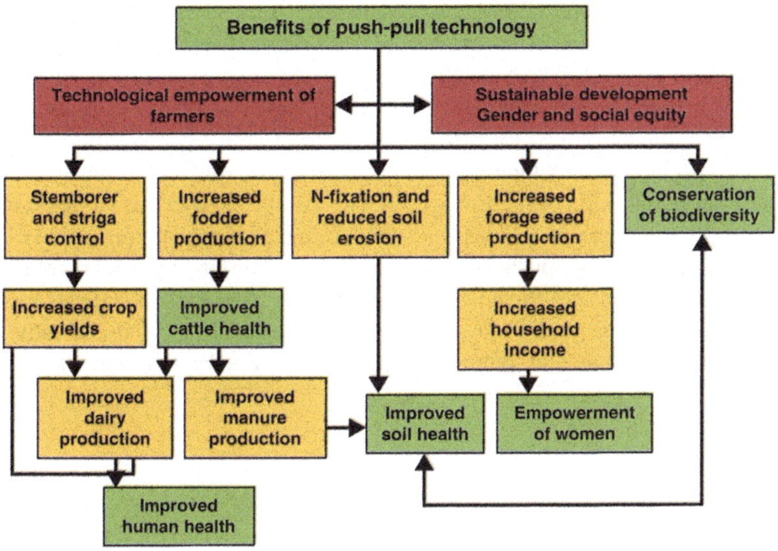

Fig. 21.13 Benefits of push–pull technology

- Antifeedants and oviposition deterrents are effectively utilized by adding pull stimuli (Raffa and Frazier 1988).
- Population reducing components' efficiency is enhanced (Martel et al. 2005b).
- Chances of pests developing resistance to pesticides are reduced, since noninsecticidal components/reduced amounts of pesticides are used (Foster et al. 2005).

21.3.2 Limitations

- Considerable research effort is required to understand the behavioral and chemical ecology of the host–pest interactions and their effects on beneficials.
- The cost of semiochemical registration is very high.
- The adoption of push–pull strategies is very much limited. At present, only two strategies have been used successfully on a commercial scale.

21.4 Integration with Other Management Strategies

Since the components used in push–pull technology involve eco-friendly and benign alternatives, they can be directly incorporated into IPM strategies. Several eco-friendly components such as biopesticides based on entomopathogenic nematodes, fungi, bacteria, and viruses, the endotoxins of *Bacillus thuringiensis* (Bt) and spinosyn (spinosad) isolated from *Saccharopolyspora spinosa* (which are commercially available as insecticides), and genetically modified crop plants expressing the Bt toxin gene can be employed for developing more effective push–pull technologies. The dead-end trap crops, which attract pests, but are unable to survive and multiply (killing either adult pests or their progeny), can also be used in this strategy. In addition, the pest predators and parasitoids which are available commercially can be utilized in improving the push–pull technologies. In order to pull parasitoids from the surrounding areas into the crop field where they are required, HIPVs including (Z)-jasmone, and sex pheromone component nepetalactone, and the lady beetle pheromones (tricosane and pentacosane) can be used which attract the aphid parasitoids.

21.5 Conclusions

The push–pull strategies for the management of insect pests in limited crops have been developed using behavioral manipulation method involving repellent/deterrent (push) and attractive/stimulant (pull) stimuli. These methods should be integrated with sustainable and environmentally sensitive components in order to reduce the use of chemical pesticides. The broad-spectrum insecticides should be replaced with innovative technologies, especially with semiochemicals, to manipulate the behavior of natural enemies for improved biological control. Adaptability of companion plants for climate resilience, capacity building and multi-stakeholder collaboration, creation of a supportive policy framework, effective transfer of technology, and improvement in input accessibility are some of the crucial aspects for further up-scaling of this technology. There is a need for developing more and more improved push–pull strategies for the management of key insect pests affecting various crops and used more widely in the future.

References

Bennison J, Maulden K, Dewhirst S, Pow EM, Slatter P, Wadhams LJ (2001a) Towards the development of a push-pull strategy for improving biological control of western flower thrips on Chrysanthemum. In: Proceedings international symposium Thrips and Tospoviruses: Thysanoptera, Reggio, Calabria, Italy, pp 199–206

Bennison J, Maulden K,Wadhams LJ, Dewhirst S (2001b) Charming the thrips from the flowers. Grower 135(23):20–22

Cook SM, Smart LE, Martin JL, Murray DA, Watts NP, Williams IH (2006) Exploitation of host plant preferences in pest management strategies for oilseed rape (*Brassica napus*). Entomol Exp Appl 119:221–229

Cook SM, Khan ZR, Pickett JA (2007) The use of push-pull strategies in integrated pest management. Annu Rev Entomol 52:375–400

Dickens JC (1999) Predator–prey interactions: olfactory adaptations of generalist and specialist predators. Agric For Entomol 1:47–54

Dickens JC, Oliver JE, Hollister B, Davis JC, Klun JA (2002) Breaking a paradigm: male-produced aggregation pheromone for the Colorado potato beetle. J Exp Biol 205:1925–1933

Duraimurugan P, Regupathy A (2005) Push-pull strategy with trap crops, neem and nuclear polyhedrosis virus for insecticide resistance management in *Helicoverpa armigera* (Hubner) in cotton. Am J Appl Sci 2:1042–1048

Foster SP, Denholm I, Thompson R, Poppy GM, Powell W (2005) Reduced response of insecticide-resistant aphids and attraction of parasitoids to aphid alarm pheromone; a potential fitness trade-off. Bull Entomol Res 95:37–46

Khan ZR, Pickett JA (2004) The 'push-pull' strategy for stemborer management: a case study in exploiting biodiversity and chemical ecology. In: Gurr GM, Wratten SD, Altieri MA (eds) Ecological engineering for pest management: advances in habitat manipulation for arthropods. CABI, Wallington, pp 155–164

Khan ZR, Ampong-Nyarko K, Chiliswa P, Hassanali A, Kimani S et al (1997) Intercropping increases parasitism of pests. Nature 388:631–632

Khan ZR, Pickett JA, van den Berg J, Wadhams LJ, Woodcock CM (2000) Exploiting chemical ecology and species diversity: stem borer and striga control for maize and sorghum in Africa. Pest Manag Sci 56:957–962

Khan ZR, Midega CAO, Hassanali A, Pickett JA, Wadhams LJ, Wanjoya A (2006a) Management of witchweed, *Striga hermonthica*, and stemborers in sorghum, *Sorghum bicolor*, through intercropping with greenleaf desmodium, *Desmodium intortum*. Int J Pest Manag 52:297–302

Khan ZR, Midega CAO, Hutter NJ, Wilkins RM, Wadhams LJ (2006b) Assessment of the potential of Napier grass (*Pennisetum purpureum*) varieties as trap plants for management of *Chilo partellus*. Entomol Exp Appl 119:15–22

Khan Z, Midega C, Pittchar J, Pickett J, Bruce T (2011) Push-pull technology: a conservation agriculture approach for integrated management of insect pests, weeds and soil health in Africa. Int J Agric Sustain 9:162–170

Kimani SM, Chhabra SC, Lwande W, Khan ZR, Hassanali A, Pickett JA (2000) Airborne volatiles from *Melinis minutiflora* P. Beauv., a non-host plant of the spotted stem borer. J Essent Oil Res 12:221–24

Kranthi K, Singh J, Iqabal A, Wu Y, Russell DA (2004) Patterns and magnitude of insectivore resistance levels in India, Pakistan and China. In: Biljon JV, Joubert D (eds) Proceedings world cotton conference III. Cape Town, South Africa, pp 1215–1229

Martel JW, Alford AR, Dickens JC (2005a) Laboratory and greenhouse evaluation of a synthetic host volatile attractant for Colorado potato beetle, *Leptinotarsa decemlineata* (Say). Agric For Entomol 7:71–78

Martel JW, Alford AR, Dickens JC (2005b) Synthetic host volatiles increase efficacy of trap cropping for management of Colorado potato beetle, *Leptinotarsa decemlineata* (Say). Agric For Entomol 7:79–86

Miller JR, Cowles RS (1990) Stimulo-deterrent diversion: a concept and its possible application to onion maggot control. J Chem Ecol 16:3197–212

Potting RPJ, Perry JN, Powell W (2005) Insect behavioural ecology and other factors affecting the control efficacy of agro-ecosystem diversification strategies. Ecol Model 182:199–216

Pyke B, Rice M, Sabine B, Zalucki MP (1987) The push-pull strategy – behavioural control of *Heliothis*. Aust Cotton Grow:7–9

Raffa KF, Frazier JL (1988) A generalized model for quantifying behavioral desensitization to antifeedants. Entomol Exp Appl 46:93–100

Smart LE, Blight MM, Pickett JA, Pye BJ (1994) Development of field strategies incorporating semiochemicals for the control of the pea and bean weevil, *Sitona lineatus* L. Crop Prot 13:127–135

Part IX
Careful Management of Farm Machinery

Chapter 22
Efficient Use of Farm Machinery

Abstract In conservation agriculture, there is a need for suitable farm machinery to carry out the operations such as direct seeding and fertilizer application without disturbing the mulch on soil surface, so as to achieve good germination with minimum soil disturbance. Special machinery is also required for weeding, pest control, harvesting, etc. These operations have to be carried out through cheaper and lighter machines with lesser horsepower to avoid soil compaction and to prevent damage to the soil. The farm machinery used for primary tillage, secondary tillage, direct seeding, and spraying is discussed.

Keywords Primary tillage • Secondary tillage • Direct seeding • Spraying • Farm machinery

22.1 Introduction

Agriculture mechanization refers to usage of any improved tool, implement, or machine by the farm workers to improve the efficiency of any operation to increase crop productivity. It also helps to reduce drudgery or stress which adversely affect human mental faculties leading to errors, imprecision and hazards, and eventually loss of efficiency if done manually. The timeliness of operations has assumed greater significance in obtaining optimal yields from different crops. For example, the sowing of wheat in Punjab is done up to the first fortnight of November. A delay beyond this period by every 1 week leads to reduction in yield of about 375 kg/ha. The timeliness is also important for farm operations like weeding, irrigation, and harvesting. To ensure timeliness of various operations, it is necessary to use farm machinery which has high output capacity. Farm mechanization has helped to increase area under cultivation and the intensity of cropping.

The quality and precision of the operations are equally significant for realizing higher yields. The various farm operations need a high degree of precision to increase the efficiency of the inputs and reduce the losses.

Farm mechanization is gaining importance because the output of work done per hour is several times more than human labor. This also entails reduction in the total labor requirement. The ever-increasing wages for the human labor, their

P.P. Reddy, *Sustainable Intensification of Crop Production*,
DOI 10.1007/978-981-10-2702-4_22

nonavailability at the time of need, and the rising cost of maintaining draft animals also speak in favor of mechanization.

22.2 Primary Tillage

Opening of the compacted soil with the help of different tools and implements preferably by plows is known as primary tillage or plowing. This is mainly done to open the hard soil. In addition to open the hard soil, the aim of primary tillage is also to achieve inversion (whenever necessary) of soil and uprooting of stubbles and weeds. Some of the types of primary tillage practices include deep tillage, subsoiling, and year-round tillage.

The major primary tillage implements used are wooden plows (country plow), soil inversion plows (moldboard, disk, and turnwrest), and special plows (subsoil plow, chisel plow, ridge plow, rotary plow or rotary hoes, and basic lister). The selection of plows depends on the suitability for different situations (soil condition and nature of weed problem) (Table 22.1).

22.2.1 Moldboard Plow

The traditional primary plowing tool is the moldboard plow (Fig. 22.1). The moldboard plow shears off a section of soil, tips it upside down, and fractures it along several planes resulting in better pulverization. During this process, any organic matter on the soil surface is buried. The animal-drawn moldboard plow is small, plow to a depth of 15 cm. Victory plow is an animal-drawn moldboard plow with a short shaft. Two moldboard plows of bigger size are attached to the tractor and plowed to a depth of 25–30 cm. Moldboard plows are used for soil inversion. When finished, this implement leaves the soil surface very rough with a series of ridges and furrows.

Table 22.1 Suitability of plows for different situations

Plow	Purpose or situation
Moldboard plow (animal drawn)	Incorporation of manure, fertilizers, and crop residues in soil
Moldboard plow (tractor drawn)	Deep plowing and inversion
Disk plow, country plow	Soil inversion and cutting of creeping or spreading grass

Fig. 22.1 The moldboard plow at work

22.2.2 *Disk Plow*

The disk plow is having a little resemblance to the common moldboard plow. A large, revolving, concave steel disk replaces the share and the moldboard. A disk plow is an implement with a series of three to ten large (30–45 cm) disks mounted on a frame at an angle to the direction of travel (Fig. 22.2). The disk turns the furrow slice to one side with a scooping action. The usual size of the disk is 60 cm in diameter and this turns a 35–30 cm furrow slice. The disk plow is more suitable for land with much fibrous growth of weeds, as the disk cuts and incorporates the weeds. The disk plow works well in soils free from stones.

22.2.3 *Chisel Plow/Subsoiler*

It is mainly used for breaking hard pans and for deep plowing (60–70 cm) with fewer disturbances to the top layers. Its body is thin with replaceable cutting edge, which has minimum disturbance to the top layers. It contains a replaceable share to shatter at the lower layers (Fig. 22.3).

A chisel plow is used for primary tillage that loosens the soil but does not invert it. Chisel plowing to eight inches leaves the soil rough with about 50–80 % residue cover. Light disking can then reduce residues to 30–50 %. Seeds are then planted through the remaining residues.

Fig. 22.2 Disk plow

Fig. 22.3 Chisel plow/subsoiler

22.3 Secondary Tillage

Finer operations that are performed on the soil after primary tillage are known as secondary tillage operations. Different types of implements like disk harrows, cultivators, plank, and rollers are used for secondary tillage. The first step is commonly accomplished with a tandem disk. Secondary tillage is usually a two-step process. First ridges left from plowing are smoothed out and large clods are broken. Then smaller lumps are pulverized and a fine seedbed is produced.

22.3.1 Tandem Disk Harrow

The typical tandem disk harrow has four gangs of disks set like the four arms of an "X." The front two gangs turn the soil inward, and the back two turn it back out. The disk harrow consists of a number of concave disks of 45–55 cm in diameter (Fig. 22.4). The disks cut through the soil and effectively pulverize the clods. Blade harrows are used for different purposes like removal of weeds and stubbles, crushing of clods, working of soil to shallow depth, covering the seeds, inter-cultivation and harvesting of groundnut, etc.

22.3.2 Spring-Tooth Harrow (Field Cultivator)

A spring-tooth harrow can also be used in secondary tillage. This implement is made of long, springy, C-shaped teeth with a shear point or broad shovel that digs into the soil (Fig. 22.5). A finishing harrow or drag is used to pulverize the soil clods into a smooth, fine surface. This is an implement used for finer operations like breaking clods and working the soil to a fine tilth for preparing seedbed. It is also called as cultivator, tiller, or tooth harrow. It is used to further loosen the previously plowed land before sowing. It is also used to destroy weeds that germinate after plowing. The main object of providing two rows and staggering the position of

Fig. 22.4 The tandem disk harrow

Fig. 22.5 Spring-tooth harrow

tynes is to provide clearance between tynes so that clods and plant residues can freely pass through without blocking. The number of tynes ranges from 7 to 13. The shares of the tynes can be replaced when they are worn-out.

22.3.3 Bed Former

Bed forming is to build a protected, well-drained area in which to grow the crop, providing improved drainage and even better seed germination (Fig. 22.6).

22.4 Direct Seeding

The direct seeding machines are developed to open uniform depth furrow, drop seeds and fertilizers uniformly at depths desired without mixing (seed and fertilizer metering), and prevent injury to seeds, by keeping proper space between seeds, by covering the seeds with soil to provide proper compaction, and by retaining previous crop residues.

22.4.1 Jab Planter

The jab planter is a manual implement used for sowing of seeds and fertilizers by small-scale farmers under no-tillage system (Fig. 22.7).

Fig. 22.6 The bed former

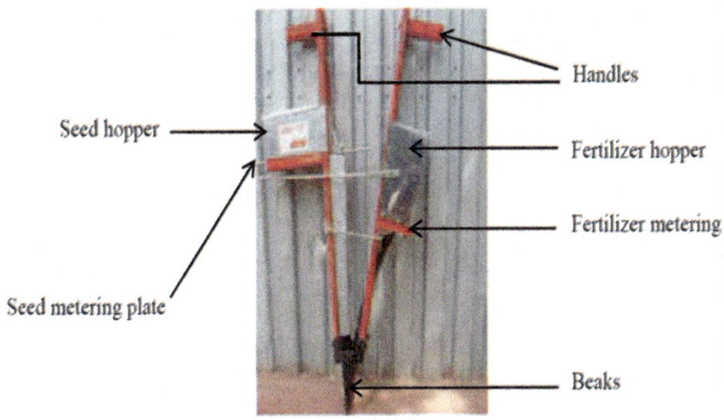

Fig. 22.7 The jab planter

22.4.2 Animal-Drawn Direct Seeder

The animal traction direct seeder is used under no-tillage conditions to plant in surface residue cover (Fig. 22.8). It has two hoppers for seed and fertilizer and a coulter (cutting disk). The trailing type of seed cum fertilizer drill is provided with rear cage wheel to support and to provide power to seed and fertilizer metering device.

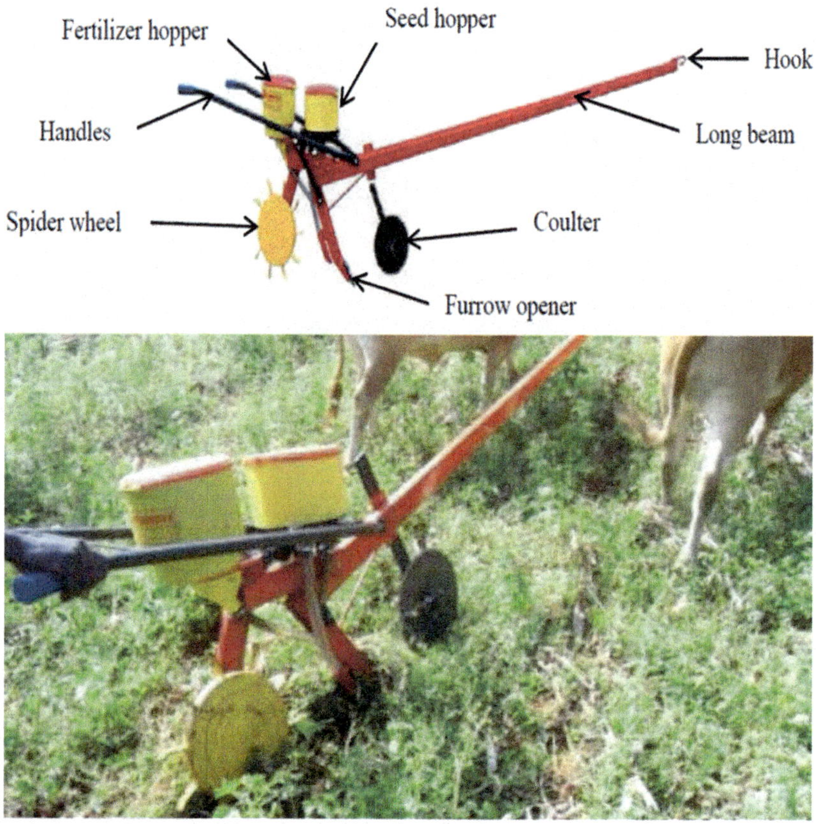

Fig. 22.8 The animal traction direct seeder

22.4.3 Tractor-Drawn Direct Seeding Machine

The trailing type of tractor-drawn direct seeding machine is supported with tractors hitch in the front and two cage wheels in the back to support the seed and fertilizer drill machine and to provide power required to move seed and fertilizer delivery by chain and gear mechanism (Fig. 22.9).

22.4.4 Rice Drum Seeder

The rice drum seeder machine places the rice seed in puddle soil for good germination instead of transplanting of rice seedlings (Fig. 22.10). This cost-effective beneficial method is practiced in regions where shortage of labor is a major problem during the time of transplanting. The rice yield under direct seeding is comparable to transplanted rice (no yield reduction).

Fig. 22.9 Tractor-drawn direct seeding machine

Fig. 22.10 Rice drum seeder

22.4.5 Transplanter

The rice transplanter is a specifically designed machine to help in transplanting of rice seedlings into paddy fields (Fig. 22.11). These machines generally use less time and labor than sowing seedlings by hand while producing higher yields.

Fig. 22.11 The rice transplanter

Fig. 22.12 Strip-till planter

22.4.6 Strip-Till Planter

With no primary tillage, a specialized implement tills a band of soil and plants seeds into the band (Fig. 22.12). Another implement sweeps residues off a strip into the middle of the rows. This operation normally leaves about 50 % of crop residue.

Fig. 22.13 Ridge-till planter

22.4.7 Ridge-Till Planter

The ridge-till system (Fig. 22.13) is used in cool and moist conditions. Seed is planted on six-inch ridges with crop residues swept into the shallow furrows. About two-thirds of crop residues remain after planting. Cultivation with special tools minimizes residue burial and rebuilds ridges for the coming year.

22.4.8 Mulch-Till Planter

Mulch-till planter is used to till the soil before planting (Fig. 22.14). The disturbance occurs to more than 30 % of the soil surface on the field, while some residue remains on the soil surface.

22.5 Weed Management Equipment

22.5.1 Manual Tools/Equipment

The above-ground weeds can be manually managed by using a slasher or a sickle or a push/pull hand weeder (Fig. 22.15) which can be easily fabricated by local artisans. The slasher or a sickle takes a long time for weeding, while the push/pull hand weeder is easy to use, effective and fast.

Fig. 22.14 Mulch-till planter

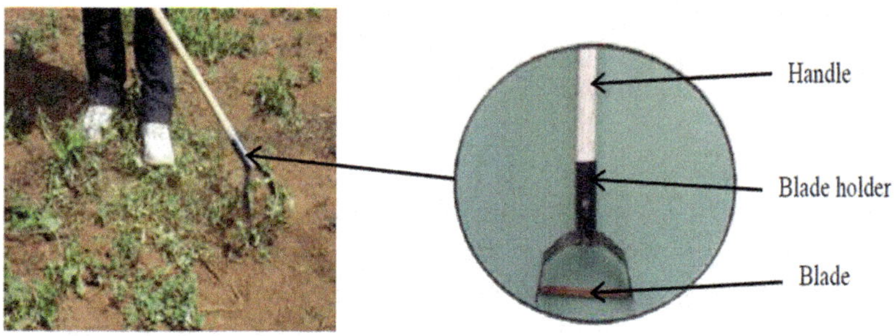

Fig. 22.15 Various parts of a weed scraper

22.5.2 Animal-Drawn Equipment

The animal-drawn weed control equipment includes the duck-foot weeder (Fig. 22.16) and shallow sweepers. The duck-foot weeder does not turn the soil over, while shallow sweepers with wide blades can make a single pass between row crops. The animal-drawn weeders are very fast and need only one-fifth of the time required for manual weeding.

22.6 Spraying Machinery

Spraying machinery is used for application of agro-chemicals such as pesticides and herbicides. The different types of spraying machinery are as follows:

- Knapsack sprayer
- Motorized knapsack sprayer
- Tractor-mounted sprayer

Fig. 22.16 Various parts of duck-foot weeder

22.6.1 Knapsack Sprayer

Knapsack sprayer (Fig. 22.17) is provided with inbuilt storage tank with varying capacity of 14, 16, and 18 L. The operator can pump with one hand and spray with the other. Uniform pressure (2–3 kg/cm^3) can be maintained by keeping the pump in continuous operation. It has swath width of 22.5–35 cm and field capacity of 0.5–1.0 ha/hr with an application rate of 250–400 L/ha. It can be used to spray row crops, small trees, and shrubs.

22.6.2 Motorized Knapsack Sprayer/Duster

It is a shoulder-mounted motorized sprayer suitable for several plant protection operations (Fig. 22.18). It blows the spray liquid as a mist and dust as fine particles. It can be used for spraying tree crops.

22.6.3 Tractor-Mounted Sprayer

It is suitable for use when the crop is young and planted at row spacing that is convenient for tractor movement (Fig. 22.19). The sprayer is mounted on tractor (on three-point linkage) and the pump is operated by PTO power. It has a 200–400 L capacity and gives wide coverage of 7–8 m with up to 20 nozzles. The sprayer has a field capacity of 0.75–L ha/hr with an application rate of 250–400 L/ha.

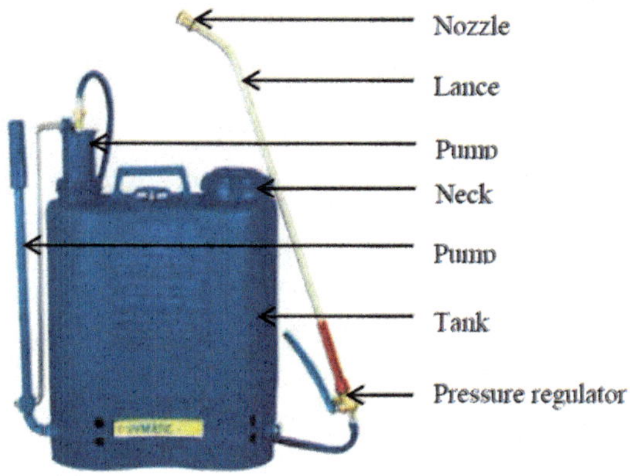

Fig. 22.17 Various parts of a knapsack sprayer: photo by KENDAT

Fig. 22.18 Motorized knapsack sprayer/duster

22.7 Conclusions

The availability of cheaper and suitable equipment for farmers is one of the main limitations for adoption of conservation agriculture practices. There is a need for equipment that farmers can use to get a good seed placement (and fertilizer if

Fig. 22.19 Tractor-mounted high volume sprayer

necessary) into the permanent soil cover in order to get better germination with minimum soil disturbance. Alternative farm machinery system is the need of the hour so that farm operations can be carried out by animal power or tractors with less horsepower. More efforts are required in the future to stimulate additional research on development of better and more fuel efficient farm machinery.

Part X
Integrated Crop-Livestock Production

Chapter 23
Integrated Crop–Livestock Farming Systems

Abstract In view of concerns regarding increasing regulation of concentrated animal feeding operations, long-term sustainability, natural resource degradation, and stability and profitability of farm income, there is a considerable interest in integrated farming systems. Integration of crop, pasture, and livestock is mutually beneficial to each other, since crop residues can be used as animal feed, while animal manure can be utilized to enhance soil tilth, fertility, and carbon sequestration that can enhance agricultural productivity. The combined system enhances soil biological activity and nutrient recycling, improves profits, increases crop yields, intensifies land use, prevents soil erosion, reduces poverty and malnutrition, and strengthens environmental sustainability. The "win–win" outcomes can be achieved by providing higher investment in research and development and favorable incentives, policies, and technologies to enable sustainable, integrated, and productive integrated farming systems.

Keywords Carbon sequestration • Integrated farming system • Crop–livestock integration • Environmental sustainability • Resource conservation • Crop residues • Soil fertility

23.1 Introduction

The serious constraints that beset livelihoods and sustainability in the current agricultural land use pattern include poor soil health and therefore low soil productivity; soil inversion tillage practices, which degrades soil porosity, organic matter, and soil biota; suboptimal soil and crop and pest management practices; inadequate crop diversification and crop residue management; and poor integration of livestock in the production system.

In view of increased urbanization and higher incomes in developing countries, there is a greater demand for nutritive diets comprising of more horticultural (fruits and vegetables) and animal-source foods (meat, milk, eggs, and fish) (FAO 2011; Otte et al. 2012). Hence, there is an immediate need for integrated crop–livestock farming systems to produce both crop and animal-source food products.

P.P. Reddy, *Sustainable Intensification of Crop Production*,
DOI 10.1007/978-981-10-2702-4_23

The agricultural system that combines both crop and livestock production is called integrated farming system in which the by-product of one system becomes the input for other. This system helps the farmer with a limited land to diversify his farm for crop production and to maintain a few heads of livestock in order to exploit unutilized resources, to improve quality and quantity of food produced, and to enhance cash income (Delgado et al. 1999).

Integrated farming systems provide primary source of livelihood for about one billion people and produce 50 % of cereals, 34 % of beef, and 30 % of milk (Herrero et al. 2009). The profitability and environmental sustainability of the farms and communities can be increased by adoption of integrated crop–livestock systems.

A sustainable approach to attain reasonably high productivity with considerable scope for recycling of resources and providing ecological soundness is called the integrated farming system.

The aims of integrated farming systems are as follows:

- Effective integration of crops and livestock by careful land use
- Enhancing productivity of both crops and livestock
- Production of a good quality and more diverse foods
- Protection of the environment using natural resources efficiently and prudently

The key aspects of integrated farming systems include (Fig. 23.1):

- Crop production
- Livestock production
- Nutrient recycling
- Crop residues

23.2 Types of Integrated Farming Systems

The three types of crop–livestock combined systems based on the use of agriculture techniques, the local agro-ecological conditions, and the degree of integration among crop and livestock are as follows (Rota 2009):

- Pasture cropping system (sowing annual cereal crops into perennial pasture)
- Nonchemical input cropping system (same as pasture cropping system without the use of fertilizers, herbicides, and pesticides)
- Holistic management system (same as pasture cropping system with inclusion of grazing by livestock into the farming system)

23.2.1 Pasture Cropping System

The system consists of direct seeding of annual cereal crops (oats, millet, or others) into perennial pasture (usually native species) that combines advantages obtained from pastures with those from crops. The land is used for growing crops and pasture

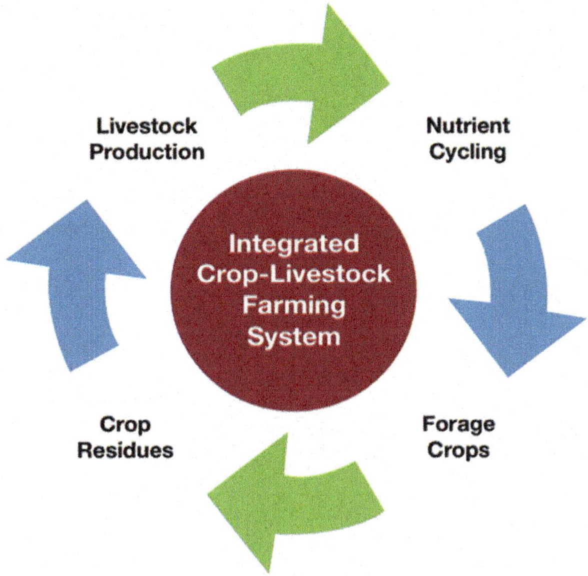

Fig. 23.1 The key aspects of integrated farming systems

symbiotically for dual purposes of both food crop production and pasture maintenance. After the harvest of the cereal crops, livestock can return to the pasture.

23.2.2 Nonchemical Input Cropping System

This very low-cost method is similar to pasture cropping system that has considerable effects on the preservation of biodiversity, since fertilizers and/or pesticides (including herbicides) are not at all used.

23.2.3 Holistic Management System

The holistic management system integrates the inclusion of grazing by livestock into the farming system in a way which regenerates the pastures.

The characteristics of the three integrated farming systems are presented in Table 23.1.

The available resources are efficiently used to the maximum extent by recycling due to interaction of crop–livestock integrated system to create a synergy. The agricultural productivity is increased by using crop residues as animal feed and livestock and livestock by-product production and processing to improve soil fertility, thereby reducing the use of chemical fertilizers (Fig. 23.2). The farmers' income can be significantly increased in several parts of the country by adopting integrated crop and livestock farming systems (Table 23.2).

Table 23.1 Characteristics of the three integrated farming systems

	Pasture cropping system	Nonchemical inputs cropping system	The holistic management system
Key principles	Direct seeding technique of sowing one annual crop into perennial pasture so as to have these crops grow symbiotically	No herbicides or pesticides used	Combination of grazing into the farming system
		No fertilizers applied	
		Limited soil disturbance	
Main environmental benefits	Improvements in soil fertility	Preservation of biodiversity	Reducing soil erosion
	Improved water use efficiency	Positive effects on soil compaction and water content	Enriched biodiversity
	Pastures enhancement		Sustainable land livestock carrying capacity
	Enriched biodiversity		
	Flexible farming system that has potential benefits for rejuvenation of native pastures		
Financial aspects	Reduced costs for inputs/technology acquisition: one annual crop is seeded and manure is used as a natural organic fertilizer	Zero costs for fertilizers and herbicides acquisition	Increased potential income generating activities (particularly due to the combination of crop and livestock production)
	No need to resow pasture		Reduced costs for inputs acquisition
Principal risks	Risks on soil fertility derive from inappropriate animal grazing	Inadequate supply of forages to animals if the grazing system is not properly managed with a long-term vision	Risks might derive from the number of animals grazing, the type and the timing of grazing
	In case of inadequate rainfall, benefits are reduced		
Impact on soil carbon sequestration	Enhanced soil carbon sequestration	Enhanced soil carbon sequestration	Enhanced soil carbon sequestration

The beef cattle production is enhanced by integrating with cropland to improve economic returns (Small and McCaughey 1999). The net profitability of about $9000 can be achieved by combing crops and beef cows compared with crops only (Anderson and Schatz 2003).

The fisheries + piggery + poultry farming system in China gave maximum economic returns (registered 48.6 % gain) as compared to the traditional rice-wheat system, generating additional employment for about 500 man days ha^{-1} yr^{-1} (Gill et al. 2005).

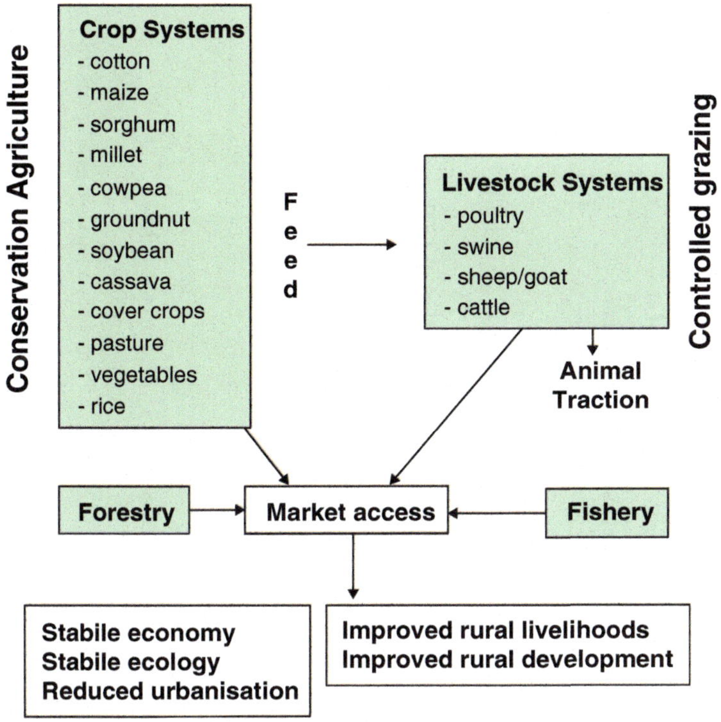

Fig. 23.2 Integrated crop–livestock farming systems

The rate of accumulation of soil C is doubled in a legume–grain crop rotation by adding cattle (Drinkwater et al. 1998). Soil C sequestration is improved by decomposing residues and manure (Soussana et al. 2004). The topsoil C increased annually to the extent of 1200 and 2300 kg ha^{-1} yr^{-1} under grazed switch grass (*Panicum virgatum*) and grazed smooth brome grass (*Bromus inermis*), respectively, over a corn–soybean–3-yr–alfalfa rotation (Al-Kaisi et al. 2005).

The highest net profit (Rs 20,581) was obtained under irrigated conditions in arable farming system with crossbred cows, followed by mixed farming with buffaloes (Rs 6218), and lowest in arable farming in Haryana (Rs 4615) (Singh et al. 1993).

The integration of coconut with vegetables, mushroom, poultry, and dairy in Goa increased the economic returns to Rs 17,518 ha^{-1} yr^{-1} as compared to the cultivation of cashew nut alone (the local traditional practice).

The higher annual productivity and economic return of Rs 1,31,118 were obtained by integration of rice with fish (400 reared in 3 ponds of 0.04 ha each), poultry (20 Babcock layer bird), pigeon (40 pairs), and goat (Tellicherry breed of 20 female and 1 male in 0.03 ha deep litter system). This integrated farming system generated the employment opportunities for additional 207 man day's yr^{-1} (Table 23.3) (Jayanthi et al. 2001). The fish were fed with poultry, pigeon, and goat droppings. Likewise, extra poultry, pigeon, and goat manure were applied to the rice crop.

Table 23.2 Effect of integrated farming system on the economic viability in different states of India (Manjunatha et al. 2014)

State	Prevailing system	Net returns (Rs)	Integrated farming systems	Net returns (Rs)	Reference
Tamil Nadu	Rice – rice – black gram	8312	Rice – rice – cotton + maize + poultry/fish	17,209	Shanmugasundaram et al. (1995)
	Rice–rice	15,229	Rice–rice–Azolla/Calotropis + fish	17,488	
	Rice – rice – rice – fallow-pulses	13,790	Rice – rice – rice –fallow – cotton + maize + duck/fish	24,117	Ganesan et al. (1990)
	Cropping alone	36,190	Cropping + fish + poultry	97,731	Jayanthi et al. (2001)
			Cropping + fish + pigeon	98,788	
			Cropping + fish + goat	1,31,118	
	Rice	22,971	Rice + fish	28,569	Balusamy et al. (2003)
			Rice + Azolla + fish	31,788	
Goa	Cashew	36,330	Coconut + forage + dairy	32,335	Manjunath and and Itnal (2003)
			Rice – brinjal (0.5 ha) + rice + cowpea (0.5 ha) + mushroom + poultry	75,360	
Madhya Pradesh	Arable farming	24,093	Mixed farming + dairy (2 cows)	37,668	Tiwari et al. (1999)
Uttar Pradesh	Sugarcane + wheat	41,017	Sugarcane + wheat + dairy	47,737	Singh (2004)
Karnataka	Rice–rice system	21,559	Rice – fish – poultry	62,977	Chnnabasavanna and Biradar (2007)

Table 23.3 Productivity and economic analysis of integrated farming system in Tamil Nadu (1998–2000)

Farming systems	Rice-equivalent yield (t/ha)	Net returns (Rs./ha)	B/C ratio	Employment generation (man days)
Cropping alone	13.0	37,153	2.43	369
Cropping + fish + poultry	29.6	97,731	3.02	515
Cropping + fish + pigeon	29.2	98,788	3.06	515
Cropping + fish + goat	37.7	1,31,118	3.36	576

Table 23.4 Economics of rice-*Azolla* fish integrated farming system

Farming systems	Gross income (Rs)			Total expenditure (Rs)	Net income (Rs)
	Crop	Fish	Total		
Rice	43,291	–	43,291	20,320	22,971
Rice + fish	39,447	11,422	50,869	22,300	28,569
Rice + azolla + fish	40,752	13,649	54,401	22,613	31,788

Table 23.5 Components of crop and livestock production

Crop production components	Livestock production components
Agro-forestry	Apiary
Azolla farming	Aquaculture
Crop husbandry	Biogas plants
Fodder production	Duck rearing
Horticulture	Fish farming
Mushroom cultivation	Goat rearing
Nursery	Pigeon rearing
Seed production	Piggery
Sericulture	Poultry
	Rabbitory

The rice + *Azolla* + fish, and rice + fish farming systems in Tamil Nadu gave net incomes of Rs 8817 and Rs.3219 ha^{-1} as compared to the rice monoculture, respectively. The fish utilized Azolla and phytoplankton as food resources at the rice bottom (Table 23.4) (Balusamy et al. 2003).

23.3 Components

Integrated farming systems mainly have two major components:

- Crop production (crops, pastures)
- Livestock production

The various components of crop and livestock production are presented in Table 23.5.

The components, outputs, outcomes, and impact of integrated farming systems are presented in Table 23.6.

23.4 Benefits and Limitations

23.4.1 Benefits

The overall benefits of the holistic crop–livestock integration can be summarized as follows:

23.4.1.1 Feed Resources

In developing countries, the crop residues such as straw, stover, and other fibrous by-products from the cereals, pulses, oil crops, roots, and tubers serve as a valuable, low-cost feed resource (major source of nutrients) for livestock production. Crop residues form the only source of feed during drought seasons in semiarid areas with low and uncertain rainfall (Valbuena et al. 2012).

23.4.1.2 Organic Soil Nutrients

The soil organic matter from animal manure supplies the nutrients required by the crops [containing 8 kg of nitrogen, 4 kg of phosphorus, and 16 kg of potassium per ton (FAO 1999)] and contributes to improve the physical, chemical, and biological properties of soil, like nutrient and water retention capacities, more biologically active soils, more productive rangeland or cropland, and controlled livestock carrying capacity. In the integrated farming systems, 23 % of the nitrogen requirement for crop production comes from livestock (Liu et al. 2010).

23.4.1.3 Draft Power

In developing countries, the animal draft power still plays a significant role to ease the drudgery and burden of hand cultivation, harvesting, and other cropping operations despite increased farm mechanization (FAO 2011). The need for foreign exchange to buy expensive tractors and fuel can be reduced by using draft animal (Jahake 1992). About 52 % of the cultivated area in the developing world (excluding China) is farmed entirely with draft animal, animal traction, bringing heavy but potentially very productive soil into production (International Food Security Treat Campaign 1984).

Table 23.6 The components, outputs, outcomes, and impact of integrated farming systems

Components	Outputs	Outcomes	Impact
Knowledge base	Current state of agropastoral systems in the participating countries documented	Informed research and development interventions in the crop–livestock system	Increased revenues and well-being of producers in agropastoral systems
	Database on current state of agropastoral system in project areas established	Improved collaboration among crop–livestock value chain actors	
	Gaps in research and development within the crop–livestock system identified		
	Framework for consultation between actors sharing common natural resources developed		
Technologies and innovations	Improved forage and soil health-enhancing crops proposed to farmers	Increased resilience of production systems to failure	
	Improved livestock rations for ruminant fattening proposed	Increased production of diversified food products	
	Innovative agropastoral rotational and other cultural practices proposed to farmers	Reduced incidents of conflicts between crop growers and livestock farmers	
	Options for reducing potential conflicts between crop growers and pastoralists identified		
Capacity strengthening	Capacity strengthening needs of farmers in agropastoral systems identified	Increased production of forage, meat, and other crop–livestock products	
	Instructional materials detailing the steps needed in improving intensive crop–livestock systems developed	Increased volumes of crop and livestock products marketed	
	Capacity of farmers to intensify crop–livestock systems strengthened		
	Capacity of actors in agropastoral systems to access inputs and services strengthened		
	Mechanism for improved flow of market information developed		

23.4.1.4 Cash Income

Animals produce livestock products (meat, milk, eggs, wool, and hides) or can be converted into prompt cash in times of need by selling the livestock. One of the significant contributions from livestock production is the cash inflow which can be reinvested in another enterprise. For example, to purchase inputs for cropping enterprises, the farmer sells goats in Zimbabwe (Homann et al. 2007). The economic benefits include reduced costs for inputs/technology acquisition, increased savings, and increased food security particularly due to the integrated crop and livestock production systems.

23.4.1.5 Environmental Benefits

The integrated system helps in the enhanced soil carbon sequestration and related reduction of the existing carbon dioxide from the atmosphere, control desertification, protection from drought, reduction of crop pests (consequent reductions in the use of pesticides), and use of cleaner water. The use of organic manures helps to minimize the requirement of chemical fertilizers which in turn reduces pollution from fertilizer manufacturing factories. The environmental pollution can be reduced through soil erosion control.

23.4.1.6 Enhanced Biodiversity

The major biodiversity benefits of the integrated systems include improved economic viability for organic production and restoration of natural wildlife habitat.

23.4.1.7 Mutual Benefits

The by-products of one enterprise are utilized as resources for the other enterprise. The crop residues and their by-products contribute to improved animal nutrition and productivity by supplementing inadequate feed supplies for animals; while the livestock manure is used to increase crop production. The livestock also transforms plant energy into useful activities like animal power that is used for plowing, water lifting for irrigation, marketing, transport, milling, road construction, and logging.

23.4.1.8 Energy Production

The animal excreta can be utilized for the production of biogas and energy for household purposes such as cooking and lighting or for rural industries like powering mills and water pumps. The use of dung cakes, charcoal, and wood as fuel can be replaced with biogas. One cubic meter of biogas can be produced by 25 kg of fresh cow dung, which can be used to provide energy for light, heat, or motive power.

23.4.2 Limitations

The limitations of integrated farming systems are as follows:

- The low digestibility and protein content of crop residues reduce their nutritional values as animal feeds. The digestibility and protein content in crop residues can be increased by using chemicals and machinery which are expensive and are not readily available (Keftasa 1988).
- Crop residues are either misapplied or disregarded, since they are primarily soil regenerators.
- The nutrients in organic manures can be lost due to intensive recycling.
- The import of production and transportation and the surpluses of animal manure are lost in the environment, if the manure nutrient use efficiencies are not improved or properly applied.
- Since the inorganic fertilizers supply the nutrients faster and are easier to apply, the farmers prefer them over organic manures.
- The enhancement of digestibility in crop residues requires costly chemicals and machinery which the farmers cannot afford.
- Bulk application of organic manures (not available) is required to fulfill the needs of crop plants.
- The transportation of bulky organic manures is not feasible and is costly.
- When applied excessively, both N and P are causes for environmental concern.

23.5 Challenges and Opportunities

23.5.1 Challenges

- Reducing pressures on natural resources
- Promoting strategies to develop crop–livestock interactions and synergies with the following aims:
  - Effective integration with careful land use
  - Productivity enhancement
  - Food production expansion
  - Environmental protection by efficient use of natural resources
- Devising sustainable manure management system without spreading contaminants and environmental losses

- Enhancing biomass digestibility by biodigesters and their large-scale dissemination

23.5.2 Opportunities

- Coordinating the crop production enterprises during the rainy season and livestock enterprises in the dry season.
- Sustainable intensification of crop–livestock integration systems.
- Sustainable crop–livestock integration can be supported by several indigenous, emerging, and improved technologies such as crop rotation with cereals and grain legumes, integrated nutrient and weed and pest management, year-round feed packages for livestock, and the development of modeling.
- Integrating crop–livestock farming systems is more lucrative than following individual enterprises.
- Development of alternatives for inorganic fertilizers and heavy farm machinery by using animal manure and draft power.

23.6 Conclusions and Ways Forward

There is considerable potential for improving the integration of crops and livestock in mixed systems to enhance the productivity of both crops and livestock to meet the global food supply and livelihoods in a more sustainable and eco-friendly manner. The by-product of one enterprise becomes the input for other in this system by mobilizing the resources effectively. The profit margin increases (30–50 %) by increasing the number of enterprises, generating employment opportunities. The new improved and innovative approach of integrated crop–livestock farming system offers sustainable production intensification for small farmers in reducing the malnutrition and poverty.

There is a need to give more emphasis for the following aspects in future for attaining sustainable production intensification:

- Development of socially acceptable and economically viable research modules of farming for different holding sizes.
- Generating the database with respect to economics, infrastructure, type of farming system, and sustainability under different agroclimatic zones.

- There is a justification for significantly more investment in research and development by providing incentives (financial assistance), policies, and technologies to encourage sustainable, integrated, and productive mixed crop–livestock systems in order to achieve win–win outcomes.
- Farming system technologies developed at research stations need to be assessed and refined under farmers' field conditions.

References

Al-Kaisi MM, Yin X, Licht MA (2005) Soil carbon and nitrogen changes as influenced by tillage and cropping systems in some Iowa soils. Agric Ecosyst Environ 105:635–647

Anderson V, Schatz B (2003) Biological and economic synergies, and methods of integrating beef cow and field crops enterprises. 2002 Unified Beef Cattle and Range Research Report, Dep. of Animal and Range Sci., Agric. Exp. Stn., North Dakota State Univ., Fargo. Available at www.ext.nodak.edu/extpubs/ansci/beef/2002/beef01.htm#biological. Accessed 5 Sept 2006; verified 22 Nov. 2006)

Balusamy M, Shanmugham PM, Baskaran R (2003) Mixed farming an ideal farming. Intensiv Agric 41(11&12):20–25

Chnnabasavanna AS, Biradar DP (2007) Relative performance of different rice-fish-poultry integrated farming system models with respect to system productivity and economics. Karnataka J Agric Sci 20(4):706–709

Delgado C, Rosegrant M, Steinfeld H (1999) Livestock to 2020: the next food revolution. IFPRI Brief 61, IFPRI (International Food Policy Research Institute), Washington, DC, USA

Drinkwater LE, Wagoner P, Sarrantonio M (1998) Legume based cropping systems have reduced carbon and nitrogen losses. Nature 396:262–265

FAO (1999) World production of animal protein and need for a new approach. FAO, Rome, pp 98–99

FAO (2011) World livestock 2011 – livestock in food security. FAO (Food and Agriculture Organization of the United Nations), Rome. www.fao.org/docrep/014/i2373e/i2373e.pdf

Ganesan G, Chinnasamy KN, Balasubramanian A, Manickasundram P (1990) Studies on rice based farming system with duck cum fish culture in deltaic region of Thanjavur district, Tamil Nadu. Farm Syst Newslett 1(2):14

Gill MS, Samra JS, Singh G (2005) Integrated farming system for realizing high productivity under shallow water-table conditions. Research bulletin, Department of Agronomy, Punjab Agricultural University, Ludhiana, 29 pp

Herrero M, Thornton PK, Notenbaert A, Msangi S, Wood S, Kruska R, Dixon J, Bossio D, van de Steeg J, Freeman HA, Li X, Parthasarathy Rao P (2009) Drivers of change in crop-livestock systems and their impacts on agro-ecosystems services and human well-being to 2030. ILRI (International Livestock Research Institute), Nairobi, Kenya. www.goo.gl/KbZxDa

Homann S, van Rooyen A, Moyo T, Nengomasha Z (2007) Goat production and marketing: baseline information for semiarid Zimbabwe. ICRISAT (International Crops Research Institute for the Semi-Arid Tropics), Bulawayo, Zimbabwe

International Food Security Treat Campaign (IFSTC) (1984) Sustainable product from livestock in Sub-Saharan Africa. International Livestock Center for Africa, Programme Planning and Funding requirement, pp 93–97

Jahake HE (1992) Livestock production systems and livestock development in tropical Africa. Kieler Wissenschaftsverlag Vauk, Kiel, pp 72–74

Jayanthi C, Rangasamy A, Mythili S, Balusamy M, Chinnusamy C, Sankaran N (2001) Sustainable productivity and profitability to integrated farming systems in low land farms. In: Singh AK,

Gangwar B, Pankaj, Pandey PS (eds) National symposium on farming system research on new millennium extended summaries. PDCSR, Modipuram, pp 79–81
Keftasa D (1988) Role of crop residues as livestock feed in ethiopian highlands. In: Proceedings of the third workshop at the International Conference Centre, International Livestock Centre for Africa (ILCA), Arusha, United Republic of Tanzania
Liu J, You L, Amini M, Obersteiner M, Herrero M, Zehnder AJB, Yang H (2010) A high-resolution assessment of nitrogen flows in cropland. Proc Natl Acad Sci U S A 107:835–840
Manjunath BL, Itnal CJ (2003) Farming system options for small and marginal holdings in different topographies of Goa. Indian J Agron 48(1):4–8
Manjunatha SB, Shivmurthy D, Satyareddi SA, Nagaraj MV, Basavesha KN (2014) Integrated farming system – an holistic approach: a review. Res Rev J Agric Allied Sci 3(4):30–38
Otte J, Costales A, Dijkman J, Pica-Ciamarra U, Robinson T, Ahuja V, Ly C, Roland-Holst D (2012) Livestock sector development for poverty reduction: an economic and policy perspective – livestock's many virtues. FAO, Rome. www.fao.org/docrep/015/i2744e/i2744e00.pdf
Rota A (2009) Integration of crop and livestock production in conservation agriculture: guidelines for project design. International Fund for Agricultural Development, Rome, Italy
Shanmugasundaram VS, Baluswamy M, Rangaswamy A (1995) Integrated farming system research in Tamil nadu. J Farming Syst Res Dev 1(1&2):1–9
Singh G (2004) Farming systems options in sustainable management of national resources. In: Proceedings of national symposium on alternative farming systems. PDCSR, Modipuram, pp 80–94
Singh CB, Renkema JA, Dhaka JP, Singh K, Schiere JB (1993) Income and employment on small farmers. In: Proceeding of international workshop on feeding of ruminants on fibrous crop residues: aspects of treatment, feeding, nutrient evaluation, research and extension. National Dairy Research Institute, Karnal, Haryana, pp 67–76
Small JA, McCaughey WP (1999) Beef cattle management in Manitoba. Can J Anim Sci 79:539–545
Soussana JF, Loiseau P, Vuichard N, Ceschia E, Balesdent J, Chevallier T, Arrouays D (2004) Carbon cycling and sequestration opportunities in temperate grasslands. Soil Use Manag 20:219–230
Tiwari SP, Ravi R, Nandeha KL, Vardia HK, Sharma RB, Rajgopal S (1999) Augmentation of economic status of Bastar tribals through integrated (crop, livestock, poultry, duck, fish) farming system. Indian J Animal Sci 69(6):448–452
Valbuena D, Erenstein O, Homann-Kee Tui S, Abdoulaye T, Claessens L, Duncan AJ, Gerard B, Rufino MC, Teufel N, van Rooyen A, van Wijk MT (2012) Conservation agriculture in mixed crop-livestock systems: scoping crop residue trade-offs in Sub-Saharan Africa and South Asia. Field Crop Res 132:175–184

Part XI
The Way Forward

Chapter 24
Past Experience and Future Scenario

Abstract Agricultural intensification practices that were followed (excessive tillage, mono-cropping, increased inputs, higher production costs, and indiscriminate use of pesticides) to achieve green revolution in the past have led to costly environmental damage and need to be modified in order to achieve greater sustainability. In order to produce more from the same area of land, there is a need to adapt a new paradigm, Sustainable Intensification of Crop Production (SICP), to conserve natural resources, reduce negative impacts on the environment, and enhance the provision of agro-ecosystem services. The chapter highlights the development of policies and institutions to support the key management practices and technologies required to implement SICP, recognizing that the appropriate combination will depend on local needs and conditions.

Keywords New paradigm • Sustainable intensification • Conserving resources • Supportive policies

24.1 Introduction

In view of increasing world population, the global demand for food is growing steadily because of urbanization, change in food habits, and increase in per capita income. It is difficult to meet the present-day food demand sustainably by using the current dominant production systems based on the so-called modern tillage-based and agro-chemically driven high-carbon footprint agricultural production systems—the dominant paradigm of the twentieth century. The climate change is also posing additional challenge to increase the food production.

The crop production and productivity are threatened by continuous soil degradation (soil erosion by water and wind and loss of soil organic matter, structure, and soil compaction), increasing water scarcity (rain water runoff), excessive tillage, mono-cropping, increased inputs, higher production costs, and indiscriminate use of pesticides, energy scarcity, and market globalization. With the present intensive production systems, it is not possible to adequately capture the required ecosystem services such as clean water, erosion control, carbon sequestration, nutrient cycling, etc. for the community.

P.P. Reddy, *Sustainable Intensification of Crop Production*,
DOI 10.1007/978-981-10-2702-4_24

24.2 Past Experience

Huggins and Reganolds (2008) opined that "The age-old practice of turning the soil before planting a new crop is a leading cause of farmland degradation. Tillage is a root cause of agricultural land degradation – one of the most serious environmental problems worldwide – which poses a threat to food production and rural livelihoods." The role of soil microorganisms in soil biological processes in mainstream production system paradigm has been totally ignored in the twentieth century.

The agricultural practices that are now adopted are causing damage to soils and reducing crop yields and other ecosystem services on a sustainable basis. The present-day practice of plowing the land before planting a crop loosens the soil and prepares a weed-free seedbed. The nutrients taken up by the plants are replaced with inorganic fertilizers. The agricultural input-supplying agencies recommend the farmers higher levels of fertilizer and pesticide applications and the use of genetically engineered seeds for increased production. These practices are responsible for degrading the soils that are not sustainable economically and environmentally.

It is predicted that the climate change shall have enormous negative impact on productivity, with reduced food availability and human well-being and increased greenhouse gas emissions in all developing countries resulting in increased global food prices ranging from 7 to 20 % by 2050 (Spielman and Pandya-Lorch 2010; Fischer et al. 2009).

24.3 Roads to Sustainable Intensification

The farmers and the scientists involved in agricultural research and development worldwide are at a crossroad in deciding which way to go forward in the twenty-first century. The farmers should take up more productive, profitable, and sustainable practices for crop production that do not harm the soil, landscape, and environment and provide higher productivity and enhanced ecosystem services. There is an urgent need for a change toward sustainable intensified crop production systems. The sustainable intensification practices are aimed at meeting the ever-growing demand for food, for reducing rural poverty, and for utilizing natural and human resources in an environmentally, socially, and financially sustainable way.

The increased production from the same piece of land while avoiding the more extreme effects of biological and environmental consequences is the challenge. Integration of traditional agricultural knowledge with new ecological knowledge into the sustainable intensification can meet this challenge. The implementation of strategies such as agro-forestry, intercropping, and mixed arable-livestock systems, integrated nutrient—organic matter management and pest management approaches—and reduction of off-site problems is receiving increasing attention as pathways to sustainable agricultural production.

The "precision agriculture" is an emerging technology that increases efficiency in application of inputs differentially across fields to match the crop demands. In order to reduce reliance on pesticides without affecting yields, improved pest management practices such as integration of crop diversification, host resistance, nutrient management, and biological control have to be adopted.

The sustainable intensification is attributed with multiple benefits such as increase in organic matter, improvement and maintenance of soil porosity (prolong the availability of plant-available soil water in times of drought), cleaner water due to less erosion, reduce weed, insect pest and disease incidence by biological means, raising agro-ecological diversity, favoring biological nitrogen fixation, providing better food security, increased profit, requirement of less farm power, less flooding, recharge of aquifers, and reduction in labor and time, resulting in both raised and stabilized yields with lower costs of production (Friedrich et al. 2009; Kassam et al. 2009) (Table 24.1).

24.4 Implementation

The biggest obstacle for no-till adoption in most parts of the world is because the plow as the symbol of agriculture is still deeply rooted in many cultures and mind-set. Erosion, soil degradation, and desertification are the consequences of tillage farming. A change has to take place in our mind-set, and we need to overcome this myth of tillage necessity for crop production.

The main barriers that need to be overcome for sustained growth of no-till to take place include:

- Know how (knowledge on how to do it)
- Nonavailability of adequate direct seeding machines
- Nonavailability of suitable herbicides
- Policy inadequacies
- Subsidies based on commodities
- Tradition and prejudice (mind-set)

The sustainable intensification practices have to be refined before their implementation that requires local adaptation to specific cropping systems and cultures across diverse agro-ecological and socioeconomic situations. In this context, farmers' field schools (FFS) can be utilized for empowering farmers through phased learning and discovery processes. The lessons already learnt by the countries which have adapted these practices should share their experiences with new countries adapting these technologies. The organizational strategies for implementation include:

- Adaption of participatory approach in research and development.
- Farmer organizations should take higher responsibilities for agricultural innovations.

Table 24.1 Effects of sustainable intensification on multiple benefits due to improved agro-ecosystem services

	Sustainable intensification component			
Ecosystem services	Mulch cover (crop residues, cover crops, green manures)	No tillage (minimal or no soil disturbance)	Legumes (for fixing N and supplying plant nutrients)	Crop rotation (for several beneficial purposes)
Simulate optimum "forest-floor" conditions	*	*		
Reduce evaporative loss of moisture from soil surface	*			
Reduce evaporative loss from upper soil layers	*	*		
Minimize oxidation of soil organic matter, CO_2 loss		*		
Minimize compactive impacts by intense rainfall, passage of feet, machinery	*			
Minimize temperature fluctuations at soil surface	*			
Provide regular supply of organic matter as substrate for soil organisms' activity	*			
Increase, maintain nitrogen levels in root zone	*	*	*	*
Increase CEC of root zone	*	*	*	*
Maximize rain infiltration, minimize runoff	*	*		
Minimize soil loss in runoff, wind	*	*		
Permit, maintain natural layering of soil horizons by actions of soil biota	*	*		
Minimize weeds	*	*		*
Increase rate of biomass production	*	*	*	*

(continued)

Table 24.1 (continued)

	Sustainable intensification component			
Ecosystem services	Mulch cover (crop residues, cover crops, green manures)	No tillage (minimal or no soil disturbance)	Legumes (for fixing N and supplying plant nutrients)	Crop rotation (for several beneficial purposes)
Speed soil porosity's recuperation by soil biota	*	*	*	*
Reduce labor input		*		
Reduce fuel-energy input		*	*	
Recycle nutrients	*	*	*	*
Reduce pest pressure of pathogens				*
Rebuild damaged soil conditions and dynamics	*	*	*	*
Pollination services	*	*	*	*

Friedrich et al. (2009) and Kassam et al. (2009)

- Networking of a multi-stakeholder knowledge management system.
- Farmers organizations should involve in capacity building.
- Payments for environmental services.
- Understand belowground processes by consulting with expert scientists.

24.5 Policies and Institutions

There is a need to rethink on policies and institutions for sustainable crop production intensification in view of unprecedented challenges to agriculture—including natural resources degradation, energy scarcity, market globalization, population growth, and climate change. These policies should focus on enabling smallholder farmers, in low-income developing economies in particular, to adopt sustainable intensification.

There is a need to give more emphasis on the following aspects while developing the policies and institutions:

- Pricing of inputs and outputs
- Regulation of seed sector
- Incentives for ecosystem services
    - Improving water, soil, and air quality
    - Preserving agricultural landscapes (cultural)

  - Preserving biodiversity and wildlife habitats
  - Reducing greenhouse gas emissions
  - Sequestering soil carbon

- Investment in agriculture
- Strengthening institutions
- Availability of key resources

  - Land
  - Plant genetic resources
  - Research

    – Increase funding
    – Strengthen research systems
    – Focus research on sustainable intensification
    – Priority to benefit smallholders
    – Learn from past experiences

  - Transfer of technology
  - Technologies and information
  - Loans for farmers
  - Productive social safety nets

- Regulation of agricultural marketing
- Loans for purchase of direct seeding equipment
- Establishment of CA producer associations

  - Common use of CA implements
  - Mutual learning

- Support for agricultural service providers
- Teaching of CA principles in agricultural schools and colleges

24.6 Conclusions

The tillage-based agriculture is considered not compatible with sustainable agriculture in view of the large-scale degradation of our land resource base and failure to optimize the crop productivity (efficiency) and output globally. In addition, tillage farming is not in a position to provide the needed range and quality of environmental services that are controlled in the soil system by ecosystem functions. The alternative to tillage-based agriculture is sustainable intensification of crop production (SICP) that is ecologically based and is universally applicable to all land-based production systems. This new innovative system covers all aspects of SICP-based production systems, including crop, nutrient, pest, water management, farm power and mechanization, public and private sector engagement, and policy and institutional support.

SICP constitutes a new production paradigm for most farmers globally that can be practiced by smallholder poor farmers, building on their own natural resource base, and can produce more from less without ecological consequences. By following SICP practices, the farmer can obtain environmental services from agricultural land use, address global challenges of climate change, and cope with the rise in food, energy, and production costs.

The early adopters face many hurdles before the full benefits of SICP can be harvested, because it is knowledge intensive and a complex system to learn and implement. It needs a dynamic support of enabling policies and institutional support to farmers and supply chain service providers for scaling up to achieve national impact.

SICP has established itself as a technology that can no longer be ignored by policy makers, scientists, institutions, extension workers, farmers as well as machine manufacturers, and other agriculture related industries.

There is a need for research into suitable innovations (technological and socio-economic), targeted financial investments and public private partnerships, active participation by smallholder famers, and, above all, political leadership in the process. The paradigm of sustainable intensification shows the way forward to the twenty-first century.

References

Fischer RA, Byerlee D, Edmeades GO (2009) Can technology deliver on the yield challenge to 2050? In: Paper presented at the FAO expert meeting: how to feed the world in 2050, FAO, Rome

Friedrich T, Kassam AH, Shaxson F (2009) Conservation agriculture. In: Agriculture for developing countries. Science and Technology Options Assessment (STOA) project. European Parliament, European Technology Assessment Group, Karlsruhe, Germany

Huggins DR, Reganold JP (2008) No-till: the quiet revolution. Sci Am 299:70–77

Kassam AH, Friedrich T, Shaxson F, Jules P (2009) The spread of conservation agriculture: justification, sustainability and uptake. Int J Agric Sustain 7(4):292–320

Spielman DJ, Pandya-Lorch R (2010) Fifty years of progress. In: Spielman DJ, Pandya-Lorch R (eds) Proven successes in agricultural development: a technical compendium to millions fed. International Food Policy Research Institute, Washington, DC, pp 1–30

Annexures

Annexure I. Glossary

Abiotic (Nonliving) Abiotic resources comprise nonliving things (e.g., land, water, air, and minerals).

Abiotic Stress Abiotic stress is defined as the negative impact of nonliving factors on the living organisms in a specific environment. Abiotic stress is essentially unavoidable.

Adaptation Adjustment of natural systems to a new or changing environment

Adaptation to Climate Change Refers to adjustment in natural systems in response to actual or expected climatic stimuli or their effects, which moderates harm or exploits beneficial opportunities. Various types of adaptation can be distinguished, including anticipatory and reactive adaptation, private and public adaptation, and autonomous and planned adaptation.

Afforestation The conversion from other land uses into forest or the increase of the canopy cover to above the 10 % threshold

Agricultural Biodiversity The component of biodiversity that is relevant to food and agriculture production. The term agro-biodiversity encompasses genetic species and ecosystem diversity.

Agricultural Intensification Refers to any practice that increases productivity per unit of land area at some cost in labor or capital inputs

Agricultural Landscape Quality Different ways in which agricultural models for land use modify the surroundings and influence ecological processes

P.P. Reddy, *Sustainable Intensification of Crop Production*,
DOI 10.1007/978-981-10-2702-4

Agro-ecological Restoration Restoration of the relationship between farming and nature by following an ecosystem approach, welcoming greater biodiversity and use of free ecological services

Agro-ecology In general it has three meanings or forms related to the application of ecology to agricultural systems: (1) a scientific discipline, (2) an agricultural practice, and (3) a social movement. While there are many different definitions, one of the broadest definitions is "the integrative study of the ecology of the entire food system, encompassing ecological, economic and social dimensions." Alternatively agro-ecology refers to "the study of purely ecological phenomena within the crop field, such as predator/prey relations, or crop/weed competition. Agro-ecology often incorporates ideas about a more environmentally and socially sensitive approach to agriculture, one that focuses not only on production, but also on the ecological sustainability of the productive system and goes well beyond the limits of the agricultural field."

Agro-ecosystem A seminatural or modified natural ecosystem managed by humans for food and agricultural production purposes

Agro-forestry Collective term for land-use systems and technologies in which woody perennials are deliberately used on the same land management unit as agricultural crops and/or animals, in some form of either spatial arrangement or temporal sequence

Allelopathy Direct influence from a chemical released from one plant on the development and growth of another one. Allelopathic substances, if present in crop varieties, may reduce the need for weed management, particularly herbicide use.

Alley Farming (Cropping) It is an agro-forestry practice that consists in planting perennial trees or shrubs on the sides of crops.

Alternative Agriculture (Agricultura Alternativa) Agricultural approach that attempts to provide a balanced environment, sustained yields and soil fertility, and natural pest control through the design of diversified agro-ecosystems and the use of self-sustaining technologies, based on ecological principles

Aquaculture The farming of aquatic organisms in inland and coastal areas, involving intervention in the rearing process to enhance production and the individual or corporate ownership of the stock being cultivated

Aquifers An aquifer is an underground layer of water-bearing permeable rock or unconsolidated materials (gravel, sand, silt, or clay) from which groundwater can be usefully extracted using a well.

Beneficial Insects Some insects have beneficial roles for nature: (1) plants reproduction (pollinators), (2) waste biodegradations (decomposers), and (3) natural resistance of agro-ecosystems/natural control of harmful species (natural enemies, predators, parasitoids). They also have beneficial roles for humans as edible insect species in nutrition, insect valuable products (e.g., silk and honey), and biomimicry among others.

Biodiversity The variability among living organisms from all sources including terrestrial, marine, and other aquatic ecosystems and the ecological complexes of which they are part; this includes diversity within and among species and diversity within and among ecosystems.

Biofuel Energy carriers that store the energy derived from biomass (e.g., food, fiber, and wood process residues from the industrial sector; energy crops, short rotation crops, and agricultural wastes from the agriculture sector; and residues from the forestry sector)

Biogeochemical Cycle The manner in which the atoms of an element critical to life (such as carbon, nitrogen, or phosphorus) move from the bodies of living organisms to the physical environment and back again

Biological Pest Control Method of controlling pests, diseases, and weeds in agriculture that relies on natural predation, parasitism, or other natural mechanisms that restrain the development of pathogenic organisms

Biomass The total mass of living organisms in a given area or volume; recently dead plant material is often included as dead biomass.

Biotic Living organisms that make up the biotic parts of ecosystems

Buffer Zone (Organic Agriculture) A clearly defined and identifiable boundary area bordering an organic production site that is established to limit the application of, or contact with, prohibited substances from an adjacent area

Carbon Cycle The flow of carbon (in various forms such as carbon dioxide) through the atmosphere, ocean, terrestrial biosphere, and lithosphere

Carbon Sequestration The process of increasing the carbon content of a carbon reservoir other than through the atmosphere

Change in Productivity Approach Economic valuation techniques that value the impact of changes in ecosystems by tracing their impact on the productivity of economic production processes

Climate Change Climate change refers to any long-term trends in climate over many years or decades, around which climate variability may be evident year on year. Hence, a single warmer or cooler year on its own is not sufficient evidence to assert that climate is changing, but systematic changes in average conditions over many years do provide evidence of a changing climate. The United Nations Framework Convention on Climate Change (UNFCCC) defined climate change as: "a change of climate which is attributed directly or indirectly to human activity that alters the composition of the global atmosphere and which is in addition to natural climate variability observed over comparable time periods." The UNFCCC thus makes a distinction between climate change attributable to human activities altering the atmospheric composition and climate variability attributable to natural causes.

Climate-Resilient Agriculture Type of agriculture that builds resilience to climate change and reduces greenhouse gas emissions through, for example, increased sequestration of carbon in soil

Compensation and Rewards for Environmental Services (CRES) Contractual arrangements and negotiated agreements among ecosystem stewards environmental service beneficiaries and/or intermediaries, for the purpose of enhancing, maintaining, reallocating, or offsetting damage to environmental services. A particular CRES contract or negotiated agreement will include a compensation or reward instrument or combination of instruments.

Compensation for Environmental Services (CES) Payments or other forms of restitution made to economic service beneficiaries or ecosystem stewards to offset foregone entitlements to environmental services or ecosystem stewardship benefits

Competition Interaction between individuals, brought about by a shared requirement for a resource, and leading to a reduction in the survivorship, growth, and/or reproduction of at least some of the competing individuals concerned

Composting Natural process of "rotting" or decomposition of organic matter by microorganisms under controlled conditions

Conservation Agriculture (CA) Conservation agriculture aims to achieve sustainable and profitable agriculture and subsequently aims at improved livelihoods of farmers through the application of the three CA principles: minimal soil disturbance, permanent soil cover, and crop rotations. To do so it promotes no-tillage to safeguard soil biodiversity, uses several organic fertilization practices such as rotations and mulching, but allows the use of genetically modified organisms (GMOs) and chemical inputs, namely, herbicides.

Conservation-Tillage Farming It is a practice used to reduce the effects of tillage on soil erosion; however, it still depends on tillage as the structure-forming element in the soil.

Conventional-Tillage or Intensive-Tillage It involves full-width tillage and may involve one, three, or perhaps up to 15 tillage passes. There is less than 15% residue on the soil surface after planting. Moldboard plowing and/or multiple tillage trips are involved.

Corridors Corridors refer to landscape elements that connect similar patches through a dissimilar matrix or aggregation of patches. The patches connected by corridors are often called nodes. A given corridor (e.g., a road) may function as a corridor for some organisms (humans) and a barrier for others (slugs).

Crop Diversification Crop diversification refers to varied crop associations and/or rotations (involving annual and/or perennial crops including trees). Crop diversification is intended to give a wider choice in the production of a variety of crops in a given area so as to expand production related activities on various crops and also to lessen risks. Crop diversification is generally viewed as a shift from traditionally grown less remunerative crops to more remunerative crops.

Crop Rotations The practice of alternating the species or families of annual and/or biannual crops grown on a specific field in a planned pattern or sequence so as to break weed, pest, and disease cycles and to maintain or improve soil fertility and organic matter content

Crop Wild Relatives Plant species that are more or less genetically related to crops but, unlike them, have not been domesticated

Deficit Irrigation Irrigation scheme in which water supply is less than the crop's full requirements and mild stress is allowed during growth stages that are less sensitive to moisture deficiency. The expectation is that any yield reduction will be limited and additional benefits are gained by diverting the saved water to irrigate other crops.

Deforestation The conversion of forest to another land use or the long-term reduction of tree canopy cover below the 10 % threshold

Disservices Some agricultural practices can generate negative effects over the environment leading to ecological disservices (e.g., decreases in water and air quality or a contribution to biodiversity loss). They reduce productivity or increase production costs.

Disturbance A cause, a physical force, an agent, or a process, causing a perturbation in an ecological component or system, relative to a specific reference state and system, defined by specific characteristics

Diversity The variety and relative abundance of different entities in a sample

Drip Irrigation Technique for achieving a low-rate, high-frequency, or long-duration water delivery through pipes to drip nozzles located near the plants

Drought The phenomenon that exists when precipitation is significantly below normal recorded levels, causing serious hydrological imbalances that often adversely affect land resources and production systems

Drought-Tolerant Crops Crops which are resilient to water shortages

Dry Land Farming Dry land farming is practiced in areas which are characterized by low and scanty rainfall with erratic distribution, leading to wide fluctuations in crop production.

Eco-functional Intensification Increase in food production, a considerable reduction in the negative impacts of agriculture on the environment, ecosystem services, and the use of nonrenewable resources and energy would become prerequisites for human survival.

Ecological Balance A state of dynamic equilibrium within a community of organisms in which genetic, species, and ecosystem diversity remain relatively stable, subject to gradual changes through natural succession

Ecological Diversity The variety of forests, deserts, grasslands, oceans, streams, lakes, and other biological communities interacting with one another and with their nonliving environment

Ecological Footprint The area of productive land and aquatic ecosystems required to produce the resources used and to assimilate the wastes produced by a defined population at a specified material standard of living, wherever on earth that land may be located

Ecological Intensification Ecological intensification refers to maximization of primary production per unit area without compromising the ability of the system to sustain its productive capacity. This entails management practices that optimize nutrient and energy flows and use local resources, including horizontal combinations (such as multiple cropping systems or polycultures), vertical combinations (such as agro-forestry), spatial integration (such as crop–livestock or crop-fish systems), and temporal combinations (rotations). A further definition is the following: An alternative approach for mainstream agriculture to meet current challenges. Ecological intensification aims to match or augment yield levels while minimizing negative impacts on the environment and ensuing negative feedbacks on agricultural productivity, by integrating the management of ecosystem services delivered by biodiversity into crop production systems.

Ecosystem The interactive system formed from all living organisms and their abiotic (physical and chemical) environment within a given area. Ecosystems cover a hierarchy of spatial scales and can comprise the entire globe, biomes at the continental scale or small, well-circumscribed systems such as a small pond.

Ecosystem-Based Practices Management practices used to buffer the impacts of climate variability and increase overall resilience (e.g., diversification, moisture conservation, protection of pollinators, and biological indicators)

Ecosystem Assessment A social process through which the findings of science concerning the causes of ecosystem change, their consequences for human well-being, and management and policy options are brought to bear on the needs of decision-makers.

Ecosystem Functions Activities, roles, or processes performed by structures. There are five main types of functions: capture (input), resources are brought into the system; production, resources are "manufactured" within the system; cycling, resources are transported within the system; storage, resources are conserved within the system; and output, resources leave the system. Structures often are involved in more than one function, and a function often requires more than one structure.

Ecosystem Health A measure of the stability and sustainability of ecosystem function or ecosystem services that depends on an ecosystem being active and maintaining its organization, autonomy, and resilience over time

Ecosystem Services The benefits people obtain from ecosystems. These include provisioning services such as food and water; regulating services such as flood and disease control; cultural services such as spiritual, recreational, and cultural benefits; and supporting services such as nutrient cycling that maintain the conditions for life on earth.

Efficient Water Management In agriculture, efficient water management means getting the right amount of water to the crops at the right time with minimum labor and expense.

Erosion The process of removal and transport of soil and rock by weathering and mass wasting and the action of streams, glaciers, waves, winds, and underground water.

Evapotranspiration (ET) ET is the measurement of moisture that plants and land lose through evaporation and transpiration processes due to heat, humidity, and wind. This amount is what should be replaced while irrigating.

Ex Situ Conservation Conservation of plant genetic resources in conditions different from those of their natural habitats

Extensive Agriculture Farming system that requires small amounts of labor and capital in relation to area of land being farmed

Fallow Land Land set aside for at least one season and sometimes for periods of up to 10 years or more with the aim of improving soil fertility and controlling pest and diseases. It may also refer to the period between the harvest of one crop and the planting of the next crop.

Farm Diversity Variety of farm activities, diverse grassland ecotypes, site-specific crop rotations with high numbers of crops in sequence, and the integration of livestock into cropping systems

Farmer Field School (FFS) FFS refers to group-based learning methodology that has been used by a number of governments, NGOs, and international agencies to promote Integrated Pest Management. It brings together concepts and methods from agro-ecology, experiential education, and community development.

Fertigation It is the application of fertilizers, soil amendments, or other water soluble products through an irrigation system.

Food Chain Food chain refers to the sum of all the processes and activities in a food system. A food system comprises multiple food chains operating at the global, national, and local levels.

Food Insecurity Situation when people lack secure access to sufficient amounts of safe and nutritious food for normal growth and development and an active and healthy life. It may be caused by the unavailability of food, insufficient purchasing power, inappropriate distribution, or inadequate use of food at the household level.

Food and Nutrition Security This exists when all people, at all times, have physical, social, and economic access to sufficient, safe, and nutritious food that meets their dietary needs and food preferences for an active and healthy life.

Food Security Situation that exists when all people, at all times, have physical, social, and economic access to sufficient, safe, and nutritious food that meets their dietary needs and food preferences for an active and healthy life.

Food System Encompasses (i) activities related to the production, processing, distribution, preparation, and consumption of food and (ii) the outcomes of these activities contributing to food security (food availability, with elements related to production, distribution, and exchange; food access, with elements related to affordability, allocation, and preference; and food use, with elements related to nutritional value, social value, and food safety). The outcomes also contribute to environmental and other securities (e.g., income).

Forage Food of any kind for animals, especially for horses and cattle, as grass, pasture, hay, corn, and oats

Fossil Fuel Fossil source fuels, that is, hydrocarbons found within the top layer of the earth's crust

Genetic Erosion Genetic erosion is a process that refers to a change in genetic diversity over time. To measure this change, some authors argue that it is important to focus on the loss of richness or evenness of alleles, while others argue that it is more important to focus on the loss of genes or genotypes.

Genetic Vulnerability The susceptibility of genetically uniform crops to damage or destruction caused by outbreaks of a disease or pest or unusually poor weather conditions or climatic change

Genetically Modified Organism (GMO) Organism in which the genetic material has been changed through modern biotechnology in a way that does not occur naturally by multiplication and/or natural recombination (e.g., a plant may be given fish genetic material that increases its resistance to frost).

Green Manuring To cover crop grown to help maintain soil organic matter and increase nitrogen availability

Greenhouse Emissions Gaseous constituents of the atmosphere, both natural and anthropogenic, that absorb and emit radiation at specific wavelengths within the spectrum of infrared radiation emitted by the earth's surface, the atmosphere, and clouds. This property causes the greenhouse effect.

Habitat The particular environment or place where an organism or species tend to live; a more locally circumscribed portion of the total environment

Home Gardens Farming system which combines different physical, social, and economic functions on the area of land around the family home

Hydrological Cycle The process of evaporation, transpiration, vertical and horizontal transport of vapor, condensation, precipitation, interception, runoff, infiltration, percolation, storage, the flow of water from continents to oceans, and return.

Humus Decomposed, dark brown, and amorphous organic matter of soils, having lost all trace of the structure and composition of the vegetable and animal matter from which it was derived. Humus hence refers to any organic matter that has reached a point of stability and which is used in agriculture to amend soil.

In Situ Conservation Conservation of plant genetic resources in their natural habitats

Incentives to Adopt Good Agricultural Practices (GAPs) The incentives for farmers to adopt GAPs include economic incentives such as increasing and/or stabilizing revenue; reducing average costs, improved market access, increased capital valuation of farm assets, and reduced vulnerability to poor agricultural practices of other farmers; regulatory or legal incentives including changes in ownership rights or tax burdens, liability rules, and subsidies; and human capital incentives including access to new skills.

Integrated Pest Management (IPM) Careful consideration of all available pest control techniques and subsequent integration of appropriate measures that discourage the development of pest populations and keep pesticides and other interventions to levels that are economically justified and reduce or minimize risks to human health and the environment. IPM emphasizes the growth of a healthy crop with the least possible disruption to agro-ecosystems and encourages natural pest control mechanisms.

Intensification Intensification in conventional agriculture is understood primarily as using a higher input of nutrient elements and of pesticides per land unit. It also means more energy (direct for machinery and indirect for inputs).

Intercropping Growing two or more crops as a mixture in the same field at the same time

Interventionist Approach In which most aspects of production are controlled by technological interventions such as soil tilling, protective or curative pest and weed control with agrochemicals, and the application of mineral fertilizers for plant nutrition

Land Cover The physical coverage of land, usually expressed in terms of vegetation cover or lack of it that is influenced by land use

Land Degradation Temporary or permanent lowering of the productive capacity of land

Land Use Land use is the total of arrangements, activities, and inputs undertaken in a certain land cover type (a set of human actions). It also refers to social and economic purposes for which land is managed (e.g., grazing, timber extraction, and conservation).

Land-Use Change A change in the use or management of land by humans, which may lead to a change in land cover

Land Race Land race refers to domesticated animals or plants adapted to the natural and cultural environment in which they live (or originated) and, in some cases, work. They often develop naturally with minimal assistance or guidance from humans using traditional breeding methods.

Landscape Landscape is an area of land that contains a mosaic of ecosystems, including human-dominated ecosystems. The term cultural landscape is often used when referring to landscapes containing significant human populations.

Landscape Composition The physical and biological characteristics of a landscape element

Landscape Contrast The degree to which adjacent landscape elements differ from each other, with respect to species composition and physical attributes

Landscape Diversity Diversity has been characterized as having three components: compositional, structural, and process: compositional diversity refers to the variation in types of landscape elements or vegetation types, their relative proportions within the landscape, their degree of rarity or commonness; structural diversity describes the variation in sizes and shapes of landscape elements, as well as diversity of pattern; and process diversity relates to the variety of landscape flows, functions, and processes present.

Living Mulch A cover crop that is interplanted with the primary crop(s) during the growing season

Matrix Most connected portion of the landscape, that is, the vegetation type that is most contiguous. This becomes increasingly less true as forests become fragmented by clear cutting; where in some cases it may not be possible to discern a matrix at all.

Mineralization The process by which organic residues in the soil are broken down to release mineral nutrients that can be utilized by plants

Mitigation (of Climate Change) An anthropogenic intervention to reduce the sources or enhance the sinks of greenhouse gases

Monoculture This refers to a cultivation system in which a single crop species covers a plot of land.

Mosaics Mosaics are evident at all scales from submicroscopic to the planet and universe. All mosaics are composed of spatial elements (patches, corridors, and matrix). Those at the landscape scale are commonly called landscape elements, and those at the regional scale are landscapes.

Mulching Application of mulch improves the water use efficiency and helps in water saving by reducing the ET losses and increased yields of a number of field crops during summer months.

Mulch-Till Full-width tillage involving one or more tillage trips which disturbs the entire soil surface and is done prior to and/or during planting. Tillage tools such as chisels, field cultivators, disks, sweeps, or blades are used. Weed control is accomplished with crop protection products and/or cultivation. It leaves more than 30 % soil covered with residues after seeding.

Multifunctionality Recognizes agriculture as a multi-output activity producing not only commodities (food, feed, fibers, agro-fuels, medicinal products, and ornamentals) but also non-commodity outputs such as environmental services, landscape amenities, and cultural heritages. It also recognizes that some of the non-commodity outputs may exhibit the characteristics of externalities or public goods, such that markets for these goods function poorly or are nonexistent.

Multiple Cropping Systems Planting two or more species in the same field during the same growing season is multiple cropping systems. It can take the form of double-cropping, in which a second crop is planted after the first has been harvested, or relay cropping, in which the second crop is started amidst the first crop before it has been harvested.

Natural Selection The process by which adaptive traits increase in frequency in a population due to the differential reproductive success of the individuals that possess the traits

Niche (Ecological) All of the interactions of a species with the other members of its community including competition, predation, parasitism, and mutualism are niche. A variety of abiotic factors, such as soil type and climate, also define a species' niche. Each of the various species that constitute a community occupies its own ecological niche.

Neglected and Underutilized Species Non-commodity crops, which are part of a larger biodiversity portfolio, once more popular and today neglected by users' groups for a variety of agronomic, genetic, economic, social, and cultural factors

On-Farm Conservation Management of genetic diversity of locally developed crop varieties (landraces) by farmers within their own agricultural, horticultural, or agrisilvicultural systems

Open Pollination The natural dispersal of pollen among all the members of a cross-pollinating crop population, resulting in the maximum degree of genetic mixing and diversity

Organic Agriculture Holistic production management system which promotes and enhances agro-ecosystem health, including biodiversity, biological cycles, and soil biological activity, is organic agriculture. It emphasizes the use of management practices in preference to the use of off-farm inputs, taking into account that regional conditions require locally adapted systems.

Organic Fertilization The use of natural organic fertilizer that helps to provide all the nutrients required by the plants and increases the quality of the soil with a natural microorganism environment

Organic Matter Plant and animal residues at various stages of decomposition, cells and tissues of soil organisms, and substances synthesized by the soil population

Participatory Plant Breeding Farmers participate in the selection of parent materials and in on-farm evaluations.

Pastoral System The use of domestic animals as a primary means for obtaining resources from habitats

Patches Areas of vegetation that are relatively homogeneous internally and that differ from what surrounds them (the matrix or other patches). In a complex landscape where a matrix is not apparent, forest stands are also patches.

Payments for Environmental Services (PES) Transparent system for the additional provision of environmental services through conditional payments to voluntary providers

Pheromones A chemical secreted by an animal, especially an insect that influences the behavior or development of others of the same species

Plant Genetic Resources Inter- and intraspecific diversity of crops, varieties, and related wild species which are central to agricultural development and improvements

Pollination The completion of the sexual phase of reproduction in some plants by the transportation of pollen

Polyculture Complex form of intercropping in which a large number of different plants maturing at different times are planted together

Pressurized Irrigation Pressurized irrigation systems are designed to deliver water with the flow rate and pressure required by the farm irrigation systems, sprinkling, or micro-irrigation and respecting the time, duration, and frequency decided by the farmers.

Provisioning Services The products obtained from ecosystems, including, for example, genetic resources, food and fiber, and freshwater.

Rain-Fed Agriculture Agricultural practice relying exclusively on rainfall as its source of water

Rangeland An area where the main land use is related to the support of grazing or browsing mammals

Reduced-Till Reduced-till systems are somewhat similar to mulch till in that they involve full-width tillage, use the same implements, and may use one to three tillage trips. Reduced-till, however, leaves 15–30% residue on the soil surface after planting. Weed control is accomplished with crop protection products and/or row cultivation.

Regulating Services The benefits obtained from the regulation of ecosystem processes, including, for example, the regulation of climate, water, and some human diseases.

Relative Humidity The amount of water vapor (vapor pressure) in a given parcel of air divided by the maximum amount of water vapor the parcel of air could contain at a given temperature (saturation vapor pressure) before it would begin to condense into water droplets.

Relay Cropping Cropping systems in which two or more crops are grown in sequence in the same field in the same year, with little or no overlap in time. Not a true form of polyculture because very little interspecies interaction usually occurs in these systems.

Resilience The capacity of a system to tolerate impacts of drivers without irreversible change in its outputs or structure

Resurgence Abnormal increase in pest populations often exceeding the economic threshold level, following the insecticide application

Rewards of Environmental Services (RES) Inducements provided to ecosystem stewards to enhance or continue to maintain environmental services

Ridge-Till The soil is left undisturbed from harvest to planting except for strips up to one-third of the row width. Planting is completed on the ridge and usually involves the removal of the top of the ridge. Planting is completed with sweeps, disk openers, coulters, or row cleaners. Residue is left on the surface between ridges. Weed control is accomplished with crop protection products (frequently banded) and/or cultivation. Ridges are rebuilt during row cultivation.

Ripper-Furrower System A farming system using a tractor-drawn ripper furrower to rip the hard pan to a depth of 60 cm and form furrows for in-field rainfall harvesting. The harvested water is concentrated in the root zone of crops, which are planted in the rip lines together with a mixture of fertilizer and manure.

Salinization The accumulation of salts in soils

Smallholder The definition of smallholders differs between countries and between agro-ecological zones. In favorable areas of smallholder subsistence agriculture with high population densities, smallholders often cultivate less than one hectare of land, whereas they may cultivate ten hectares or more in semiarid areas or manage up to ten heads of livestock.

Social Costs and Benefits These are costs and benefits as seen from the perspective of society as a whole. These differ from private costs and benefits in being more inclusive and in being valued at social opportunity cost rather than market prices, where these differ.

Soil Health The capacity of soil to function as a living system

Soil Organic Matter (SOM) Soil organic matter is any material produced originally by living organisms (plant or animal) that is returned to the soil and goes through the decomposition process. At any given time, it consists of a range of materials from the intact original tissues of plants and animals to the substantially decomposed mixture of materials known as humus.

Soil Structure Soil structure is defined by the way individual particles of sand, silt, and clay are assembled. Single particles when assembled are called aggregates. Aggregation of soil particles can occur in different patterns, resulting in different soil structures.

Strip-Till The soil is left undisturbed from harvest to planting except for strips up to one-third of the row width (strips may involve only residue disturbance or may include soil disturbance). Planting or drilling is accomplished using disk openers, coulter(s), row cleaners, in-row chisels, or rototillers. Weed control is accomplished primarily with crop protection products.

Subsoiling or Ripping Soil preparation treatment done with tined implements to break up hard pans without turning the soil upside down

Supplemental Irrigation The process of providing additional water to stabilize or increase yields under site conditions where a crop can normally be grown under direct rainfall, the additional water being insufficient to produce a crop. The concept consists in making up rainfall deficits during critical stages of the crops in order to increase yields.

Supporting Services Ecosystem services that is necessary for the production of all other ecosystem services. Some examples include biomass production, production of atmospheric oxygen, soil formation and retention, nutrient cycling, water cycling, and provisioning of habitat.

Sustainability A characteristic or state whereby the needs of the present and local population can be met without compromising the ability of future generations or populations in other locations to meet their needs

Sustainable Development Development that meets the needs of the present without compromising the ability of future generations to meet their own needs

Sustainable Intensification of Crop Production (SICP) System which produces more from the same area of land while conserving resources, rescuing negative impacts on the environment, and enhancing natural capital and the flow of ecosystem services

Threshold A point or level at which new properties emerge in an ecological, economic, or other system, invalidating predictions based on mathematical relationships that apply at lower levels. For example, species diversity of a landscape may decline steadily with increasing habitat degradation to a certain point and then fall sharply after a critical threshold of degradation is reached. Human behavior, especially at group levels, sometimes exhibits threshold effects. Thresholds at which irreversible changes occur are especially of concern to decision-makers.

Transgenic Plants Transgenic plants are plants possessing a single or multiple genes, transferred from a different species.

Uncertainty An expression of the degree to which a future condition is unknown

Upscaling The process of aggregating or extrapolating information collected at a fine resolution to a courser resolution or greater extent

Utility In economics, the measure of the degree of satisfaction or happiness of a person

Vermicomposting The use of earthworms for composting organic residues

Vulnerability Vulnerability is the degree to which a system is susceptible to or unable to cope with, adverse effects of climate change, including climate variability and extremes. Vulnerability is a function of the character, magnitude, and rate of climate variation to which a system is exposed, its sensitivity, and its adaptive capacity.

Water-Use Efficiency Carbon gain in photosynthesis per unit water lost in evapotranspiration. It can be expressed on a short-term basis as the ratio of photosynthetic carbon gain per unit transpirational water loss or on a seasonal basis as the ratio of net primary production or agricultural yield to the amount of available water.

Zero Tillage No-till farming (sometimes called zero tillage) is a way of growing crops from year to year without disturbing the soil through tillage.

Annexure II. Acronyms

ACT	African Conservation Tillage Network
AF	Agro-forestry
AMF	Arbuscular mycorrhizal fungi
Bt	*Bacillus thuringiensis*
CA	Conservation agriculture
CBC	Conservation biological control
CEC	Cation exchange capacity
CGMC	Cover/green manure crop
CIMMYT	International Maize and Wheat Improvement Center
CP	Cultural practices
CRM	Crop residue management
CT	Conventional tillage
CULTAN	Controlled Uptake Long-Term Ammonium Nutrition
DI	Deficit irrigation
ECAF	European Conservation Agriculture Federation
EPM	Ecological pest management
ET	Evapotranspiration
FAO	Food and Agriculture Organization of the United Nations
FDP	Fertilizer Deep Placement
FFS	Farmer Field School
FUE	Fertilizer use efficiency
GHG	Greenhouse gas
GM	Genetically modified
GMO	Genetically modified organism
HIPVs	Herbivore-induced plant volatiles
IPM	Integrated pest management
IAASTD	International Assessment of Agricultural Knowledge, Science and Technology for Development
ICARDA	International Center for Agricultural Research in the Dry Areas
IC-LS	Integrated crop–livestock systems
ICRAF	World Agro-forestry Center
INM	Integrated nutrient management
INRM	Integrated Natural Resources Management

IRRI	International Rice Research Institute
ISFM	Integrated soil fertility management
LER	Land Equivalent Ratio
MDGs	Millennium Development Goals
MT	Mulch Till
NCP	North China Plain
NCPAH	National Committee on Plasticulture Applications in Horticulture
NGO	Non-Governmental Organization
NOx	Nitrogen oxides
NPV	Nuclear Polyhedrosis Virus
NSC	National Seeds Corporation
NSE	(Seed related)
NSKE	Neem seed kernel extract
NT	No-till
NUE	Nutrient use efficiency
ODP	Oviposition-deterring pheromone
OM	Organic matter
PGR	Plant genetic resources
PRD	Partial root-zone drying
RDI	Regulated deficit irrigation
RT	Ridge till
RWH	Rainwater harvesting
SAU	State Agricultural University
SDI	Subsurface drip irrigation
SI	Sustainable intensification
SI	Supplemental irrigation
SICP	Sustainable intensification of crop production
SOC	Soil organic carbon
SOM	Soil organic matter
SMB	Soil microbial biomass
SRC	Short rotation coppice
ST	Strip till
TKW	Thousand kernel weight
TPS	True potato seed
UNEP	United Nations Environment Program
WP	Water productivity
WUE	Water use efficiency

Index

P.P. Reddy, *Sustainable Intensification of Crop Production*,
DOI 10.1007/978-981-10-2702-4